Lecture Notes in Networks and Systems 990

Series Editor

Janusz Kacprzyk ⓘ, *Systems Research Institute, Polish Academy of Sciences, Warsaw, Poland*

Advisory Editors

Fernando Gomide, *Department of Computer Engineering and Automation—DCA, School of Electrical and Computer Engineering—FEEC, University of Campinas— UNICAMP, São Paulo, Brazil*

Okyay Kaynak, *Department of Electrical and Electronic Engineering, Bogazici University, Istanbul, Türkiye*

Derong Liu, *Department of Electrical and Computer Engineering, University of Illinois at Chicago, Chicago, USA*

Institute of Automation, Chinese Academy of Sciences, Beijing, China

Witold Pedrycz, *Department of Electrical and Computer Engineering, University of Alberta, Alberta, Canada*

Systems Research Institute, Polish Academy of Sciences, Warsaw, Poland

Marios M. Polycarpou, *Department of Electrical and Computer Engineering, KIOS Research Center for Intelligent Systems and Networks, University of Cyprus, Nicosia, Cyprus*

Imre J. Rudas, *Óbuda University, Budapest, Hungary*

Jun Wang, *Department of Computer Science, City University of Hong Kong, Kowloon, Hong Kong*

The series "Lecture Notes in Networks and Systems" publishes the latest developments in Networks and Systems—quickly, informally and with high quality. Original research reported in proceedings and post-proceedings represents the core of LNNS.

Volumes published in LNNS embrace all aspects and subfields of, as well as new challenges in, Networks and Systems.

The series contains proceedings and edited volumes in systems and networks, spanning the areas of Cyber-Physical Systems, Autonomous Systems, Sensor Networks, Control Systems, Energy Systems, Automotive Systems, Biological Systems, Vehicular Networking and Connected Vehicles, Aerospace Systems, Automation, Manufacturing, Smart Grids, Nonlinear Systems, Power Systems, Robotics, Social Systems, Economic Systems and other. Of particular value to both the contributors and the readership are the short publication timeframe and the world-wide distribution and exposure which enable both a wide and rapid dissemination of research output.

The series covers the theory, applications, and perspectives on the state of the art and future developments relevant to systems and networks, decision making, control, complex processes and related areas, as embedded in the fields of interdisciplinary and applied sciences, engineering, computer science, physics, economics, social, and life sciences, as well as the paradigms and methodologies behind them.

Indexed by SCOPUS, INSPEC, WTI Frankfurt eG, zbMATH, SCImago.

All books published in the series are submitted for consideration in Web of Science.

For proposals from Asia please contact Aninda Bose (aninda.bose@springer.com).

Álvaro Rocha · Hojjat Adeli ·
Gintautas Dzemyda · Fernando Moreira ·
Aneta Poniszewska-Marańda
Editors

Good Practices and New Perspectives in Information Systems and Technologies

WorldCIST 2024, Volume 6

 Springer

Editors
Álvaro Rocha
ISEG
Universidade de Lisboa
Lisbon, Portugal

Hojjat Adeli
College of Engineering
The Ohio State University
Columbus, OH, USA

Gintautas Dzemyda
Institute of Data Science and Digital
Technologies
Vilnius University
Vilnius, Lithuania

Fernando Moreira
DCT
Universidade Portucalense
Porto, Portugal

Aneta Poniszewska-Marańda
Institute of Information Technology
Lodz University of Technology
Łódz, Poland

ISSN 2367-3370   ISSN 2367-3389 (electronic)
Lecture Notes in Networks and Systems
ISBN 978-3-031-60327-3   ISBN 978-3-031-60328-0 (eBook)
https://doi.org/10.1007/978-3-031-60328-0

Preface

This book contains a selection of papers accepted for presentation and discussion at the 2024 World Conference on Information Systems and Technologies (WorldCIST'24). This conference had the scientific support of the Lodz University of Technology, Information and Technology Management Association (ITMA), IEEE Systems, Man, and Cybernetics Society (IEEE SMC), Iberian Association for Information Systems and Technologies (AISTI), and Global Institute for IT Management (GIIM). It took place in Lodz city, Poland, 26–28 March 2024.

The World Conference on Information Systems and Technologies (WorldCIST) is a global forum for researchers and practitioners to present and discuss recent results and innovations, current trends, professional experiences, and challenges of modern Information Systems and Technologies research, technological development, and applications. One of its main aims is to strengthen the drive toward a holistic symbiosis between academy, society, and industry. WorldCIST'23 is built on the successes of: WorldCIST'13 held at Olhão, Algarve, Portugal; WorldCIST'14 held at Funchal, Madeira, Portugal; WorldCIST'15 held at São Miguel, Azores, Portugal; WorldCIST'16 held at Recife, Pernambuco, Brazil; WorldCIST'17 held at Porto Santo, Madeira, Portugal; WorldCIST'18 held at Naples, Italy; WorldCIST'19 held at La Toja, Spain; WorldCIST'20 held at Budva, Montenegro; WorldCIST'21 held at Terceira Island, Portugal; WorldCIST'22 held at Budva, Montenegro; and WorldCIST'23, which took place at Pisa, Italy.

The Program Committee of WorldCIST'24 was composed of a multidisciplinary group of 328 experts and those who are intimately concerned with Information Systems and Technologies. They have had the responsibility for evaluating, in a 'blind review' process, the papers received for each of the main themes proposed for the conference: A) Information and Knowledge Management; B) Organizational Models and Information Systems; C) Software and Systems Modeling; D) Software Systems, Architectures, Applications and Tools; E) Multimedia Systems and Applications; F) Computer Networks, Mobility and Pervasive Systems; G) Intelligent and Decision Support Systems; H) Big Data Analytics and Applications; I) Human-Computer Interaction; J) Ethics, Computers & Security; K) Health Informatics; L) Information Technologies in Education; M) Information Technologies in Radiocommunications; and N) Technologies for Biomedical Applications.

The conference also included workshop sessions taking place in parallel with the conference ones. Workshop sessions covered themes such as: ICT for Auditing & Accounting; Open Learning and Inclusive Education Through Information and Communication Technology; Digital Marketing and Communication, Technologies, and Applications; Advances in Deep Learning Methods and Evolutionary Computing for Health Care; Data Mining and Machine Learning in Smart Cities: The role of the technologies in the research of the migrations; Artificial Intelligence Models and Artifacts for Business Intelligence Applications; AI in Education; Environmental data analytics; Forest-Inspired

Computational Intelligence Methods and Applications; Railway Operations, Modeling and Safety; Technology Management in the Electrical Generation Industry: Capacity Building through Knowledge, Resources and Networks; Data Privacy and Protection in Modern Technologies; Strategies and Challenges in Modern NLP: From Argumentation to Ethical Deployment; and Enabling Software Engineering Practices Via Last Development Trends.

WorldCIST'24 and its workshops received about 400 contributions from 47 countries around the world. The papers accepted for oral presentation and discussion at the conference are published by Springer (this book) in four volumes and will be submitted for indexing by WoS, Scopus, EI-Compendex, DBLP, and/or Google Scholar, among others. Extended versions of selected best papers will be published in special or regular issues of leading and relevant journals, mainly JCR/SCI/SSCI and Scopus/EI-Compendex indexed journals.

We acknowledge all of those that contributed to the staging of WorldCIST'24 (authors, committees, workshop organizers, and sponsors). We deeply appreciate their involvement and support that was crucial for the success of WorldCIST'24.

March 2024

Álvaro Rocha
Hojjat Adeli
Gintautas Dzemyda
Fernando Moreira
Aneta Poniszewska-Marańda

Organization

Conference

Honorary Chair

Hojjat Adeli   The Ohio State University, USA

General Chair

Álvaro Rocha   ISEG, University of Lisbon, Portugal

Co-chairs

Gintautas Dzemyda   Vilnius University, Lithuania
Sandra Costanzo   University of Calabria, Italy

Workshops Chair

Fernando Moreira   Portucalense University, Portugal

Local Organizing Committee

Bożena Borowska   Lodz University of Technology, Poland
Łukasz Chomątek   Lodz University of Technology, Poland
Joanna Ochelska-Mierzejewska   Lodz University of Technology, Poland
Aneta Poniszewska-Marańda   Lodz University of Technology, Poland

Advisory Committee

Ana Maria Correia (Chair)   University of Sheffield, UK
Brandon Randolph-Seng   Texas A&M University, USA

Chris Kimble   KEDGE Business School & MRM, UM2,
                                   Montpellier, France
Damian Niwiński   University of Warsaw, Poland
Eugene Spafford   Purdue University, USA
Florin Gheorghe Filip   Romanian Academy, Romania
Janusz Kacprzyk   Polish Academy of Sciences, Poland
João Tavares   University of Porto, Portugal
Jon Hall   The Open University, UK
John MacIntyre   University of Sunderland, UK
Karl Stroetmann   Empirica Communication & Technology
                                   Research, Germany
Marjan Mernik   University of Maribor, Slovenia
Miguel-Angel Sicilia   University of Alcalá, Spain
Mirjana Ivanovic   University of Novi Sad, Serbia
Paulo Novais   University of Minho, Portugal
Sami Habib   Kuwait University, Kuwait
Wim Van Grembergen   University of Antwerp, Belgium

Program Committee Co-chairs

Adam Wojciechowski   Lodz University of Technology, Poland
Aneta Poniszewska-Marańda   Lodz University of Technology, Poland

Program Committee

Abderrahmane Ez-zahout   Mohammed V University, Morocco
Adriana Peña Pérez Negrón   Universidad de Guadalajara, Mexico
Adriani Besimi   South East European University, North
                                   Macedonia
Agostinho Sousa Pinto   Polytechnic of Porto, Portugal
Ahmed El Oualkadi   Abdelmalek Essaadi University, Morocco
Akex Rabasa   University Miguel Hernandez, Spain
Alanio de Lima   UFC, Brazil
Alba Córdoba-Cabús   University of Malaga, Spain
Alberto Freitas   FMUP, University of Porto, Portugal
Aleksandra Labus   University of Belgrade, Serbia
Alessio De Santo   HE-ARC, Switzerland
Alexandru Vulpe   University Politechnica of Bucharest, Romania
Ali Idri   ENSIAS, University Mohamed V, Morocco
Alicia García-Holgado   University of Salamanca, Spain

Almir Souza Silva Neto	IFMA, Brazil
Álvaro López-Martín	University of Malaga, Spain
Amélia Badica	Universiti of Craiova, Romania
Amélia Cristina Ferreira Silva	Polytechnic of Porto, Portugal
Amit Shelef	Sapir Academic College, Israel
Ana Carla Amaro	Universidade de Aveiro, Portugal
Ana Dinis	Polytechnic of Cávado and Ave, Portugal
Ana Isabel Martins	University of Aveiro, Portugal
Anabela Gomes	University of Coimbra, Portugal
Anacleto Correia	CINAV, Portugal
Andrew Brosnan	University College Cork, Ireland
Andjela Draganic	University of Montenegro, Montenegro
Aneta Polewko-Klim	University of Białystok, Institute of Informatics, Poland
Aneta Poniszewska-Maranda	Lodz University of Technology, Poland
Angeles Quezada	Instituto Tecnologico de Tijuana, Mexico
Anis Tissaoui	University of Jendouba, Tunisia
Ankur Singh Bist	KIET, India
Ann Svensson	University West, Sweden
Anna Gawrońska	Poznański Instytut Technologiczny, Poland
Antoni Oliver	University of the Balearic Islands, Spain
Antonio Jiménez-Martín	Universidad Politécnica de Madrid, Spain
Aroon Abbu	Bell and Howell, USA
Arslan Enikeev	Kazan Federal University, Russia
Beatriz Berrios Aguayo	University of Jaen, Spain
Benedita Malheiro	Polytechnic of Porto, ISEP, Portugal
Bertil Marques	Polytechnic of Porto, ISEP, Portugal
Boris Shishkov	ULSIT/IMI - BAS/IICREST, Bulgaria
Borja Bordel	Universidad Politécnica de Madrid, Spain
Branko Perisic	Faculty of Technical Sciences, Serbia
Bruno F. Gonçalves	Polytechnic of Bragança, Portugal
Carla Pinto	Polytechnic of Porto, ISEP, Portugal
Carlos Balsa	Polytechnic of Bragança, Portugal
Carlos Rompante Cunha	Polytechnic of Bragança, Portugal
Catarina Reis	Polytechnic of Leiria, Portugal
Célio Gonçalo Marques	Polytenic of Tomar, Portugal
Cengiz Acarturk	Middle East Technical University, Turkey
Cesar Collazos	Universidad del Cauca, Colombia
Cristina Gois	Polytechnic University of Coimbra, Portugal
Christophe Guyeux	Universite de Bourgogne Franche Comté, France
Christophe Soares	University Fernando Pessoa, Portugal
Christos Bouras	University of Patras, Greece

Christos Chrysoulas	London South Bank University, UK
Christos Chrysoulas	Edinburgh Napier University, UK
Ciro Martins	University of Aveiro, Portugal
Claudio Sapateiro	Polytechnic of Setúbal, Portugal
Cosmin Striletchi	Technical University of Cluj-Napoca, Romania
Costin Badica	University of Craiova, Romania
Cristian García Bauza	PLADEMA-UNICEN-CONICET, Argentina
Cristina Caridade	Polytechnic of Coimbra, Portugal
Danish Jamil	Malaysia University of Science and Technology, Malaysia
David Cortés-Polo	University of Extremadura, Spain
David Kelly	University College London, UK
Daria Bylieva	Peter the Great St. Petersburg Polytechnic University, Russia
Dayana Spagnuelo	Vrije Universiteit Amsterdam, Netherlands
Dhouha Jaziri	University of Sousse, Tunisia
Dmitry Frolov	HSE University, Russia
Dulce Mourato	ISTEC - Higher Advanced Technologies Institute Lisbon, Portugal
Edita Butrime	Lithuanian University of Health Sciences, Lithuania
Edna Dias Canedo	University of Brasilia, Brazil
Egils Ginters	Riga Technical University, Latvia
Ekaterina Isaeva	Perm State University, Russia
Eliana Leite	University of Minho, Portugal
Enrique Pelaez	ESPOL University, Ecuador
Eriks Sneiders	Stockholm University, Sweden; Esteban Castellanos ESPE, Ecuador
Fatima Azzahra Amazal	Ibn Zohr University, Morocco
Fernando Bobillo	University of Zaragoza, Spain
Fernando Molina-Granja	National University of Chimborazo, Ecuador
Fernando Moreira	Portucalense University, Portugal
Fernando Ribeiro	Polytechnic Castelo Branco, Portugal
Filipe Caldeira	Polytechnic of Viseu, Portugal
Filippo Neri	University of Naples, Italy
Firat Bestepe	Republic of Turkey Ministry of Development, Turkey
Francesco Bianconi	Università degli Studi di Perugia, Italy
Francisco García-Peñalvo	University of Salamanca, Spain
Francisco Valverde	Universidad Central del Ecuador, Ecuador
Frederico Branco	University of Trás-os-Montes e Alto Douro, Portugal
Galim Vakhitov	Kazan Federal University, Russia

Gayo Diallo	University of Bordeaux, France
Gabriel Pestana	Polytechnic Institute of Setubal, Portugal
Gema Bello-Orgaz	Universidad Politecnica de Madrid, Spain
George Suciu	BEIA Consult International, Romania
Ghani Albaali	Princess Sumaya University for Technology, Jordan
Gian Piero Zarri	University Paris-Sorbonne, France
Giovanni Buonanno	University of Calabria, Italy
Gonçalo Paiva Dias	University of Aveiro, Portugal
Goreti Marreiros	ISEP/GECAD, Portugal
Habiba Drias	University of Science and Technology Houari Boumediene, Algeria
Hafed Zarzour	University of Souk Ahras, Algeria
Haji Gul	City University of Science and Information Technology, Pakistan
Hakima Benali Mellah	Cerist, Algeria
Hamid Alasadi	Basra University, Iraq
Hatem Ben Sta	University of Tunis at El Manar, Tunisia
Hector Fernando Gomez Alvarado	Universidad Tecnica de Ambato, Ecuador
Hector Menendez	King's College London, UK
Hélder Gomes	University of Aveiro, Portugal
Helia Guerra	University of the Azores, Portugal
Henrique da Mota Silveira	University of Campinas (UNICAMP), Brazil
Henrique S. Mamede	University Aberta, Portugal
Henrique Vicente	University of Évora, Portugal
Hicham Gueddah	University Mohammed V in Rabat, Morocco
Hing Kai Chan	University of Nottingham Ningbo China, China
Igor Aguilar Alonso	Universidad Nacional Tecnológica de Lima Sur, Peru
Inês Domingues	University of Coimbra, Portugal
Isabel Lopes	Polytechnic of Bragança, Portugal
Isabel Pedrosa	Coimbra Business School - ISCAC, Portugal
Isaías Martins	University of Leon, Spain
Issam Moghrabi	Gulf University for Science and Technology, Kuwait
Ivan Armuelles Voinov	University of Panama, Panama
Ivan Dunđer	University of Zagreb, Croatia
Ivone Amorim	University of Porto, Portugal
Jaime Diaz	University of La Frontera, Chile
Jan Egger	IKIM, Germany
Jan Kubicek	Technical University of Ostrava, Czech Republic
Jeimi Cano	Universidad de los Andes, Colombia

Jesús Gallardo Casero	University of Zaragoza, Spain
Jezreel Mejia	CIMAT, Unidad Zacatecas, Mexico
Jikai Li	The College of New Jersey, USA
Jinzhi Lu	KTH-Royal Institute of Technology, Sweden
Joao Carlos Silva	IPCA, Portugal
João Manuel R. S. Tavares	University of Porto, FEUP, Portugal
João Paulo Pereira	Polytechnic of Bragança, Portugal
João Reis	University of Aveiro, Portugal
João Reis	University of Lisbon, Portugal
João Rodrigues	University of the Algarve, Portugal
João Vidal de Carvalho	Polytechnic of Porto, Portugal
Joaquin Nicolas Ros	University of Murcia, Spain
John W. Castro	University de Atacama, Chile
Jorge Barbosa	Polytechnic of Coimbra, Portugal
Jorge Buele	Technical University of Ambato, Ecuador; Jorge Gomes University of Lisbon, Portugal
Jorge Oliveira e Sá	University of Minho, Portugal
José Braga de Vasconcelos	Universidade Lusófona, Portugal
Jose M. Parente de Oliveira	Aeronautics Institute of Technology, Brazil
José Machado	University of Minho, Portugal
José Paulo Lousado	Polytechnic of Viseu, Portugal
Jose Quiroga	University of Oviedo, Spain
Jose Silvestre Silva	Academia Military, Portugal
Jose Torres	University Fernando Pessoa, Portugal
Juan M. Santos	University of Vigo, Spain
Juan Manuel Carrillo de Gea	University of Murcia, Spain
Juan Pablo Damato	UNCPBA-CONICET, Argentina
Kalinka Kaloyanova	Sofia University, Bulgaria
Kamran Shaukat	The University of Newcastle, Australia
Katerina Zdravkova	University Ss. Cyril and Methodius, North Macedonia
Khawla Tadist	Morocco
Khalid Benali	LORIA - University of Lorraine, France
Khalid Nafil	Mohammed V University in Rabat, Morocco
Korhan Gunel	Adnan Menderes University, Turkey
Krzysztof Wolk	Polish-Japanese Academy of Information Technology, Poland
Kuan Yew Wong	Universiti Teknologi Malaysia (UTM), Malaysia
Kwanghoon Kim	Kyonggi University, South Korea
Laila Cheikhi	Mohammed V University in Rabat, Morocco
Laura Varela-Candamio	Universidade da Coruña, Spain
Laurentiu Boicescu	E.T.T.I. U.P.B., Romania

Lbtissam Abnane	ENSIAS, Morocco
Lia-Anca Hangan	Technical University of Cluj-Napoca, Romania
Ligia Martinez	CECAR, Colombia
Lila Rao-Graham	University of the West Indies, Jamaica
Liliana Ivone Pereira	Polytechnic of Cávado and Ave, Portugal
Łukasz Tomczyk	Pedagogical University of Cracow, Poland
Luis Alvarez Sabucedo	University of Vigo, Spain
Luís Filipe Barbosa	University of Trás-os-Montes e Alto Douro
Luis Mendes Gomes	University of the Azores, Portugal
Luis Pinto Ferreira	Polytechnic of Porto, Portugal
Luis Roseiro	Polytechnic of Coimbra, Portugal
Luis Silva Rodrigues	Polytencic of Porto, Portugal
Mahdieh Zakizadeh	MOP, Iran
Maksim Goman	JKU, Austria
Manal el Bajta	ENSIAS, Morocco
Manuel Antonio Fernández-Villacañas Marín	Technical University of Madrid, Spain
Manuel Ignacio Ayala Chauvin	University Indoamerica, Ecuador
Manuel Silva	Polytechnic of Porto and INESC TEC, Portugal
Manuel Tupia	Pontifical Catholic University of Peru, Peru
Manuel Au-Yong-Oliveira	University of Aveiro, Portugal
Marcelo Mendonça Teixeira	Universidade de Pernambuco, Brazil
Marciele Bernardes	University of Minho, Brazil
Marco Ronchetti	Universita' di Trento, Italy
Mareca María Pilar	Universidad Politécnica de Madrid, Spain
Marek Kvet	Zilinska Univerzita v Ziline, Slovakia
Maria João Ferreira	Universidade Portucalense, Portugal
Maria José Sousa	University of Coimbra, Portugal
María Teresa García-Álvarez	University of A Coruna, Spain
Maria Sokhn	University of Applied Sciences of Western Switzerland, Switzerland
Marijana Despotovic-Zrakic	Faculty Organizational Science, Serbia
Marilio Cardoso	Polytechnic of Porto, Portugal
Mário Antunes	Polytechnic of Leiria & CRACS INESC TEC, Portugal
Marisa Maximiano	Polytechnic Institute of Leiria, Portugal
Marisol Garcia-Valls	Polytechnic University of Valencia, Spain
Maristela Holanda	University of Brasilia, Brazil
Marius Vochin	E.T.T.I. U.P.B., Romania
Martin Henkel	Stockholm University, Sweden
Martín López Nores	University of Vigo, Spain
Martin Zelm	INTEROP-VLab, Belgium

Mazyar Zand	MOP, Iran
Mawloud Mosbah	University 20 Août 1955 of Skikda, Algeria
Michal Adamczak	Poznan School of Logistics, Poland
Michal Kvet	University of Zilina, Slovakia
Miguel Garcia	University of Oviedo, Spain
Mircea Georgescu	Al. I. Cuza University of Iasi, Romania
Mirna Muñoz	Centro de Investigación en Matemáticas A.C., Mexico
Mohamed Hosni	ENSIAS, Morocco
Monica Leba	University of Petrosani, Romania
Nadesda Abbas	UBO, Chile
Narasimha Rao Vajjhala	University of New York Tirana, Tirana
Narjes Benameur	Laboratory of Biophysics and Medical Technologies of Tunis, Tunisia
Natalia Grafeeva	Saint Petersburg University, Russia
Natalia Miloslavskaya	National Research Nuclear University MEPhI, Russia
Naveed Ahmed	University of Sharjah, United Arab Emirates
Neeraj Gupta	KIET group of institutions Ghaziabad, India
Nelson Rocha	University of Aveiro, Portugal
Nikola S. Nikolov	University of Limerick, Ireland
Nicolas de Araujo Moreira	Federal University of Ceara, Brazil
Nikolai Prokopyev	Kazan Federal University, Russia
Niranjan S. K.	JSS Science and Technology University, India
Noemi Emanuela Cazzaniga	Politecnico di Milano, Italy
Noureddine Kerzazi	Polytechnique Montréal, Canada
Nuno Melão	Polytechnic of Viseu, Portugal
Nuno Octávio Fernandes	Polytechnic of Castelo Branco, Portugal
Nuno Pombo	University of Beira Interior, Portugal
Olga Kurasova	Vilnius University, Lithuania
Olimpiu Stoicuta	University of Petrosani, Romania
Patricia Quesado	Polytechnic of Cávado and Ave, Portugal
Patricia Zachman	Universidad Nacional del Chaco Austral, Argentina
Paula Serdeira Azevedo	University of Algarve, Portugal
Paula Dias	Polytechnic of Guarda, Portugal
Paulo Alejandro Quezada Sarmiento	University of the Basque Country, Spain
Paulo Maio	Polytechnic of Porto, ISEP, Portugal
Paulvanna Nayaki Marimuthu	Kuwait University, Kuwait
Paweł Karczmarek	The John Paul II Catholic University of Lublin, Poland

Pedro Rangel Henriques	University of Minho, Portugal
Pedro Sobral	University Fernando Pessoa, Portugal
Pedro Sousa	University of Minho, Portugal
Philipp Jordan	University of Hawaii at Manoa, USA
Piotr Kulczycki	Systems Research Institute, Polish Academy of Sciences, Poland
Prabhat Mahanti	University of New Brunswick, Canada
Rabia Azzi	Bordeaux University, France
Radu-Emil Precup	Politehnica University of Timisoara, Romania
Rafael Caldeirinha	Polytechnic of Leiria, Portugal
Raghuraman Rangarajan	Sequoia AT, Portugal
Radhakrishna Bhat	Manipal Institute of Technology, India
Raiani Ali	Hamad Bin Khalifa University, Qatar
Ramadan Elaiess	University of Benghazi, Libya
Ramayah T.	Universiti Sains Malaysia, Malaysia
Ramazy Mahmoudi	University of Monastir, Tunisia
Ramiro Gonçalves	University of Trás-os-Montes e Alto Douro & INESC TEC, Portugal
Ramon Alcarria	Universidad Politécnica de Madrid, Spain
Ramon Fabregat Gesa	University of Girona, Spain
Ramy Rahimi	Chungnam National University, South Korea
Reiko Hishiyama	Waseda University, Japan
Renata Maria Maracho	Federal University of Minas Gerais, Brazil
Renato Toasa	Israel Technological University, Ecuador
Reyes Juárez Ramírez	Universidad Autonoma de Baja California, Mexico
Rocío González-Sánchez	Rey Juan Carlos University, Spain
Rodrigo Franklin Frogeri	University Center of Minas Gerais South, Brazil
Ruben Pereira	ISCTE, Portugal
Rui Alexandre Castanho	WSB University, Poland
Rui S. Moreira	UFP & INESC TEC & LIACC, Portugal
Rustam Burnashev	Kazan Federal University, Russia
Saeed Salah	Al-Quds University, Palestine
Said Achchab	Mohammed V University in Rabat, Morocco
Sajid Anwar	Institute of Management Sciences Peshawar, Pakistan
Sami Habib	Kuwait University, Kuwait
Samuel Sepulveda	University of La Frontera, Chile
Sara Luis Dias	Polytechnic of Cávado and Ave, Portugal
Sandra Costanzo	University of Calabria, Italy
Sandra Patricia Cano Mazuera	University of San Buenaventura Cali, Colombia
Sassi Sassi	FSJEGJ, Tunisia

Seppo Sirkemaa — University of Turku, Finland

Sergio Correia — Polytechnic of Portalegre, Portugal

Shahnawaz Talpur — Mehran University of Engineering & Technology Jamshoro, Pakistan

Shakti Kundu — Manipal University Jaipur, Rajasthan, India

Shashi Kant Gupta — Eudoxia Research University, USA

Silviu Vert — Politehnica University of Timisoara, Romania

Simona Mirela Riurean — University of Petrosani, Romania

Slawomir Zolkiewski — Silesian University of Technology, Poland

Solange Rito Lima — University of Minho, Portugal

Sonia Morgado — ISCPSI, Portugal

Sonia Sobral — Portucalense University, Portugal

Sorin Zoican — Polytechnic University of Bucharest, Romania

Souraya Hamida — Batna 2 University, Algeria

Stalin Figueroa — University of Alcala, Spain

Sümeyya Ilkin — Kocaeli University, Turkey

Syed Asim Ali — University of Karachi, Pakistan

Syed Nasirin — Universiti Malaysia Sabah, Malaysia

Tatiana Antipova — Institute of Certified Specialists, Russia

TatiannaRosal — University of Trás-os-Montes e Alto Douro, Portugal

Tero Kokkonen — JAMK University of Applied Sciences, Finland

The Thanh Van — HCMC University of Food Industry, Vietnam

Thomas Weber — EPFL, Switzerland

Timothy Asiedu — TIM Technology Services Ltd., Ghana

Tom Sander — New College of Humanities, Germany

Tomasz Kisielewicz — Warsaw University of Technology

Tomaž Klobučar — Jozef Stefan Institute, Slovenia

Toshihiko Kato — University of Electro-communications, Japan

Tuomo Sipola — Jamk University of Applied Sciences, Finland

Tzung-Pei Hong — National University of Kaohsiung, Taiwan

Valentim Realinho — Polytechnic of Portalegre, Portugal

Valentina Colla — Scuola Superiore Sant'Anna, Italy

Valerio Stallone — ZHAW, Switzerland

Verónica Vasconcelos — Polytechnic of Coimbra, Portugal

Vicenzo Iannino — Scuola Superiore Sant'Anna, Italy

Vitor Gonçalves — Polytechnic of Bragança, Portugal

Victor Alves — University of Minho, Portugal

Victor Georgiev — Kazan Federal University, Russia

Victor Hugo Medina Garcia — Universidad Distrital Francisco José de Caldas, Colombia

Victor Kaptelinin — Umeå University, Sweden

Contents

4th Workshop on Open Learning and Inclusive Education Through Information and Communication Technology

1st Workshop on Environmental Data Analytics

1st Workshop on AI in Education

1st Workshop on Artificial Intelligence Models and Artifacts for Business Intelligence Applications

1st Workshop on The Role of the Technologies in the Research of the Migrations

12nd Workshop on Special Interest Group on ICT for Auditing and Accounting

2nd Workshop on Data Mining and Machine Learning in Smart Cities

**2nd Workshop on Enabling Software Engineering Practices Via Last
Development Trends**

1st Workshop on Data Privacy and Protection in Modern Technologies

GDPR-Compliant Data Breach Detection: Leveraging Semantic Web and Blockchain

Kainat Ansar[1]([✉]), Mansoor Ahmed[2], Muhammad Irfan Khalid[3],
and Markus Helfert[2]

[1] Department of Computer Science, COMSATS University, Islamabad, Pakistan
`kainat.ansar@gmail.com`
[2] ADAPT Centre, Innovation Value Institute, Maynooth University, Maynooth,
Ireland
`{mansoor.ahmed,markus.helfert}@mu.ie`
[3] Faculty of Computing and Information Technology, Department of Information
Technology, University of Sialkot, Sialkot, Pakistan
`irfanse6235@gmail.com`

Abstract. Insider attacks are becoming common and have a significant financial impact on organizations. Insider threats come from within the targeted organization, and insider attacks are carried out by users who have been granted access to the target organization's network, apps, or databases. An attacker with administrative capabilities can edit logs and login records to remove traces of the attack, making insider attacks difficult to detect. Such data breaches may severely negatively influence the life of the Data Owner. Creating a mechanism for quickly identifying data breaches is still essential and difficult. The General Data Protection Regulation (GDPR) has established processes and guidelines to address data privacy issues. Due to this, when a data breach occurs, the Data Controller is required under the GDPR implementation to notify the Data Protection Authority. To address these problems, this article proposes a GDPR-compliant data breach detection system with a severity assessment mechanism using the semantic web and blockchain technology. The suggested method can generate alert notifications for each data breach. Consequently, with the help of the severity assessment mechanism, the proposed model conducts a breach assessment to indicate the data breach's severity level.

Keywords: Data Breach Detection · GDPR · Blockchain · Semantic Web

1 Introduction

For every organization, a data breach is a serious issue. Any incident in which data is seen, deleted, altered, or transferred by an unauthorized party or an authorized person unintentionally or intentionally is referred to as a data breach [1]. A data breach can be caused by a variety of factors, including hardware

Á. Rocha et al. (Eds.): WorldCIST 2024, LNNS 990, pp. 3–11, 2024.
https://doi.org/10.1007/978-3-031-60328-0_1

issues, software crashes, phishing, malware, ransomware, distributed denial-of-service, human error, misplaced or lost data storage devices (such as USB drives, laptops, portable drives, and so on), malicious insiders, and external issues such as power outages [2]. However, our research focuses on malicious insider threats.

Insider data breaches are growing more common and have a higher financial impact on organizations. A recent report states that insider threats are responsible for 60% of data breaches [3]. An insider is typically someone who has allowed access to company resources and intentionally or unintentionally damages the company. Current or former employees, contractors, partners, or employees who have access to an organization's systems or data may pose a threat to them [4].

Since the General Data Protection Regulation (GDPR) went into effect in May 2018, there has been a paradigm shift in data privacy [5]. The GDPR specifies processes and rules to address the challenges of insider threat and data protection. As a result, when a data breach occurs, the Data Controller (DC) must notify the Data Protection Authority (DPA) and the affected Data Owner (DO). He will face significant fines if he fails to notify the breaches within a particular time frame. According to GDPR, organizations that experience a data breach may be fined up to 4% of their annual revenue, or €20 million, whichever is greater [6]. Such a system that can detect data breaches is required to avoid severe penalties.

On the other hand, in contrast to traditional internet technology, which simply provides a "network of information," blockchain is a cutting-edge technology that provides a "network of value" [7]. Ethereum blockchain utilizes particular languages such as solidity [8] to become fully programmable, allowing the building of modern decentralized applications. These decentralized applications use smart contracts. Smart contracts are coding scripts that enable users to execute transactions without the possibility of fraud or third-party interference [9].

To address the issue of data breaches, this article developed a GDPR-compliant detection system that takes advantage of the semantic web, smart contracts, and blockchain technologies. The processes of the proposed system methodology and the functioning of smart contracts are detailed in the subsequent sections.

2   Related Work

Malicious insider threats have recently been identified as one of the organizations' most harmful breach attacks. Data breaches are security incidents that occur when an attacker gains access to a company's network, application, or database and performs malicious activity. Numerous studies have been conducted to solve this issue.

In [10], the authors presented a data leakage prevention system. Authors employed document semantic signatures to detect breaches. When the semantic signatures of the outgoing document match those of the original document, the system detects data leaks. A sensitive file, however, can avoid detection if an attacker encrypts it and sends it via email. In this circumstance, the detection system cannot recognize the encrypted data as a sensitive file. As a result, sensitive data may be leaked. In [11], an anomaly detection model is proposed for database protection. The Hidden Markov Model (HMM) was utilized for prediction, and the authors achieved minimal false-positive rates. The HMM-based system, on the other hand, is dependent on the training dataset. If the training dataset is insufficient, the system may generate false-positive alarms.

The authors suggested a three-tiered data protection strategy in [12] in response to the information leakage concern created by cloud indexing. However, the process requires a pre-defined data classification. Data that has been misclassified may be leaked. The authors of [13] presented a data leak prevention strategy based on Named Entity Recognition (NER). However, the approach did not use semantic technologies to provide meaning to entities. As a result, spelling errors and related words could impact NER.

To detect insider attacks in relational database systems, the authors proposed a blockchain-based framework in [14]. However, the authors' solution only addresses the private data and centralized control system, in which a private blockchain network is built in a privately controlled environment with no democratic participants. Furthermore, regardless of whether a network is built on blockchain technology, attackers can manipulate any data or network within an organization. Storing all proof within the same centralized controls or system can increase the attack risk. The authors employed a private blockchain network, meaning anyone with access to that company can modify the private blockchain network even if the entire organization is compromised.

In [15], a blockchain-based event-driven data alteration detection system is presented. However, the model described in the study does not clarify how the framework will function technically. Furthermore, the paper lacks any practical application examples or solutions. It is important to note that storing any data in any structure mandates using a smart contract, yet this approach does not share any knowledge of the smart contract. Furthermore, this paper does not specify how to keep data evidence or fingerprints on the blockchain network. Due to the lack of an appropriate structure, this approach will be ambiguous and impractical. Furthermore, existing research lacks GDPR-compliant practical methods for data leak detection that Data Protection Authorities and Data Controllers can use to determine if it is required to notify affected Data Owners.

To summarize the above discussion, we find that existing blockchain-based data breach detection solutions have several limitations. As a result, developing a system capable of addressing the issues mentioned above is challenging. Considering the limitations of previous studies, we present a novel Personal Data Breach Detection (PDBD) technique in this paper. The following are the main contributions of our work.

i A GDPR-compliant PDBD model is developed. It will enable the DC to quickly determine the necessary mitigation measures for data breach events.
ii Semantic Web Rule Language (SWRL) rules are developed for the Data Breach Severity Assessment (DBSA) mechanism. This will result in providing the DC with a computable tool to assess the severity of data breaches. It will also help the DC in the process of notifying about breaches accordingly to the data protection authorities and the affected DO.

iii Severity level detection ontology is developed to calculate breach severity index score. Also, ontology will indicate breach severity level using SWRL rules.
iv Hash Variance Algorithm (HVA) is introduced to reduce the computational overhead of both DBSA and Ethereum.

3 Use Case Scenario

The use case scenario for the health industry is discussed in this part to show how the system performs. In current hospitals, collecting and processing personal data from patients has become mandatory. Almost every hospital department handles protected health information and personally identifiable information about patients. It is hard to recover privacy or restore psychosocial damage when an insider attack discloses a patient's private information. Furthermore, compromised information can interfere with hospital operations and negatively impact the health and well-being of the patient. If immediate treatment is not received, this condition may result in death or permanent disability due to these operating delays.

The use case scenario assumes that John, the data processor, is a medical specialist who frequently requests patients' medical records for operational needs, and Michael, the Data Owner, is the patient. Michael, the patient, agrees with having her medical data preserved on the blockchain. John can get the required patient data from the patient database by submitting a request to Robert, the Data Controller. Robert uses our proposed system for tasks like data verification and consent validation. Before providing John with any data, our recommended approach allows Robert to detect any alterations to the database record and confirm its authenticity.

3.1 System Design and Methodology

Figure 1 illustrates the proposed data breach detection model and its components with operation flow on DBSA and Ethereum layers. The main components of the proposed model are as follows.

Data Consumer:
Supposedly trusted third parties or data consumers are important entities of the proposed model that request Data Owners' personal information. For instance, a surgeon who frequently seeks patients' medical records for operation purposes. (as discussed in the previous section)

Ethereum:
Ethereum is a blockchain-based platform. Blockchain technology is the collection of blocks containing transaction data linked to each other in a chain. It is a digital ledger that is secure, cryptography-based, and distributed across a network. And this ledger is such of a kind that allows your transactions to be secure, anonymous, fast, and without any central authority. We have used the Ethereum

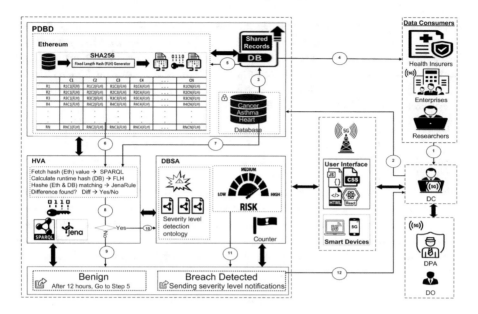

Fig. 1. Proposed System Model

(ETH) network with shared database records in this proposed model. Intending not to store all the data on the blockchain, we create a Cell Signature (CS) against each data table cell and only store that on the blockchain. These cell signatures are generated using the SHA256 [16] for each cell in a table. SHA256 is a cryptographic hash function. As such, it is practically impossible to reverse it and find a message or data that hashes to a given digest. For each row in the table, we generate cell pointer CnRn. For example, in Table X, row 1 (R1) has N columns, and the Cell Pointer (CP) will be generated as shown in Eq. 1.

$$TableX\_CP\_R1 = R1C1(FLH), R1C2(FLH), ..., R1CN(FLH) \qquad (1)$$

The sequence of CP with cell signature is depicted in Fig. 2. These CPs are then stored on a blockchain using a private key. Any modifications to a CP get logged on the blockchain with a new cell signature of the respective row. Any previous CPs of the modified data cell are also preserved in the blockchain.

HVA:

In the previous phase, we created a cell signature for each data cell and stored this CP on the blockchain. The next phase is the Hash Variance Algorithm (HVA) phase. Cell signatures created using SHA256 in the previous (ethereum phase) will serve as inputs to this phase. The function of the HVA mechanism is shown in Algorithm 1. This can reduce the computational overhead of both DBSA and Ethereum and increase the system throughput. This phase mainly utilizes semantic web technologies such as SPARQL, SWRL rules, reasoning engine, and Jena framework to fetch calculated cell signatures from the previous phase and calculate runtime cell signatures of CPs by fetching shared records from the shared part of database applications by using the SPARQL query. In other words, we need to calculate the difference between CPs (Ethereum and shared database) and then compare them one by one according to the Fixed-Length Hash (FLH) threshold set in advance. If a difference is found, it will be considered as an attack and modified record.

DBSA: The above HVA phase has described the basic structure and mechanism of data breach detection. Based on the above methodology, the calculated output value of the HVA phase can be forwarded. It is necessary to forward the final output of HVA to the DBSA phase to calculate the severity score. In this phase, severity level detection ontology is developed to calculate the breach severity index score. The presented DBSA mechanism in this synopsis uses severity assessment methodology [17] provided by the European Union Agency for Network and Information Security (ENISA). ENISA introduced a severity level assessment formula [17] to calculate the overall severity score, which is shown below in Eq. 2.

$$Severity\_level\_score = DPF * ER + SB \qquad (2)$$

where DPF is a data processing factor, ER is the ease of recognition, and SB denotes a breach situation. Furthermore, DPF includes classified breached data as simple, behavioral, financial, and sensitive. ER evaluates how easily a certain person is identified using breached data. The ER can be negligible, limited, significant, or maximum. Whereas SB includes malicious intents and security loss in terms of confidentiality, integrity, and availability.

The methodology, as mentioned above, is implemented in the DBSA phase. Severity level detection ontology is developed to apply these guidelines using the recommended methodology [17]. The main classes and subclasses are shown in Fig. 2. In addition, SWRL rules are developed to indicate the data breach's severity level. However, two rules are modeled for the proof of concept, as shown below.

Algorithm 1. Hash Variance Algorithm

1: Input Data:- Shared data_CP, Cell Signature, Ethereum_CP, Cell Signature
2: **function** HashVariance(CP, Cell SignatureLength), Diff Status
3: Fetch CP Ethereum
4: Fetch Cell signature Ethereum $->$ CS(ETH)
5: Fetch Shared data CP
6: **for** i=0; i<= Cell Signature.Length i++ **do**
7:     Calculate Cell signature Ethereum $->$ CS(DB)
8:     Calculate difference CS(ETH), CS(DB) $->$ Diff Status
9:     **if** CS(ETH) != CS(DB) and Diff status == "True" **then**
10:         Forward Diff status to DBSA
11:         Return Severity Level Index Score
12:     **end if**
13: **end for**

Fig. 2. Classes of Severity Level Detection Ontology

*Rule1: Affected_DO(?ado), Breach_Detected(?bd)(?ado dc: hasSB Min)
(?ado dc:hasER Negligible)(?ado dc: hasDPF Simple) — >
(?bd dc: setFlag Low)*

*Rule2: Affected_DO(?ado), Breach_Detected(?bd)(?ado dc: hasSB Confidential-
ity_loss)(?ado dc:hasER Maximum)(?ado dc: hasDPF Sensitive) — >
(?bd dc: setFlag High)*

4   Conclusion and Future Works

Data Controllers are obliged to implement measures that will facilitate compliance with GDPR and notify the Data Protection Authorities and every affected
party (data owner) in case of any data breaches or possible risk of data privacy
violation with undue delay (72 h). Failure to issue a breach notification within
time can result in a heavy fine. However, the ability to effectively detect a data
breach is still a critical issue and challenging task. Thus, the Data Controllers
must have an efficient system for detecting data breaches within time, along with
severity level, and in an appropriate way to manage the personal information
within organizations and smart devices. This paper presented a novel semantic-
blockchain-based model for rapid data breach detection to protect personal data

from breaches and reduce direct and indirect data damage that prevents direct and indirect personal data damage. The proposed model generates alerts against data breaches by taking into account severity assessment details and grading the breach incident according to the Data Owner's impact and the significance of the breach. In the future, we will implement this system using semantic web and blockchain technologies.

Funding Information. This research was conducted with the financial support of Science Foundation Ireland under Grant Agreement Nos. [13/RC/2106_P2] and [20/SP/8955] at the ADAPT SFI Research Centre at Maynooth University. ADAPT, the SFI Research Centre for AI-Driven Digital Content Technology is funded by Science Foundation Ireland through the SFI Research Centres Programme

References

1. Firman, A., et al.: Why does medical confidentiality matter during the Covid-19 pandemic? A case study from regulations in Indonesia. J. Legal Ethical Regul. Issues **24**(1), 13 (2021)
2. What Constitutes a GDPR Data Breach? Definition & Meaning. https://sectigostore.com/blog/what-constitutes-a-gdpr-data-breach-definition-meaning/. Accessed Oct 2022
3. Insider Threats Are Becoming More Frequent and More Costly. https://www.idwatchdog.com/insider-threats-and-data-breaches/. Accessed Oct 2022
4. 2022 Ponemon Cost of Insider Threats Global Report. https://www.proofpoint.com/us/resources/threat-reports/cost-of-insider-threats. Accessed Oct 2022
5. General Data Protection Regulation (GDPR). https://en.wikipedia.org/wiki/General_Data_Protection_Regulation. Accessed Apr 2021
6. Article 83 European Union General Data Protection Regulation (GDPR): General conditions for imposing administrative fines. http://www.privacy-regulation.eu/en/article-83-general-conditions-forimposing-administrative-fines-GDPR.htm#5. Accessed Apr 2021
7. Farhan, H.K.: Blockchain: Transforming the Fourth Industrial Revolution. Global Foundation for Cyber Studies and Research (2020)
8. Dannen, C.: Bridging the blockchain knowledge gap. In: Introducing Ethereum and Solidity. Apress, Berkeley, Springer (2019). https://doi.org/10.1007/978-1-4842-2535-6_1
9. Alotaibi, S.J.: Using blockchain for smart contracts. In: Innovative and Agile Contracting for Digital Transformation and Industry 4.0, pp. 208–221. IGI Global (2021)
10. Alhindi, H., Traore, I., Woungang, I.: Preventing data leak through semantic analysis. Internet Things **14**, 100073 (2021)
11. Fadolalkarim, D., Bertino, E., Sallam, A.: An anomaly detection system for the protection of relational database systems against data leakage by application programs. In: 36th International Conference on Data Engineering (ICDE). IEEE (2020)
12. Squicciarini, A., Sundareswaran, S., Lin, D.: Preventing information leakage from indexing in the cloud. In: 2010 IEEE 3rd International Conference on Cloud Computing. IEEE (2010)

13. Gómez-Hidalgo, J.M., et al.: Data leak prevention through named entity recognition. In: 2010 IEEE Second International Conference on Social Computing. IEEE (2010)
14. Srivastava, S.S., et al.: Verity: blockchains to detect insider attacks in DBMS. arXiv preprint arXiv:1901.00228 (2019)
15. Srivastava, S., Kumar, A., Jha, S.K., Dixit, P., Prakash, S.: Event-driven data alteration detection using block-chain. Secur. Privacy 4(2), e146 (2021)
16. Handschuh, H., van Tilborg, H.C.A.: SHA Family (Secure Hash Algorithm) (2005)
17. Manson, C.G., Gorniak, S.: Recommendations for a methodology of the assessment of severity of personal data breaches. ENISA (European Union Agency for Network and Inform. Security) Working Document, v1. 0 (2013)

Leveraging Blockchain Technologies for Secure and Efficient Patient Data Management in Disaster Scenarios

Muhammad Irfan Khalid[1]([envelope]), Mansoor Ahmed[2], Kainat Ansar[3],
and Markus Helfert[2]

[1] Faculty of Computing and Information Technology, Department of Information
Technology, University of Sialkot, Sialkot, Pakistan
irfanse6235@gmail.com
[2] ADAPT Center, Innovation Value Institute, Maynooth University,
Maynooth, Ireland
{mansoor.ahmed,markus.helfert}@mu.ie
[3] Department of Computer Science, COMSATS University, Islamabad, Pakistan

Abstract. This paper offers a proof-of-concept system for patient drug
prescriptions, utilizing blockchain technology to minimize the sharing of
patient data between hospitals and pharmacies. Ensuring security and
privacy guarantees is crucial for adopting healthcare management systems, with specific data protection requirements mandated by regulations like the General Data Protection Regulation (GDPR). This work
emphasizes the drawbacks of traditional centralized electronic healthcare systems. To minimize data sharing while maintaining patient data
confidentiality, we leverage the Hyperledger Fabric blockchain to establish private data channels and incorporate client identity query features.
Our approach demonstrates effective monitoring, prevention, and management of potential disasters arising from centralized systems, which
pose a single point of failure risk for servers housing sensitive patient
data and can compromise the confidentiality of such data.

Keywords: Blockchain Technologies · Data Management · Disaster
Management · Decentralized Systems · Drug Prescriptions ·
Information Sharing

1 Introduction

Blockchain has not been limited to ordinary concepts; it is now transformed
into a blend of numerous replicas, like networking, mathematics, cryptography,
and distributed consensus algorithms. Inherently in the blockchain, some blocks
hold certain information depending upon the type of data one will store in
these blocks. Thus, these blocks make blockchain the pool of proceedings [1].
Blockchain was developed from a Bitcoin paper published by Nakamoto in 2008.
It is a peer-to-peer network where all participants (peers) serve as a node, and

A. Rocha et al. (Eds.): WorldCIST 2024, LNNS 990, pp. 12–21, 2024.
https://doi.org/10.1007/978-3-031-60328-0_2

all the nodes hold the same information. In the existing research, many efforts have been intended to research and explore the adoption of blockchain technology in the healthcare sector to address the issues of patient healthcare data and its sharing among other entities [2]. State-of-the-art tools aim to digitize the healthcare sector so that innovation and automation should be included in the presently opted systems, i.e., currently, healthcare and medical data are being stored in a centralized server, and blockchain technology is one of those innovative technologies disrupting the world. Apart from that, there should be mechanisms where patients control their health information instead of storing them on the organization's servers (i.e., already using centralized servers), which can be hacked. The current research is on how healthcare information is saved, used, and forwarded to any organization/institution. Identity management and confidential data protection will be crucial in blockchain technology applications [3]. In Fig. 1, we have elaborated on the centralized systems currently being used where a patient shares her sensitive data with the hospital (doctors), and the doctor shares that sensitive data and prescriptions with the pharmacy.

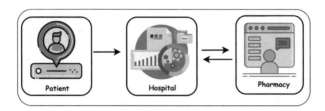

Fig. 1. The traditional patient drug prescription system having a centralized architecture.

2   Our Contributions

During the COVID-19 times, many solutions have been developed to monitor and diagnose patients remotely, but still, there have been serious concerns from the patients about their confidential data exposure via online software systems. Various techniques have been used to confirm the confidentiality of the patient data. Still, a big challenge is building a system or drafting a design that can ascertain that patient's data is being used according to the consent and that their confidential details have not been shared with unidentified parties. In this work, we contributed in the following ways:

- We study the implications of private blockchain to only allow known entities to the network where the patient-sensitive data is shared in a minimum amount.
- We validated hypothetical findings by writing chain codes that have demonstrated the conformance of confidentiality of the patient data. The detailed

code for this implementation can be seen here: Blockchain-based-Patient-Drug-Prescription-System
- We discovered the major challenges while deploying blockchain-based patient data analysis and drug prescriptions and gave their practical solutions using the built-in features of Hyperledger Fabric blockchain.

3 State of the Art Related Studies

Healthcare and medical data are currently stored in centralized servers, posing challenges to seamless sharing among public healthcare organizations and hospitals. The limitations stem from the framework of these centralized systems. Blockchain technology offers a promising solution for sectors handling sensitive data, providing enhanced security, transparency, and immutability [4]. Issues with patient health information exchange persist, with hospitals and government institutions holding patient data without explicit consent. Authors in [5] emphasize the benefits of blockchain in healthcare, highlighting its endorsement and security for patient information. The immutable and transparent nature of blockchain addresses the growing concerns surrounding patient data integrity. In contrast to traditional data storage methods, researchers explore the potential of blockchain technologies.

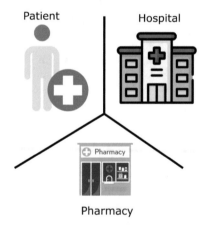

Fig. 2. Typical actors in a drug prescription system.

For instance, in another research, authors discuss patient data storage on a permissioned blockchain, introducing the Public Healthcare Data Gateway (HDG) [6]. Researchers explore permission-less blockchain applications for storing personal health records, while some advocate for private blockchain technology, exemplified by the MedChain system, to securely share medical data among entities [7]. The authors stress the importance of automated and secure public healthcare systems, particularly in pandemics. The need for systems replacing physical human movement is evident, with a call for fully automated public healthcare implementations [8]. Suggestions include enabling hospital staff, including doctors and nurses, to work securely from home using blockchain's security, privacy, and decentralization features [9]. Private blockchain technology with known nodes is favored for developing enterprise blockchain-based public healthcare and record-keeping systems, enhancing security and privacy during the pandemic. Blockchain's increasing attention during the pandemic is notable, with discussions on its potential to transform public healthcare and medicine sectors through features like immutability, decentralization, and time stamping [10]. Advancements in IoT, sensors, and actuators drive the shift of the public healthcare sector towards a virtual mode, streamlining operations without physical presence. The sheer volume of healthcare data underscores the pivotal role of blockchain in efficiently managing and maintaining this information.

4   Proposed Solution

This section will precisely illustrate our proposed solution along with the involved actors, their interactions, and implementations. The depiction of these actors can be seen in Fig. 2.

4.1   System Model

Our model essentially consists of three actors that are:

1. Patient
2. Hospital
3. Pharmacy

All three physical actors interact with each other to perform the required tasks. Typically, patients share their sensitive data with the hospital, including age, sex, genetic data, and disease symptoms. Then, doctors within a hospital can analyze data and carry out a diagnosis. Eventually, doctors will prescribe some drugs to the patients, and the pharmacy must be aware of it to deliver drugs correctly. The idea of this work is to share a minimum amount of data by the patient with the hospital (doctors) and the pharmacy so that the confidentiality of patient data can be guaranteed effectively.

4.2   System Model Bringing Blockchain into the Use Case

To tackle the issue of sharing patient-sensitive data with the Hospital and then with the Pharmacy, we have used the Hyperledger Fabric blockchain. From the intrinsic features of the Hyperledger Fabric blockchain perspective, we organized the use-case actors into the following organizations:

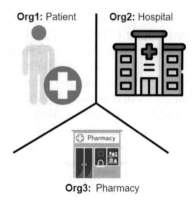

Fig. 3. Organizations listing on Hyperledger Fabric network.

1. Organization "Org1": represents the Patient who wants to share her data with the Hospital so doctors can perform medical evaluations.
2. Organization "Org2": represents the Hospital, and therefore the doctors, who have to access the Patient's data for assigning drugs.
3. Organization "Org3": represents the Pharmacy, which may access only the drug assignments data without accessing the rest of the patient data.
   Figure 3 shows three organizations hosted on the Hyperledger Fabric network.

Figure 3 shows that three organizations are hosted on the Hyperledger Fabric blockchain. Essentially, the patient is sharing her sensitive data, like age, gender, disease information, etc., with the hospital; the hospital is the organization where doctors will look into the information provided by the patient and then will share that evaluation and the final data to the pharmacy that the pharmacy organization may give those accurate drugs to that particular patient. However, a few challenges need to be addressed to perform the required operations effectively. We don't want to disclose sensitive information to the hospital and pharmacy. Instead, we want to share precise information so that doctors and pharmacies may do what is needed. Below are the few crucial challenges that we have tackled in this work:

– Prevent the Pharmacy from reading Patient's sensitive data while preserving the correctness of drug assignments.
– Disallow the Patient to assign drugs.
– Design a Blockchain based system that can welcome new patients.

We will precisely elaborate on each challenge solution and how we solved each using the intrinsic features of the Hyperledger Fabric blockchain.

5   Challenges (1/3)

5.1   Preventing the Pharmacy from Reading Patient's Sensitive Data While Preserving the Correctness of Drug Assignments:

To solve this challenge, we have utilized one of the essential functionalities of the Hyperledger Fabric platform, which is about having Private Data Collections. This can be obtained by using Hyperledger Fabric Private Data functionality. We have developed two private collections, *MedicalCollection* and *DrugsCollection*. In the "*MedicalCollection*," we have stored the sensitive data shared by the patient to organization 2, which is a hospital. The private collection *MedicalCollection* can only be accessed by organization 1, the patient, and organization 2, the hospital, where doctors check the provided patient information for drug prescription. Likewise, the second private data functionality we have developed, *DrugsCollection*, is used to store the results from the doctor's prescription after carefully inspecting the sensitive data sent by the organization of one patient. Apart from that, *DrugsCollection* private data collection is accessible to all three organizations, i.e., patient, hospital, and Pharmacy, only in read-only mode, and no one can make any changes to that collection. Notice that all three entities are hosted on the Hyperledger Fabric network. The detailed depiction of this solution and practical implementation can be seen in the screenshots of the code snippets in Fig. 4.

Fig. 4. Prevent the pharmacy to read patient's sensitive data, while still preserving the correctness of drugs assignment using Hyperledger Fabric private data functionality.

We built the collection *config.json* document such that we obtain the following private data collections:

Medical Collection:

- Stores sensitive data belonging to the Patient.
- Accessible to Org1 (Patient) and Org2 (Hospital).

Drugs Collection

– Stores the result of doctor's prescriptions, so the assigned drugs.
– Accessible to Org 1 (Patient) in read-only mode.
– Accessible to Org 2 (Hospital) in read-only mode. Accessible to Org 3 (Pharmacy) in read-only mode.

The details of these simulations and the source code available at (source code available at Blockchain based Patient Drug Prescription System, where especially the collection *config.json* file pertains the logic for tackling the above issue.

5.2   Disallow the Patient to Assign Drugs

This is one of the known issues in the security and privacy paradigms where actors may try to forge their data in various ways. In this case, a patient may act as a second organization and try to assign drugs in the light of her disease information. This process may end up with the wrong drug prescription that can be very harmful to the Patient, and secondly, it is also likely that the Patient may act as one of the adversaries to compromise the confidentiality and integrity of the entire Hyperledger network. To tackle this issue, we have used one of the intrinsic features of the Hyperledger Fabric Client: the Identity Query functionality provided by the API (Fig. 6).

Fig. 5. Disallowing the Patient to assign drugs by using Hyperledger Fabric Client Identity Query functionality provided by the API.

Here, we can assert that the requesting actor is managed by an allowed organization's membership service provider, MSP. So, we disallow the Org1 members (Patients) to assign drugs to themselves. The depictions of this solution can be seen in Fig. 5, while the code snippets to simulate this solution in Hyperledger Fabric blockchain are available in Fig. 7.

```
// verifyClientOrgMatchesPeerOrg checks that the client is from the same org as the
peer
async verifyClientOrgMatchesPeerOrg(ctx){
    const ClientMSPID = await ctx.clientIdentity.getMSPID();
    if (!ClientMSPID && ClientMSPID === '') {
        throw new Error ("Failed getting the client's MSPID.");
    }
    const peerMSPID = await ctx.stub.getMspID();
    if (!peerMSPID && peerMSPID === '') {
        throw new Error("Failed getting the peer's MSPID.");
    }
    if (ClientMSPID !== peerMSPID) {
        throw new Error(`Client from org ${ClientMSPID}`);
    }
}
```

Fig. 6. Disallowing the Org1 to assign drugs by using Hyperledger Fabric Client Identity Query functionality provided by the API.

5.3 Design a Blockchain Based System that Can Welcome New Patients:

One of the significant issues in the current blockchain ecosystems is their scalability and availability as the network grows exponentially. Our implementation stands good only if there is one Patient from Organization 1, and then that Patient interacts with the one hospital that is Organization 2, and consequently, that one hospital assesses the Patient's information and sends it to the pharmacy. But the question is, what if we need more patients, which is a real problem as any hospital may have thousands of patients simultaneously? So, we devised another workaround to accomplish this challenge by instantiating the chain code for each new Patient joining the organization 1. Each chain code must be named uniquely and match the Private Data Collection assigned to its Patient. Adding new patients is accomplished by instantiating the chain code for each new Patient, as shown in Fig. 7.

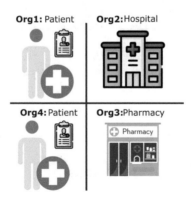

Fig. 7. Making blockchain scalable to add more patients.

```
const {Contract} = require('fabric-contract-api');
const MedicalCollection = 'medicalCollection0001'
const DrugsCollection = 'drugsCollection0001'
class HealthCare extends Contract {
    async InitLedger(ctx) {
    // creates empty ledger
    }
    // transaction invoked by patient
    async CreateMedicalData(ctx, id, age, weight, symptoms) {
    // check if client invoking the transaction belongs to
       // Patient Organization
    }
    // transaction invoked by hospital or patient
    async ReadMedicalData(ctx, id) {
    // check if client invoking the transaction belongs to
       // Hospital or Patient Organization
    }
    // transaction invoked by hospital
    async AssignDrugs(ctx, id, drugs) {
    // check if client invoking the transaction belongs to
       // Hospital organization
    }
    // transaction invoked by pharmacy
    async ReadDrugs(ctx, id) {
    // check if client invoking the transaction belongs to
       // Patient or Pharmacy or Patient Organization
    }
```

Fig. 8. The entire chain code including *CreateMedicalData* transaction, *ReadMedical-Data* transaction, requesting the transaction *AssignDrugs* and requesting the *Read-Drugs* transaction.

5.4   Hyperledger Fabric Network Topology and Configuration

1. Starting from the test network.
2. Add Org 3.
3. Have one peer per each Organization.

5.5   Overview and Workflow of the Chaincode:

- Patient belonging to Org1 requests the *CreateMedicalData* transaction.
- An asset is created inside *MedicalCollection*.
- Doctor belonging to Org2 reads the asset using *ReadMedicalData transaction*.
- Doctor belonging to Org2 prescribes some drugs to the patient by requesting the transaction *AssignDrugs*.
- An asset is created inside *DrugsCollection*.
- Pharmacist belonging to Org 3 requests the assigned drugs by requesting the *ReadDrugs* transaction.

A complete overview of the chain code used to solve the above issues can be seen in the code snippets in Fig. 8.

6   Conclusion and Future Works

This paper explores the application of Hyperledger Fabric Blockchain in managing sensitive patient data, aiming to minimize data sharing while ensuring effective communication between patients, doctors, and pharmacies. Our focus is on achieving a balance between providing doctors with the necessary information and safeguarding patient confidentiality. Through simulations, we decentralized all actors using Hyperledger Fabric, leveraging features like private data and client identity queries to address potential challenges and prevent disasters in patient data management. We assessed the risks associated with centralized data storage systems and proposed decentralized solutions to manage emergencies such as patient data leakage, ensuring confidentiality, output integrity, and overall system immutability, availability, and scalability through permission blockchains. Future implementations will incorporate privacy-preserving technologies like differential privacy and homomorphic encryption, allowing patients to share encrypted data, enhancing security, and preventing the extraction of unnecessary information from patient data.

Acknowledgments. This research was conducted with the financial support of Science Foundation Ireland under Grant Agreement Nos. [13/RC/2106_P2] and [20/SP/8955] at the ADAPT SFI Research Centre at Maynooth University. ADAPT, the SFI Research Centre for AI-Driven Digital Content Technology is funded by Science Foundation Ireland through the SFI Research Centres Programme.

References

1. Mozumder, M.A.I., Sumon, R.I., Uddin, S.M.I., Athar, A., Kim, H.C.: The metaverse for intelligent healthcare using XAI, blockchain, and immersive technology. In: 2023 IEEE International Conference on Metaverse Computing, Networking and Applications (MetaCom), pp. 612–616 (2023)
2. Gangothri, B.N., Satamraju, K.P., Malarkodi, B.: Sensor-based ambient healthcare architecture using blockchain and internet of things, pp. 543–546. Institute of Electrical and Electronics Engineers Inc. (2023)
3. Agarwal, A., Joshi, R., Arora, H., Kaushik, R.: Privacy and security of healthcare data in cloud based on the blockchain technology, pp. 87–92. Institute of Electrical and Electronics Engineers Inc. (2023)
4. Mukherjee, A., Halder, R., Chandra, J., Shrivastava, S.: HealthChain: a blockchain-aided federated healthcare management system. Institute of Electrical and Electronics Engineers Inc. (2023)
5. Ramar, K., Gopirajan, P.V., Shanmugasundaram, H., Andraju, B.P., Baskar, S.: Digital healthcare using blockchain, pp. 651–655. Institute of Electrical and Electronics Engineers Inc. (2022)
6. Global IT Research Institute: new cyber security risks for enterprise amidst COVID-19 pandemic. In: The 25th International Conference on Advanced Communications Technology: Phoenix Pyeongchang, Korea (South) (2023)
7. Marry, P., Yenumula, K., Katakam, A., Bollepally, A., Athaluri, A.: Blockchain based smart healthcare system, pp. 1480–1484. Institute of Electrical and Electronics Engineers Inc. (2023)
8. Iftikhar, M., Tahir, S., Alquayed, F., Hamid, B.: Blockchain and LDP Based Smart Healthcare IoT Applications: Architecture, Challenges and Future Works. Institute of Electrical and Electronics Engineers Inc. (2023)
9. Mukesh, V.D.K., Kiran, S.S., Rahul, K.A.U., Vanitha, V.: Blockchain in healthcare data. Institute of Electrical and Electronics Engineers Inc. (2023)
10. Baskar, S., Gopirajan, P.V.: Application of blockchain in digital healthcare, pp. 591–595. Institute of Electrical and Electronics Engineers Inc. (2023)

Oracles in Blockchain Architectures: A Literature Review on Their Implementation in Complex Multi-organizational Processes

Xavier Gutierrez[✉] and José Herrera

School of Systems Engineering and Computer Science, Universidad Nacional Mayor de San Marcos, Lima, Peru
{xavier.gutierrez,jherreraqu}@unmsm.edu.pe

Abstract. Blockchain is an innovative approach for recording and sharing data among storage nodes. Each node replicates an exact copy of the blockchain, but blockchains cannot trust external information without validation from their own nodes. To ensure integrity and security, information in the blockchain must come from trusted sources and be validated by nodes.

Oracles, conceived during Bitcoin's development, allows the implementation of complex processes that use external information to blockchains. However, using oracles poses challenges in accuracy, reliability, and timely updates. This literature review study examines oracle characteristics, their impact on trust in external information, their application in complex organizational processes, and proposed solutions. PRISMA guidelines were followed, selecting relevant articles for quality assessment.

Findings highlight the need for oracles from trusted sources, decentralized architectures, and data validation. The study concludes that proper use of oracles and development of specific techniques and protocols are crucial for data integrity and security in blockchains.

Keywords: Blockchain · Oracles · Interoperability

1 Introduction

The peculiarity of blockchains lies in their reliable internal state thanks to the process of shared and distributed computing that keeps them running. However, this also implies that blockchains cannot trust external information that does not come from their own network of nodes. In other words, to ensure the integrity and security of the information stored in the blockchain, it is necessary for all information to come from trusted sources and be validated by the blockchain nodes [3].

The idea that the blockchain might require external information was conceived from the early stages of the first blockchain proposed by S. Nakamoto. The pioneering developers of Bitcoin and its blockchain platform identified the need for escrow transactions, where a designated third party could decide whether the money involved in a transaction could be refunded or released upon the request of a chargeback. However, in practice,

Á. Rocha et al. (Eds.): WorldCIST 2024, LNNS 990, pp. 22–31, 2024.
https://doi.org/10.1007/978-3-031-60328-0_3

due to computational costs, the use of non-monetary transactions that occupied space in the blockchain was not well-received by the Bitcoin community, leading to the creation of alternative chains like Ethereum, designed to host smart contracts that could integrate with third parties [4].

A notable challenge is ensuring the accuracy and reliability of external information incorporated into the blockchain without compromising its security. This dilemma is known as the "Oracle problem". The solution to this issue is crucial to ensure the integrity and accuracy of the data stored in the blockchain and may require the implementation of specific techniques and protocols to validate and certify the authenticity of such information [3].

Validation oracles exist precisely to evaluate conditions that cannot be expressed within the blockchain and must be obtained from external servers. When a transaction depends on an external factor, the validation oracle is invoked, and the progress of the transaction is blocked until the condition regarding the external state is verified [5]. If the validation is automated, the arbiter must extract the variable from the smart contract's storage area. In cases where validation cannot be automated, a human arbiter can be called upon for approval [5].

Interorganizational processes require the exchange of data between the blockchain and external systems when multiple organizations collaborate. This is achieved using oracles, which are trusted entities that facilitate the collection, validation, and transmission of data from external sources, but despite the importance of oracles in designing and implementing blockchain use cases, research on their use does not address fundamental aspects as the origin of data, methods of integration into the blockchain, and types of data transfer [6].

In summary, oracles play a crucial role in the blockchain by enabling the incorporation of verified external information. However, challenges exist regarding the accuracy, reliability, and delays in updating external data in the blockchain. The proper use of oracles and the development of specific techniques and protocols are essential to ensure the integrity and security of the data stored in the blockchain.

2 Methodology

2.1 Systematic Review Design

The design of the systematic review followed the guidelines provided by the PRISMA method. In this study, the main objective is to identify the fundamental characteristics of oracles in Blockchain architectures necessary for successfully implementing complex processes involving multiple organizations. It analyzes the role of oracles as trusted intermediaries in providing external information to the blockchain, within the context of Blockchain technology. The study examines existing literature to identify the essential attributes that ensure the integrity, transparency, and veracity of the data provided by oracles. Furthermore, it explores proposed solutions to address the challenges associated with implementing complex processes in multi-organizational environments. By providing a comprehensive overview of best practices and leading approaches in the field of oracles in Blockchain architectures, the article aims to offer valuable guidance for future research and development of Blockchain-based systems. To achieve these objectives, four research questions were formulated, each addressing a different aspect of oracles in blockchain-based applications:

- *RQ1.* What are the essential characteristics that oracles in Blockchain architectures must have to facilitate the implementation of complex multi-organizational processes?
- *RQ2.* How do these characteristics affect the integrity, transparency, and accuracy of the external information provided to the Blockchain?
- *RQ3.* Are there significant differences in the characteristics of oracles based on the type of process or the nature of the organizations involved?
- *RQ4.* What are the most common challenges and proposed solutions in the literature for implementing complex processes in Blockchain architectures involving multiple organizations?

2.2 Identification of Sources

A comprehensive search was conducted in multiple academic databases, such as ACM Digital Library, Science Direct, and Scopus. Additionally, due to the scarcity of academic literature on the use of oracles in complex processes, gray literature was obtained from sources like Google Scholar and ResearchGate. In total, 2655 potential articles were initially obtained from these sources. For these databases, the research was conducted up until June 13, 2023.

The following query was created for each academic search system:

Blockchain AND oracles AND "smart contracts" AND (interoperability OR interface) AND (Inter-organizational OR "Cross-organizational" OR inter-company OR collaborative).

3 Results

In this section, we consolidate the results from the Systematic Literature Review (SLR) and our bibliometric analysis, each shedding light on various aspects of the blockchain oracles literature. Our results reveal the status, trends, and trajectory of this specific field of study, offering a comprehensive view of its time and geographic distribution. Subsequently, we delve deeper into the specific outlets, authors, and fields of analysis that have significantly contributed to this domain. Each research question posed in the SLR is addressed systematically in this section, with corresponding results presented respectively.

3.1 Types of Oracles

In this section, as shown in Table 1, we present an exploration into the versatile world of blockchain oracles, providing a detailed analysis of the distinct types and their specific roles. Oracles can be categorized based on their source of information, degree of centralization, and direction of information flow. These categories further branch out into various types like software and hardware oracles based on their information sources, or centralized and decentralized oracles depending on the degree of control involved.

Similarly, inbound, and outbound oracles are classified based on their information flow direction.

We delve deeper into specialized types of oracles such as feed oracles, sensor oracles, and human oracles that each hold a unique place within the oracle ecosystem. Other intriguing types include prediction oracles, consensus oracles, and aggregated oracles that use specific mechanisms for information verification. Furthermore, categories like voting oracles and reputation oracles highlight the democratic and trust-based elements within certain oracle structures. The expansive range of oracles reflects the adaptability and complexity of blockchain technology and its potential for integration with diverse external data sources.

Table 1. – Classification of Oracles

Type of oracle	Sub-type of oracle	Reference	Frequency
Source of information			
	Hardware	[8–13]	6
	Software	[10–13]	4
Degree of centralization			
	Centralized	[7–15]	9
	Decentralized	[4, 7–17]	12
	Consensus	[9, 12, 15]	3
	Voting	[11, 15, 16]	3
	Reputation	[9, 11, 13, 16]	4
Direction of Information Flow			
	Inbound	[8, 11–13, 16]	5
	Outbound	[8, 11–13, 16]	5
	Event	[3]	1
	Time	[18]	1
Special Types			
	Feed	[3, 4, 11, 19, 20]	5
	Sensor	[8, 11, 21]	3
	Human	[8, 9, 11–13, 15, 21]	7
	Prediction	[11, 14–16]	4
	Aggregated	[9, 15, 18]	3
	Other types	[12, 19]	2

3.2 Limitations of Oracles

In this section, we have identified various limitations inherent to blockchain oracles. A significant concern is their dependence on centralized sources, which potentially exposes them to risks of manipulation or censorship. Additionally, a lack of consensus within oracles may lead to disagreement on data validity, undermining the trust in the data source. Subjectivity and bias, particularly in human oracles and consensus oracles, also pose challenges to the accuracy and neutrality of the data provided. From a security standpoint, these oracles may be vulnerable to various threats, including external attacks, which could lead to errors or system failures. Furthermore, the operational costs associated with maintaining and running oracles, especially decentralized ones, can be considerably high. Finally, the lack of transparency in some oracles, especially those that rely on proprietary data sources or algorithms, can result in an inability to fully verify, or validate the data they provide. These challenges underscore the need for continued research and development in the field of blockchain oracles.

3.3 Strengths of Oracles

Our bibliographic review reveals several strengths of blockchain oracles, which contribute to their growing significance in the blockchain ecosystem. The ability to provide reliable information from multiple external sources stands as a primary advantage, promoting the use of real-time information and fostering the integration of the physical world into the digital realm of blockchain. The use of cryptographic techniques enhances the security of data transmission, ensuring data integrity and confidentiality. Furthermore, certain types of oracles can utilize expert knowledge and human judgment, adding a valuable dimension to decision-making processes. The inherent flexibility and adaptability of blockchain oracles allow them to accommodate a diverse array of use cases, from identity validation to smart contract execution. Moreover, the incorporation of consensus mechanisms in some oracle designs aids in mitigating the risk of misinformation and manipulation. These strengths highlight the integral role that blockchain oracles play in bridging the gap between blockchain technology and the real world.

4 Discussion

From classifying oracles based on source of information, degree of centralization, and direction of information flow to identifying their various strengths and limitations, our results present a multifaceted view of this rapidly evolving field. Our analysis also unearths crucial criteria for assessing the reliability of blockchain oracles, an area of growing significance considering the escalating integration of these systems into diverse blockchain applications. Moreover, we delve into potential strategies for improving the efficiency and security of oracles—a critical step towards enhancing the robustness and reliability of these indispensable conduits of information. As we traverse through these various dimensions, we hope to stimulate further scholarly inquiry and inspire potential practical applications of oracles in blockchain ecosystems.

4.1 Essential Characteristics of Blockchain Oracles for Facilitating Complex Multi-organizational Processes.

Oracles play a pivotal role in expanding the functionality of blockchain by providing trusted and reliable external information. The trustworthiness of these oracles, underpinned by the integrity and authenticity of the data they import to the system, forms the backbone of their usability [3, 11, 13].

Equally important is the security of oracles, necessitating robust mechanisms to verify the authenticity of the provided data and protect against tampering or manipulation [11]. Flexibility, another key attribute, allows oracles to adapt to different queries and data sources, ensuring smooth handling of a variety of external information [13].

In addition, for organizations aiming to easily include external data into their blockchain-focused operations, it's crucial that oracles are capable of seamlessly interfacing with a range of blockchain platforms and smart contract systems - this compatibility is a basic requirement [11, 13]. At the same time, scalability is paramount, enabling oracles to accommodate the increasing data volume and transactions inherent in complex multi-organizational processes [11].

Privacy considerations are essential, with oracles expected to prioritize data encryption, anonymization, and ensure sensitive information is not exposed to unauthorized parties [11]. Interoperability is another feature, facilitating data exchange and collaboration between different blockchain networks [11, 13].

Additionally, the development of a categorization framework is vital, enabling an assessment of oracle mechanisms' applicability and the understanding of their strengths and limitations [3]. Contextual considerations should also be factored in, given that the truth of the imported data can vary depending on the question's context [3].

The interaction of oracles with smart contracts necessitates the capability to send, fetch, and respond to data requests from smart contracts [11]. An analysis of existing blockchain oracles would also offer insights into their design, development requirements, and usage, guiding the creation of more efficient systems [11]. Lastly, computational efficiency is critical, with a need for minimizing computational costs and ensuring faster response times [11].

These characteristics, while not exhaustive, lay the groundwork for the design and implementation of oracles in blockchain architectures to support complex multi-organizational processes. However, the specifics may vary depending on the unique requirements and use-cases involved, underlining the need for further research and analysis.

4.2 Impact of Oracle Characteristics on Blockchain Information Integrity.

In terms of integrity, the quality of external information within the blockchain is highly contingent on the trustworthiness of the oracles [3]. The reliability of data stored within the blockchain is directly influenced by the level of trust attributed to an oracle [21]. Distinct oracle mechanisms offer varying levels of data quality and authenticity [3], and the utilization of trusted human verifiers may not always yield optimal results due to potential corruption or collusion [9]. Therefore, establishing a formalized trust model and an incentive structure can contribute significantly to ensuring data integrity on the blockchain [21].

Transparency in blockchain systems is largely facilitated by oracles. Acting as a bridge between the blockchain and the external world, oracles enable external data to be recorded in a transparent manner on the blockchain [3]. By nature, blockchain systems lack the capability to directly access data outside their networks [3], hence necessitating the role of oracles as trusted information sources. The transparent and consistent flow of data from oracles underpins the transparency of the entire blockchain system [3]. However, the underlying trust assumptions of any computational system integrating external information in a trust-free environment must be scrutinized [3].

Veracity, or the accuracy and reliability of external information provided by oracles, is a crucial factor in the effectiveness of blockchain systems. Oracles are responsible for capturing instances of external events [9], and the ability to convey accurate and trustworthy information may differ between various oracle mechanisms [3]. To assess the accuracy of information introduced to the blockchain, a robust framework that considers the nature and context of the queries is necessary [3]. The design of oracle mechanisms and models should aim to ensure a faithful representation of real-world events on the blockchain [9].

To sum up, the essential attributes of oracles directly influence the integrity, transparency, and veracity of the external data provided to the blockchain. Crucial elements including trustworthiness, structured trust models, clear information dissemination, and meticulous assessment of trust assumptions are all instrumental in upholding the integrity, transparency, and veracity of external data incorporated into blockchain systems.

4.3   Differential Oracle Characteristics: Process Type and Organization Influence.

Some authors underscore the necessity of categorizing oracle inquiries based on the spectrum of entities competent to respond [3]. They argue that current literature often lacks precision in detailing the capability of proposed oracle mechanisms, and generally, significant assumptions are made [3]. This contention points to the idea that the characteristics of oracles might differ depending on the kind of questions they are designed to answer.

Additionally, the same sources elaborate on the importance of scrutinizing trust assumptions inherent in any computational system, including blockchain oracles [3]. The authors highlight that all computational systems contain elements of trust at various stages such as hardware, network stack, and application level. This suggests that oracle characteristics may diverge based on the trust assumptions and levels related to different types of processes or the organizations involved.

Other authors provide a historical perspective of the development of blockchain oracles within Bitcoin [4] and the evolution of early oracle protocols, outlining the specific challenges encountered [4]. This indicates that the characteristics of oracles might be subject to change depending on the historical backdrop and technological progressions within various processes or organizations.

Although based on these sources, it can be inferred that oracle characteristics may indeed vary subject to the nature of the process, or the organizations involved. Nonetheless, further research is warranted for a more in-depth analysis to ascertain the specific variances in characteristics, this being possibly one of the most overlooked points in academic literature.

4.4 Challenges and Solutions in Deploying Complex Processes in Multi-organizational Blockchain Architectures.

One of the primary challenges is establishing trust and ensuring the reliability of oracles. The potential centralization of oracles exposes them to the risk of manipulation, which could lead to false or compromised information being introduced to the blockchain [11, 21]. In addition, blockchain oracles themselves can pose a risk to the decentralized nature of the blockchain ecosystem due to potential centralization, collusion, and Sybil attacks. It is therefore a significant design challenge to maintain decentralization and reliability, even when the number of oracles is limited [13].

Another challenge is scalability, as the processing and verification of massive data streams from multiple oracles can cause scalability issues for blockchain systems. Therefore, the challenge of scaling blockchain to handle large volumes of data from various oracles must be addressed [13]. Furthermore, oracles operating in external environments can trigger undesired events or compromise the performance objectives of decentralized applications, thus raising security issues. Enhancing the security of oracles to prevent unauthorized access or data manipulation is a vital challenge [13].

Various solutions have been proposed in the literature to address these challenges. Implementing mechanisms to ensure responsible behavior of oracles on the blockchain network can include integrating decentralized identity management and registration services to make oracles more accountable and reliable [13]. The reliability and decentralization of the system can be improved by enabling multiple oracles to report the same data streams or events based on on-chain consensus mechanisms, albeit this approach may increase the economic cost of the system [11].

Architectural enhancements are also proposed to bolster the security and credibility of oracles. For instance, secure triple-trusting architectures, secure elements for cryptographic software protection, and oracle reputation systems using smart contracts have been put forth [8]. These challenges and solutions underscore the importance of addressing concerns regarding trust, decentralization, scalability, and security when implementing complex processes in Blockchain architectures that involve multiple organizations using oracles. Further research and development are necessary to overcome these challenges and ensure the successful integration of oracles in complex Blockchain system.

5 Conclusions

In this research article, we conducted a comprehensive analysis of the literature on blockchain oracles. Through a systematic literature review and bibliometric analysis, we gained valuable insights into the status, trends, and trajectory of this field of study.

Our findings shed light on the various types of oracles, their strengths, limitations, and the criteria to assess their reliability. Additionally, we explored strategies to improve the efficiency and security of blockchain oracles.

The analysis revealed that blockchain oracles play a crucial role in bridging the gap between blockchain technology and the real world. Their ability to provide reliable information from multiple external sources, use cryptographic techniques for secure data transmission, and incorporate consensus mechanisms contributes to their growing significance in the blockchain ecosystem. Furthermore, the flexibility and adaptability of oracles make them suitable for diverse use cases, ranging from identity validation to smart contract execution.

However, we also identified several limitations inherent to blockchain oracles. These include their dependence on centralized sources, lack of transparency, potential security vulnerabilities, and high operational costs. Subjectivity and bias, as well as the lack of consensus among oracles, pose additional challenges to data accuracy and neutrality. These limitations emphasize the need for continued research and development in the field of blockchain oracles.

To assess the reliability of blockchain oracles, we identified key criteria such as cryptographic verification, reputation assessment, security evaluation, auditability, consensus of multiple oracles, and verification from multiple sources. Each of these criteria plays a crucial role in establishing trust and confidence in the data provided by oracles.

To improve the efficiency and security of blockchain oracles, we proposed several strategies. These include implementing robust verification and validation mechanisms, utilizing multiple oracles within a decentralized consensus protocol, employing advanced encryption and security technologies, establishing incentives for accurate data provision, conducting regular audits and monitoring, aligning with evolving security standards, defining service level agreements, and implementing backup and redundancy mechanisms.

In conclusion, our research provides a comprehensive understanding of the field of blockchain oracles. The findings highlight the strengths, limitations, and criteria for assessing their reliability, while also presenting strategies to enhance their efficiency and security. By addressing these aspects, researchers and practitioners can contribute to the advancement of blockchain oracles and their effective integration into various applications, fostering trust, and enabling the seamless flow of reliable data between the physical and digital realms.

References

1. Imran. B.: Mastering Blockchain, 3rd edn. Packt Publishing (2020)
2. Nakamoto, S.: Bitcoin: a peer-to-peer electronic cash system (2008). www.bitcoin.org
3. Bartholic, M., Laszka, A., Yamamoto, G., Burger, E.W.: A taxonomy of blockchain oracles: the truth depends on the question. In: IEEE International Conference on Blockchain and Cryptocurrency, ICBC 2022, Institute of Electrical and Electronics Engineers Inc. (2022). https://doi.org/10.1109/ICBC54727.2022.9805555
4. Caldarelli, G.: From Reality Keys to Oraclize. A Deep Dive into the History of Bitcoin Oracles (2023)

5. Xu, X., et al.: The blockchain as a software connector. In: Proceedings - 2016 13th Working IEEE/IFIP Conference on Software Architecture, WICSA 2016, Institute of Electrical and Electronics Engineers Inc., July 2016, pp. 182–191 (2016). https://doi.org/10.1109/WICSA. 2016.21

6. Mammadzada, K., Iqbal, M., Milani, F., García-Bañuelos, L., Matulevičius, R.: Blockchain Oracles: A Framework for Blockchain-Based Applications (2020). http://www.springer.com/series/7911

7. Sober, M., Scaffino, G., Spanring, C., Schulte, S.: A Voting-based Blockchain Interoperability Oracle (2021)

8. Al Sadawi, A., Hassan, M.S., Ndiaye, M.: On the integration of blockchain with IoT and the role of oracle in the combined system: the full picture. IEEE Access (2022). https://doi.org/10.1109/ACCESS.2022.3199007

9. Albizri, A., Appelbaum, D.: Trust but verify: yhe oracle paradox of blockchain smart contracts. J. Inf. Syst. **35**(2), 1–16 (2021). https://doi.org/10.2308/ISYS-19-024

10. Murimi, R.M., Wang, G.G.: On elastic incentives for blockchain oracles. J. Database Manage. **32**(1), 1–26 (2021). https://doi.org/10.4018/JDM.2021010101

11. Pasdar, A., Dong, Z., Lee, Y.C.: Blockchain oracle design patterns (2021). http://arxiv.org/abs/2106.09349

12. Beniiche, A.: A Study of Blockchain Oracles (2020). http://arxiv.org/abs/2004.07140

13. Al-Breiki, H., Rehman, M.H.U., Salah, K., Svetinovic, D.: Trustworthy blockchain oracles: review, comparison, and open research challenges. IEEE Access **8**, 85675–85685 (2020). https://doi.org/10.1109/ACCESS.2020.2992698

14. Nelaturu, K., et al.: On public crowdsource-based mechanisms for a decentralized blockchain oracle. IEEE Trans. Eng. Manag. **67**(4), 1444–1458 (2020). https://doi.org/10.1109/TEM.2020.2993673

15. Lo, S.K., Xu, X, Staples, M., Yao, L.: Reliability analysis for blockchain oracles. Comp. Electr. Eng. 83 (2020). https://doi.org/10.1016/j.compeleceng.2020.106582

16. Almi'ani, K., Lee, Y.C., Alrawashdeh, T., Pasdar, A.: Graph-based profiling of blockchain oracles. IEEE Access (2023). https://doi.org/10.1109/ACCESS.2023.3254535

17. Goswami, S., Danish, S.M., Zhang, K.: Towards a middleware design for efficient blockchain oracles selection (2022)

18. Tjiam, K., Wang, R., Chen, H., Liang, K.: Your smart contracts are not secure: investigating arbitrageurs and oracle manipulators in ethereum. In: CYSARM 2021 - Proceedings of the 3rd Workshop on Cyber-Security Arms Race, co-located with CCS 2021. Association for Computing Machinery, Inc, Nov. 2021, pp. 25–35 (2021). https://doi.org/10.1145/3474374.3486916

19. Laatikainen, G., Li, M., Abrahamsson, P.: A system-based view of blockchain governance. Inf. Softw. Technol. **157** (2023). https://doi.org/10.1016/j.infsof.2023.107149

20. Bekemeier, F.: Deceptive assurance? A conceptual view on systemic risk in decentralized finance (DeFi). In: ACM International Conference Proceeding Series. Association for Computing Machinery, Dec. 2021, pp. 76–87 (2021). https://doi.org/10.1145/3510487.3510499

21. Caldarelli, G.: Understanding the blockchain oracle problem: a call for action. Information (Switzerland) **11**(11), 1–19 (2020). https://doi.org/10.3390/info11110509

22. Caldarelli, G.: Formalizing Oracle Trust Models for blockchain-based business applications. An example from the supply chain sector (2022)

1st Workshop on Railway Operations, Modeling and Safety

Cost Effective Predictive Railway Track Maintenance

Sri Harikrishnan[1]([✉]), Verena Dorner[1]([✉]), and Shahrom Sohi[2]

[1] Institute of Digital Economics, Vienna University of Business and Economics,
Welthandelsplatz 1, 1020 Wien, Austria
{sri.harikrishnan,verena.dorner}@wu.ac.at
[2] Institute of Data, Process and Knowledge Management, Vienna University
of Business and Economics, Welthandelsplatz 1, 1020 Wien, Austria

Abstract. Efficient railway track maintenance is crucial due to the increasing demand for rail transportation. Reducing costs by optimizing maintenance procedures and operations is essential, and predictive maintenance approaches are necessary to achieve this. We summarized previous literature on track geometry and developed a predictive model for maintenance cost management. We test random forest and SVM predictions on a simulated dataset This information can assist railway companies in improving maintenance practices, planning costs, and enhancing network performance and efficiency.

Keywords: predictive maintenance · rail operations · data modelling

1 Introduction

In recent years, the railway industry has experienced a remarkable surge in passenger and cargo transportation, highlighting a consistent upward trend in railway usage [21]. This growth continued in 2021 and is expected to continue [29]. The increasing demand for railway services has created significant pressure on the railway network, driving the need for faster, heavier, and more resilient trains [3]. This expansion creates a substantial increase in resources allocated annually to construct and maintain railway tracks. Effectively utilizing these funds has become a formidable challenge, requiring the integration of advanced technologies into railway engineering [11].

In railway infrastructure, the track is critical for safe and efficient operations and is under stress due to increased demand. Inadequate track maintenance risks track quality deterioration and ultimately train derailments with severe consequences up to and including human casualties, environmental harm, damage to assets and economic losses [10]. Historical data shows that track defects are a leading cause of train accidents [23]. Maintaining continuous, reliable, and safe railway operations requires high-level track maintenance, albeit at substantial

This work was supported by Österreichische Bundesbahnen(ÖBB).

costs [28]. In the United States, over half of railroad maintenance expenditures are attributed to track maintenance [19]. Average unit costs and volumes for track renewals have been rising [20]. In 2018, railway maintenance expenditure in the European Union was estimated at 20.6 billion euros [24]. As a result, railway providers have developed a great interest in optimizing maintenance operations and costs. Our research aims to contribute to this area of research. Specifically, we have identified a need for more research on how critical track substructure geometry factors such as ballast fouling, drainage and sub-grade failures affect maintenance costs. We propose a prediction model to address this need and test it on a simulated dataset.

The goal of this research is to analyse the track geometry factors that have the most substantial impact on track maintenance costs and assess their magnitude. Additionally, the study aims to develop a data-driven predictive model to forecast future maintenance costs based on these identified geometry factors. Extracting valuable insights from these essential factors and implementing cost-effective strategies will enable railway companies to optimize maintenance practices and improve their future cost planning, ultimately enhancing performance and cost efficiency across their railway networks.

The paper is structured as follows: Sect. 2 presents the background and related work, while Sect. 3 presents our research question and describes the data analysis, Sect. 4 presents the results, and finally, Sect. 5 discusses our findings.

2 Theoretical Background

Ballast. Made of crushed hard rock materials like granite and basalt, is placed on top of the substructure layer and directly contacts the sleepers [14]. It provides the necessary support for the rail-fastener-tie track panel, reduces wheel/rail forces, facilitates surfacing and lining operations, ensures proper drainage, and offers resilience and damping for dynamic wheel/rail forces [14]. The ballast bed undergoes deformation, settling, and lateral flow during train service [9]. Ballast fouling happens when fine materials mix with fresh ballast under heavy train loads, causing a loss of efficiency in the ballast's function [12]. If contamination reaches the sleeper's bottom side, it can lead to the failure of the track substructure [9]. Maintenance is essential to restore the track to its original position when the railway track undergoes settlement beyond the allowable limits, as differential settlement can cause derailment [12].

Sub-ballast. A vital layer located between the ballast and the sub-grade. It is made up of graded gravel and sand and plays a crucial role in distributing loads, reducing stress on the sub-grade, and protecting against frost [14]. It also acts as a barrier that prevents sub-grade materials from mixing with the ballast and protects against abrasion [14]. The sub-ballast's primary function is to ensure proper drainage while directing water away from the sub-grade, which is critical to maintaining a well-protected and well-drained track foundation [14].

Sub-grade. The foundation of the railway track structure. It needs to be strong enough to handle heavy loads, especially with the introduction of heavier trains [14]. The sub-grade can be made of naturally occurring soil or man-made materials like railway embankments. It is critical to the satisfactory performance of a railway track and receives less stress compared to the layers of ballast and sub-ballast above it [14]. However, the uppermost zone experiences the most critical sub-grade stress [8]. Historically, the sub-grade has received little attention, and minimal efforts have been made to understand its characteristics [14,15].

2.1   The Effect of the Substructure on Maintenance Costs

Proper support of the track substructure is crucial for optimal track performance and impacts other components and passing vehicles. Strong resistance to plastic deformation and uniform elastic deflection characterizes good support, while soft and variable support causes damage to track components with each cycle of stress and strain [8,14].

Ballast Fouling. The accumulation of debris and contaminants in the ballast is a primary cause of track geometry degradation over time [8]. Ballast fouling can lead to a critical stage where the support of the track becomes problematic due to changes in pressure distribution beneath the sleepers [13]. Proper maintenance of the ballast layer is necessary to prevent sleeper cavities from developing and to avoid flotation of sleepers over the ballast layer.

Drainage Capability. Water is often part of a normal and functional track system. However, rainfall and ballast degradation can cause water to penetrate the track system, leading to reduced permeability and increased maintenance costs [14]. Excessive water can also reduce track stability, deteriorate track components, and transport harmful chemicals towards the sub-ballast [8]. Moisture in the layers of sub-ballast and sub-grade can improve their overall performance, but the fine particles produced by ballast degradation decrease void volume, trap moisture, and contribute to wear and tear [8,14].

Sub-grade Failures. Issues can be caused by load-related factors, soil-related factors and environmental factors including soil moisture and temperature [15]. Fine-grained soils are more likely to cause sub-grade problems due to their lower strength and reduced permeability [2]. High moisture content is a common issue that affects sub-grade performance. Moisture can infiltrate the sub-grade through various sources such as surface water and groundwater [14]. Groundwater becomes a significant problem if it is within approximately 6.1 m (20 ft) of the ground surface [14]. The freezing and thawing of the soil can also weaken the sub-grade [8].

Ballast fouling, drainage capabilities and sub-grade failures affect the track performance in significant ways. Research has revealed that track geometry deteriorates most rapidly in areas where fouled ballast and drainage problems coexist [14]. Poor drainage, fouled ballast, and sub-grade failures cause the most advanced deterioration. Fouled ballast sites experience the highest rates of track geometry degradation when water is retained due to a lack of drainage [14,27].

3   Research and Data Model

As mentioned above, in the realm of current literature, the influence of the substructure on maintenance costs has not received adequate attention. Our research aims to address this gap by formulating the following key research question:

- To which extent can geometry factors associated with the track substructure, specifically ballast fouling, drainage ability and sub-grade failures, predict cost maintenance costs for railway tracks?

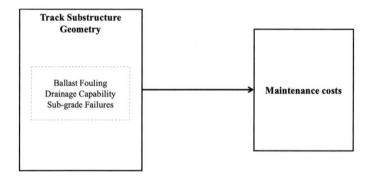

Fig. 1. Research Model

To construct our data model, we will treat ballast fouling, drainage capabilities, and sub-grade failures as independent variables, with maintenance costs as the dependent variable.

3.1   Data Simulation

Since access to real-life data on track geometry and maintenance costs proved not possible at this point, we simulated a dataset using insights gained from previous studies. The simulated data strikes a balance between real-world complexity and controllable parameters, making it a strong foundation for modelling and analysis. We quantified the relationships between the independent and dependent variables using real-world insights gained from the literature. Ballast fouling is usually measured using various methods including fouling indices and sieve analysis [14]. In our research, we simulate ballast fouling using a scale of 0 to 40, with 0 indicating fresh ballast and 40 indicating fouled ballast.

Drainage capabilities are usually measured by calculating a drainage coefficient based on field measurements, lab tests or mathematical models [14]. Due to a lack of published information on general distributions or properties of ballast draining capabilities, we use a uniform distribution between 0 and 10, with 0 indicating poor drainage.

Sub-grade failures are usually measured by visual inspections and numeric modelling [14]. We use a binomial distribution to model the presence (1) or absence (0) of sub-grade failures.

Based on previous studies, we assume the following correlation structure [14]. Ballast fouling has a negative correlation with drainage capabilities and a positive correlation with maintenance costs. Drainage capabilities have a negative correlation with sub-grade failures and maintenance costs. The presence of sub-grade failures has a positive correlation with ballast fouling and maintenance costs.

We simulated an 11-year dataset at 3-month intervals. We generated 100 replicates to provide a robust data set for analysis, simulating track geometry factors and maintenance costs for every 3-month interval. Maintenance costs were calculated based on the following mathematical model that factors in the interactions between track geometry and maintenance. We stored the data for each variable (ballast fouling, drainage, sub-grade failures, and maintenance costs) for each replicate, and aggregated it into separate data frames for each time interval.

We model maintenance costs as a linear combination of the three substructure factors:

$$
\begin{aligned}
\text{maintenance costs} = {} & \text{base cost} + k_1 \times \text{fouling} + k_2 \times \text{drainage} \\
& + k_3 \times \text{subgrade failures} + 0.02 \times \text{time component} \\
& + 0.5 \times \text{seasonal component} + \text{noise factor} \times \text{noise}
\end{aligned}
$$

We varied key parameters to encompass a broad spectrum of scenarios. For the coefficients determining relationships (k_1, k_2, k_3), we drew random values from normal distributions with means 55.0, 54.3, and 54.0, respectively, and standard deviations of 5.0. This introduced variability in the strength and direction of relationships across replicates. To simulate temporal trends, a time series component was introduced with a linear increase over time. Seasonal patterns were generated using a sine function based on the month of the year. Both components contributed to the dynamic nature of the dataset. For the introduction of random noise, we adopted a two-fold strategy. Firstly, to reflect the inherent uncertainty in time series trends, we added noise with a variable standard deviation. This variability was achieved by drawing standard deviation values from a normal distribution with a mean of 500 and a standard deviation of 50. Secondly, to account for the unpredictable nature of external influences on maintenance costs, we introduced additional random noise to the seasonal patterns.

This aimed to mimic the impact of unaccounted-for factors such as weather conditions and other external influences on maintenance costs. The noise was generated from a normal distribution with a mean of 0 and a standard deviation of 500.

3.2 Prediction Model

The second goal of our research is to test prediction models on the simulated dataset to see how well they predict maintenance costs under different conditions.

There have been various studies that review the applicability of the available methods for predicting railway track degradation, including traditional and machine learning-based methods such as probabilistic methods, Artificial Neural Networks, Linear Regression, Support Vector Machine and Grey Model [17,26].

We choose random forest and SVM. Random forest is an ensemble machine learning algorithm that uses decision trees as its base classifier. It solves classification and regression problems by combining multiple independent decision trees, using either voting or taking the average value [1,6]. Random forest has been applied successfully to several prediction problems in railway maintenance, such as predicting railcar remaining useful life [16], assessing the impact of extreme rainfall on railway disasters [18], and geo-defect forecasting [25].

SVM on the other hand, maps data into a high-dimensional space and separates them with wide decision boundaries [25]. SVM has been equally successfully applied in the railway domain, such as track geometry degradation forecasting [4], rail fastener problem inspection [5], railroad track condition monitoring [22] and in a geo-defect forecasting study [25].

Due to the highly variable and unpredictable nature of railway track data [8,14], several steps were taken to ensure that both models performed their best to both seen and unseen data. To identify the key contributors to maintenance costs, a feature importance analysis was performed using a random forest model. Features were assigned importance scores based on their contribution to predictive accuracy. To further enhance the predictive performance of the models, a meticulous process of hyperparameter tuning was executed. This involved utilizing a grid search approach and systematically exploring various combinations of hyperparameters to identify the configuration and combinations that maximize predictive accuracy. For the random forest model, parameters such as the number of trees, maximum depth, minimum samples split, and minimum samples leaf were optimized. Similarly, the SVM hyperparameters, the choice of kernel, regularization parameter (C), and gamma, were fine-tuned. The exhaustive nature of this search ensured that the models were fine-tuned and optimized for accurate predictions, enhancing their generalization capabilities to unseen data.

4   Results

For the random forest model, feature importance analysis was conducted to discern the significant contributors to maintenance costs. The selected features,

Table 1. Descriptive statistics.

	Ballast Fouling	Drainage Capabilities	Subgrade Failures	Maintenance Costs
count	4400.00	4400.00	4400.00	4400.00
mean	20.12	5.03	0.49	6410.41
std	14.26	3.58	0.50	766.97
min	0.00	0.00	0.00	5000
25%	7.29	1.7	0.00	5760.85
50%	20.53	5.076	0.00	6423.04
75%	33.39	8.27	1.00	7063.45
max	40.00	10.00	1.000	8350.74

Table 2. Correlation Matrix.

	Ballast Fouling	Drainage Capabilities	Sub-grade Failures	Maintenance Costs
Ballast Fouling	1.000000	−0.375187	0.025061	0.935413
Drainage Capabilities	−0.375187	1.000000	−0.264681	−0.137440
Sub-grade Failures	0.025061	−0.264681	1.000000	−0.002516
Maintenance Costs	0.935413	−0.137440	−0.002516	1.000000

including 'Ballast Fouling' and 'Drainage Capabilities' were then optimized through hyperparameter tuning. The SVM model underwent a similar process. Features deemed relevant, such as 'Ballast Fouling' and 'Drainage Capabilities,' were standardized and subjected to hyperparameter tuning. The in-sample performance of the models is summarized in Table 3. SVM produced a moderate MSE of 35,662.41 and an R-squared value of 0.78 for the in-sample results. While the R-squared value indicates that the SVM model provides a decent fit to the data, the random forest model performed exceptionally well, exhibiting an MSE of 45,104.62 and an R-squared value of 0.91 for the in-sample results. The results suggest that the random forest model accounts for approximately 91% of the variance, indicating a high degree of model accuracy. The out-of-sample results, presented in Table 3, reveal that the random forest model maintained a relatively lower MSE of 297,045.07 compared to SVM's 279,385.38. Although both models exhibit lower R-squared values on out-of-sample data, the random forest model's better performance in minimizing MSE suggests its better adaptability to unseen data.

Table 3. Data Analysis Results

	In-sample		Out-of-sample	
	Random Forest	SVM	Random Forest	SVM
MSE	45104.62	35662.41	297045.07	279385.38
R-squared	0.91	0.78	0.69	0.64

5   Discussion

Effective maintenance planning is essential for infrastructure managers looking to prioritize their spending. Track geometry measurements are often overlooked as a factor in maintenance costs, even when most maintenance efforts are spent on the track substructure, meaning that this area of the track should not be ignored. There is currently a lack of focus both in practice and in the literature on how track geometry factors related to the track substructure can affect future maintenance costs. By considering this issue more carefully, infrastructure managers can ensure that they are investing their maintenance budget in the most effective way possible. To test our research question, we simulated a dataset and tested two prediction models, SVM and random forest. Random forest performed better than SVM in predicting maintenance costs.

5.1   Limitations and Future Research

The main limitation of our study is the usage of a simulated dataset since we did not have access to a real-life dataset of track substructure data and maintenance costs. In simulating the dataset, we leveraged insights from the literature to create a realistic simulation, including time series trends to capture seasonalities and random noise. In the future, we plan to conduct sensitivity analyses by simulating many different scenarios. By quantifying the differences in data quality and relationships, we aim to develop a tool for railway companies that helps them decide which data to gather and which models to use. We also intend to enrich the dataset with supplementary data sources to improve predictions of maintenance costs.

The unpredictable nature of railway tracks can make it difficult for traditional supervised machine learning models to accurately predict maintenance needs. To overcome this clustering algorithms can be used to group tracks with similar attributes to identify latent patterns and categorize them into specific health states, leading to a more accurate understanding of track maintenance needs. Other mathematical and data science methodologies such as Bayesian models, Gaussian models and Markov models can also be explored to augment prediction capabilities based on this data.

5.2   Concluding Remarks

We found that there is little research on how railway track substructure data influences overall maintenance. This is a problem because the track substructure has a significant impact on maintenance costs, a third of the maintenance budget for ballasted tracks is allocated to the track substructure [7,8]. We address this research gap by implementing two models and testing their ability to predict future maintenance costs. In the future, there should be a more thorough focus on the track substructure in overall maintenance strategies and use to develop predictive models that lead to more reliable railway track operations.

References

1. Breiman, L.: Random forests. Mach. Learn. **45**, 5–32 (2001)
2. Chen, D., Trivedi, K.S.: Optimization for condition-based maintenance with semi-markov decision process. Reliab. Eng. Syst. Saf. (2005). https://doi.org/10.1016/j.ress.2004.11.001
3. Connolly, D.P., Alves Costa, P., Kouroussis, G., Galvin, P., Keith Woodward, P., Laghrouche, O.: Large scale international testing of railway ground vibrations across Europe. Soil Dyn. Earthquake Eng. **71**, 1–12 (2015)
4. Cárdenas-Gallo, I., Sarmiento, C.A., Morales, G.A., Bolívar, M.A., Akhavan-Tabatabaei, R.: An ensemble classifier to predict track geometry degradation. Reliab. Eng. Syst. Saf. (2017). https://doi.org/10.1016/j.ress.2016.12.012
5. Gibert, X., Patel, V.M., Chellappa, R.: Robust fastener detection for autonomous visual railway track inspection. In: 2015 IEEE Winter Conference on Applications of Computer Vision (2015). https://doi.org/10.1109/wacv.2015.98
6. Guo, G., Cui, X., Bowen, D.: Random-forest machine learning approach for high-speed railway track slab deformation identification using track-side vibration monitoring. Appl. Sci. **11**(11), 4756 (2021)
7. Indraratna, B., Heitor, A., Vindod, J.: Geotechnical Problems and Solutions – Practical Perspective. Taylor and Francis Group, London (2020)
8. Indraratna, B., Rujikiatkamjorn, C., Salim, W.: Advanced Rail Geotechnology – Ballasted Track, edn. 2nd. CRC Press (2023)
9. Ionescu, D.: Ballast degradation and measurement of ballast fouling. Null (2005). https://doi.org/null
10. Khajehei, H., Ahmadi, A., Soleimanmeigouni, I., Nissen, A.: Allocation of effective maintenance limit for railway track geometry. Struct. Infrastructure Eng. (2019). https://doi.org/10.1080/15732479.2019.1629464
11. Khouy, I.A., Larsson-Kråik, P.-O., Nissen, A., Kumar, U.: Cost-effective track geometry maintenance limits. Null (2016). https://doi.org/10.1177/0954409714542859
12. Kumara, J.J., Hayano, K.: Deformation characteristics of fresh and fouled ballasts subjected to tamping maintenance. Soils and Foundations (2016). https://doi.org/10.1016/j.sandf.2016.07.006
13. Köllő, S.A., Puskás, A., Köllő, G.: Influence of the maintenance work and support conditions on the optimization process of railway concrete sleepers. Procedia Eng. (2017). https://doi.org/10.1016/j.proeng.2017.02.364
14. Li, D., Hyslip, J., Sussmann, T., Chrismer, S.: Railway Geotechnics. Taylor & Francis Group, LLC (2016)
15. Li, D., Selig, E.T.: Evaluation of railway subgrade problems. Transp. Res. Rec. **1489**, 17 (2016)
16. Li, Z., Cats, O., He, Q.: Prediction of railcar remaining useful life by multiple data source fusion. IEEE Trans. Intell. Transp. Syst. (2015). https://doi.org/10.1109/tits.2015.2400424
17. Liao, Y., Han, L., Wang, H., Zhang, H.: Prediction models for railway track geometry degradation using machine learning methods: a review. Sensors **22**(19), 7275 (2022)
18. Liu, K., Wang, M., Zhou, T.: Increasing costs to Chinese railway infrastructure by extreme precipitation in a warmer world. Transp. Res. Part D: Transp. Environ. **93**, 102797 (2021)

19. Lopez-Pita, A., Teixeira, P.F., Casas, C., Bachiller, A., Ferreira, P.A.: Maintenance costs of high-speed lines in Europe state of the art. Transp. Res. Rec. **2043**(1), 13–19 (2008)
20. Network Rail. Cost benchmarking of network rail's maintenance and renewals expenditure: annual report - year 2 of control period 6. Technical report (2021)
21. Newman, P., Kenworthy, J., Glazebrook, G.: Peak car use and the rise of global rail: why this is happening and what it means for large and small cities. J. Transp. Technol. (2013)
22. Park, S., Inman, D.J., Lee, J.-J., Yun, C.B.: Piezoelectric sensor-based health monitoring of railroad tracks using a two-step support vector machine classifier. J. Infrastructure Syst. (2008). https://doi.org/10.1061/(asce)1076-0342(2008)
23. Peng, F., et al.: Track maintenance production team scheduling in railroad networks. Transp. Res. Part B: Methodol. (2012)
24. Sedghi, M., Kauppila, O., Bergquist, B., Vanhatalo, E., Kulahci, M.: A taxonomy of railway track maintenance planning and scheduling: a review and research trends. Reliab. Eng. Syst. Saf. **215**, 107827 (2021)
25. Sharma, S.S., Cui, Y.H., He, Q., Mohammadi, R.K., Li, Z.: Data-driven optimization of railway maintenance for track geometry. Transp. Res. Part C: Emerging Technol. **90**, 34–58 (2018). https://doi.org/10.1016/j.trc.2018.02.019
26. Soleimanmeigouni, I., Ahmadi, A., Kumar, U.: Track geometry degradation and maintenance modelling: a review. Proc. Instit. Mech. Eng. Part F: J. Rail Rapid Transit **232**(1), 73–102 (2018)
27. Sussmann, T.R.: Track geometry and deflection from unsprung mass acceleration data. In: Railroad Engineering 2007, London, UK (2007)
28. Xie, J., Huang, J., Zeng, C., Jiang, S.-H., Podlich, N.: Systematic literature review on data-driven models for predictive maintenance of railway track: implications in geotechnical engineering. Geosciences **10**(11), 425 (2020). https://doi.org/10.3390/geosciences10110425
29. Österreichische Bundesbahnen. Anzahl der fahrgäste der Österreichischen bundesbahnen (Öbb) von 2012 bis 2021 (in millionen) (2022)

3rd Workshop on Digital Marketing and Communication, Technologies, and Applications

The Impact of Using Digital Platforms and Sharing Online Experiences on the Reputation of a Company

Beatriz Pereira[1], Gabriela Brás[1], Elvira Vieira[1,3,2] (ID), Ana Pinto Borges[1,4(✉)] (ID), Bruno Miguel Vieira[1] (ID), and Manuel Fonseca[2,3]

[1] ISAG – European Business School and Research Center in Business Sciences and Tourism (CICET-FCVC), Porto, Portugal
{elvira.vieira,anaborges,bruno.miguel}@isag.pt
[2] Applied Management Research Unit (UNIAG), Instituto Politécnico de Bragança, Bragança, Portugal
[3] IPVC—Polytechnic Institute of Viana do Castelo, Viana do Castelo, Portugal
[4] Research Centre in Organizations, Markets and Industrial Management (COMEGI), Porto, Portugal

Abstract. The purpose of this paper is to evaluate the use of digital platforms and the sharing of online experiences concerning the sociodemographic profile of the consumer, the perception of the importance of the brand on digital platforms, and the significance of digital content in purchasing decisions. This article is pioneering for simultaneously addressing these three issues. The data is gathered using an online questionnaire distributed from April to May 2023. The analysis, including main component analysis and multiple linear regression models, was conducted using SPSS (version 21), AMOS (version 21), and STATA (version 14). The results indicate that age, level of education and the quality of digital content in the purchasing decision process are key elements in the use of digital platforms and sharing experiences. Companies that comprise the importance of these key elements can develop more effective marketing strategies, creating a positive online presence and cultivating lasting relationships with consumers.

Keywords: Digital Platforms · Consumer behavior · Digital marketing · Brand Building · Brand Reputation · Social media

1 Introduction

In recent years, the business landscape has undergone significant transformations, largely driven by the rise of digital marketing. In this context, building the brand and reputation of companies has become a strategic priority, directly influencing success in highly competitive markets.

The shift from traditional to digital marketing marks a paradigmatic change in brand-building strategies. Online presence has become crucial for companies as consumers seek information and interact with brands primarily through digital channels [1–3]. The

Á. Rocha et al. (Eds.): WorldCIST 2024, LNNS 990, pp. 47–57, 2024.
https://doi.org/10.1007/978-3-031-60328-0_5

integration of social media, SEO (Search Engine Optimization), and content marketing emerges as an essential triad for building and consolidating the digital presence of brands [4–6].

The central role of social networks in brand building is also noteworthy [7, 8]. Platforms such as Instagram, Facebook, and LinkedIn offer unique opportunities for companies to connect directly with their target audience, fostering engagement and building strong relationships. The sharing of relevant and authentic content becomes a crucial strategy for creating a positive brand image in the minds of consumers and for building online reputation.

Online reputation, often measured through customer reviews and evaluations, becomes a key indicator of the effectiveness of digital marketing strategies. Companies must actively monitor review platforms and social networks to understand and respond to consumer perceptions [9–11]. Proactive management of online reputation is vital to mitigate potential crises and strengthen consumer trust.

Due to the relevance and timeliness of the topic, this article analyzes the use of digital platforms and the sharing of online experiences regarding the sociodemographic profile of the consumer, the perception of the importance of the brand on digital platforms, and the significance of digital content in purchasing decisions. The article contributes to the literature by offering a consumer perspective.

The study adopts survey, encompassing sections on respondents' sociodemographic characteristics, assessment of brand presence and digital content importance and platforms' use using a 5-point Likert scale. To gather data, the questionnaire was distributed online through email and social media platforms from April to May 2023. The analysis employed main component analysis for the 10 statements on digital platform use, applying varimax rotation. Additionally, the Kaiser-Meyer-Olkin (KMO) test and Bartlett's test of sphericity were conducted to affirm the factorability and adequacy of the analysis. Econometric models, specifically multiple linear regression models, were executed using SPSS (version 21), AMOS (version 21), and STATA (version 14) to analyze the use and strengths of digital platforms and sharing experiences. Through the results, it will be possible to identify trends, explore integrated strategies, and highlight the importance of continuous adaptation by companies in the digital marketing landscape.

2   Literature Review

The emergence of digital platforms, the relationship between consumers, and the presence of brands online have undergone significant transformations. It is important to highlight the sociodemographic profile of users of digital platforms, the sharing of online experiences, the importance of digital content in purchasing decisions, and the contribution to brand reputation. Regarding the influence of sociodemographic profiles on the interaction and use of digital platforms, factors such as age, gender, level of education, and income can shape consumer preferences, affecting how consumers perceive and interact with brands online [12–14].

As for gender, the literature does not provide unanimous results. Some authors have observed that men are more likely to participate online than women [16], while more recent research found no gender differences in online participation [15].

In terms of age, it is argued that there is a curvilinear relationship between age and online participation [15]. Two effects are highlighted: life cycle effects and cohort effects [12]. As to life cycle effects, people go through various phases (e.g., child, student, worker, parent, retiree) and life experiences (e.g., raising children, working from home, etc.) that may stimulate their online participation and consequently the use of digital platforms [12, 17, 18]. Regarding cohort effects, especially those going through the same social scenario, the example of the "internet generation" consists of digital natives who are highly involved with technology and have a high level of online participation [17, 19]. However, merely being born into a technological world does not make the younger generation a homogeneous group of technology-savvy individuals, as adults may assume [20–22]. Furthermore, when considering socioeconomic status (education and income), it is observed that people with higher levels of education and higher income participate more online [20].

The growth of digital platforms and social networks has transformed how consumers interact, share experiences, and make purchasing decisions. In this context, the sharing of online experiences plays a crucial role in building a brand's reputation. Sharing online experiences refers to the practice of sharing experiences, opinions, and evaluations on digital platforms. Consumers increasingly trust the experiences of other users, considering this information more authentic than traditional marketing messages [23–25]. Sharing online experiences influences not only purchasing decisions but also the formation of virtual communities around common interests [26, 27].

Digital content, such as product reviews, blogs, videos, and social media posts, plays a central role in purchasing decisions. It is noted that most of the consumers research online before making a significant purchase [28–31]. The quality and authenticity of this digital content directly influence the consumer's perception of a product or service, affecting their purchasing choices [32]. Analyzing this digital content is crucial to understanding consumer preferences and adjusting marketing strategies.

A brand's reputation is largely built through online interactions. Positive sharing of experiences contributes to the construction of a strong reputation, while negative reviews can have detrimental effects [33]. Establishing a positive digital presence, proactively responding to consumer comments, and creating engaging content are effective strategies for shaping a brand's reputation [34, 35]. Companies that actively monitor online platforms, respond quickly to feedback, and cultivate positive relationships with consumers can create a high brand reputation. Therefore, it is relevant to assess the profile and perspective of the consumer in the context of the use of digital platforms and the sharing of online experiences in purchasing decisions and the development of the reputation of organizational brands.

3   Methodology and Data Analysis

Survey

To achieve our research goal, we designed a questionnaire comprising the following sections: i) the sociodemographic characteristics of the respondents (gender, age, marital status, academic qualifications, work status and monthly net income); ii) the importance given to the presence of brands on digital platforms and to digital contents in purchasing decisions, assessed on a 5-point Likert scale (1 - Not at all important, 2 - Slightly Important, 3 – Important, 4 - Fairly Important and 5 - Very Important); and iii) the level of agreement regarding the use of digital platforms (10 statements), evaluated on a 5-point Likert scale (1 - Completely disagree, 2 - Disagree, 3 - Neither agree or disagree, 4 - Agree and 5 - Completely agree) [37–40]. In the data collection process, the questionnaire was disseminated online (via email and social media platforms) between April and May of 2023.

Data analysis
In the evaluation of the 10 statements related to the use of digital platforms, we employed the main component analysis, with varimax rotation. The Kaiser-Meyer-Olkin (KMO) test of sampling adequacy and Bartlett's test of sphericity were usedto confirm the factorability and the adequacy of the analysis. For the econometric models, we ran a multiple linear regression model to explain the use and strengths of digital platforms and sharing experiences. For the analyses, we used SPSS (version 21), AMOS (version 21), and STATA (version 14) as software.

4   Results

From the description of the sample used for the present study, it is highlighted that the majority of the respondents are female, aged between 18 and 24, are single, have a higher level of education (at least a bachelor's degree), are active in the labor market (almost half working for others) and with an income level of up to 1000 EUR (Table 1).

When analyzing the statements concerning the level of agreement regarding the use of digital platforms (see table 2), we observed high average levels of agreement with the statements related to digital platforms being an easy-to-use tool (ES4), contributing to a more comprehensive perspective, a broader scope in research/information search (ES2), as they are easy to navigate (ES6), and consequently improve the quality of one's research (ES1). The lowest level of agreement, although above 3.6 out of 5, was registered in the statement "I am happy to share my good research experience with other people" (ES8). Also in Table 2, the outcomes of the factor analysis reveal that all statements exhibited significant loadings on their respective constructs, and no statement loaded onto more than one, reinforcing the independence of each construct. The constructs resulting from the data analysis were: 1- Use and strengths of digital platforms and 2 - Sharing experiences on digital platforms. The construct 1 explained 73.375% of the variance and construct 2 9.684%. These two factors measure perception and satisfaction regarding the use of digital platforms, especially in the context of research and user experience. The user experience on digital platforms is not just a secondary consideration, but a significant competitive differentiator for the reputation of brands.

Table 3 displays the two econometric models elucidating the use and strengths of digital platforms (Model 1) and the sharing experiences on these platforms (Model 2). In Model 1, it becomes evident that age emerges as a significant variable for the age range between 35 and 44 years old. This indicates a negative impact on the use and strengths of digital platforms compared to the younger age group (i.e. the interval between 18 and 24 years old). These results are in line with those found in the literature review which indicates that digital natives use digital platforms more [17, 19]. In relation to

Table 1. Sample description (n = 304)

Variable	Description	%
Gender	Female	62.2
	Male	37.8
Age	Between 18 and 24 years old	60.5
	Between 25 and 34 years old	14.1
	Between 35 and 44 years old	6.9
	Between 45 and 54 years old	13.2
	Superior to 55 years old	5.3
Marital status	Single	71.4
	Married/Civil union	24.0
	Divorced	4.6
Academic qualifications	Elementary studies	3.6
	Secondary studies	37.8
	Bachelor	46.1
	Master	11.5
	PhD	1.0
Work status	Unemployed	4.9
	Self-employed	10.5
	Paid employment	48.4
	Student/ student worker	36.2
Monthly net income (average)	Equal or Less than 530 EUR	35.2
	Between 531 EUR and 1000 EUR	30.3
	Between 1001 EUR and 1500 EUR	18.8
	Greater or equal than 15001 EUR	15.8

Source: Own elaboration. Note:* For the age variable, the average was applied

marital status, we observed that divorced and widowed respondents have a negative impact on the use of digital platforms compared to single women. As for the level of education, it is observed that it has a positive impact on the use of digital platforms in relation to respondents with a lower level. This result is integral with the argument that people with a higher level of education and income participate more online [20, 36] and sometimes use information more efficiently. Regarding the importance of digital content in the purchasing decision, we observed that as the degree of importance increases, the positive impact on the use of digital platforms also increases.

Table 2. Description of level of agreement regarding the use of digital platforms and Factorial Analysis

Statement	Loading* Use and strengths of digital platforms	Loading* Sharing experiences on digital platforms	Mean
ES1 - Using digital platforms improved the quality of my research	0.758		4.01
ES2 - The use of digital platforms contributed to a broader perspective of my research	0.748		4.05
ES3 - The use of digital platforms proves to be fundamental and useful, making my purchasing decision easier	0.774		3.98
ES4 - Digital platforms are an easy-to-use tool	0.891		4.11
ES5 - I easily find the information I'm looking for on digital platforms	0.847		4.00
ES6 - It's easy to navigate digital platforms	0,870		4.04
ES7 - On digital platforms, I can express satisfaction with my research experience		0.635	3.80
ES8 - I am happy to share my good research experience with others		0.872	3.64
ES9 - I like to help others by sharing my positive experiences		0.880	3.67
ES10 - I want to give others the opportunity to have a good research experience		0.858	3.77
Eigenvalues/Rotation Sums Squared Loadings	7.338	0.968	
Variance (%)	73.375	9.684	

(*continued*)

Table 2. (*continued*)

Statement	Loading* Use and strengths of digital platforms	Loading* Sharing experiences on digital platforms	Mean
KMO test	0.933		
Kaiser-Meyer-Olkin Measure of Sampling Adequacy (sig.)	3437.666 (0.000)		

Source: Own elaboration

In Model 2, focusing on sharing experiences on digital platforms, it's emphasized that the age range of 35 to 54 years old exhibits a negative impact compared to younger respondents, specifically those aged between 18 and 24 years old. Furthermore, respondents with a PhD are less likely to share experiences in the digital context. We also observed a positive effect of high levels of importance of digital content in the purchasing decision on sharing experiences on digital platforms. The link between online experience sharing and the influence of digital content on purchasing decisions for building and solidifying a brand's reputation is evident in the results of this model [37–40].

Table 3. Econometric models to explain the profile of use and strengths of digital platforms and sharing experiences on digital platforms

Variable	Model 1 – Use and strengths of digital platforms Coefficient	Model 2 – Sharing experiences on digital platforms Coefficient
Gender		
Female	-	-
Male	0.0601	-0.0692
Age		
Between 18 and 24 years old	-	-
Between 25 and 34 years old	-0.1552	-0.3194
Between 35 and 44 years old	-0.6132**	-0.7420**
Between 45 and 54 years old	-0.0990	-0.9936***
Superior to 55 years old	0.2058	-0.6159
Marital status		
Single	-	-
Married	0.3264	0.3181
Divorced or Widow	-0.8651**	-0.0918

(*continued*)

Table 3. (*continued*)

Variable	Model 1 – Use and strengths of digital platforms Coefficient	Model 2 – Sharing experiences on digital platforms Coefficient
Academic qualifications		
Elementary studies	-	-
Secondary studies	0.7195**	0.2825
Bachelor	0.9179***	0.0931
Master	1.0656***	0.2475
PhD	1.7445***	-1.4273**
Work status		
Unemployed	-	-
Self-employed	0.2385	-0.0064
Paid employment	0.1722	-0.2972
Student/ student worker	0.1036	-0.4052
Monthly net income (average)		
Equal or Less than 530 EUR	-	-
Between 531 EUR and 1000 EUR	-0.0911	-0.2034
Between 1001 EUR and 1500 EUR	-0.2500	-0.0459
Greater or equal than 15001 EUR	-0.3420	0.2502
Level of importance of brands' presence on digital platforms		
Not at all important	-	-
Slightly Important		
Important	0.4828	0.4525
Fairly Important	0.5893	0.0806
Very Important	1.1204	0.0767
Level of importance regarding digital content in the purchasing decision		
Not at all important	-	-
Slightly Important	1.2451**	-0.4255
Important	0.7393	0.4241
Fairly Important	0.8873**	0.7039

(*continued*)

Table 3. (*continued*)

Variable	Model 1 – Use and strengths of digital platforms Coefficient	Model 2 – Sharing experiences on digital platforms Coefficient
Very Important	1.0425**	0.6991***
Constant	2.5941***	0.3897***
Number of Observations	304	
R^2	0.2731	0.1699

Notes: significant at: * $p < 0.10$ level;** $p < 0.05$ level;*** $p < 0.01$
Source: Own elaboration

5 Conclusions

As technologies evolve and consumer expectations transform, in-depth understanding and effective application of digital marketing strategies become imperative for long-term business success. This article makes the connection, through consumer perception, between the sharing of online experiences and the influence of digital content on purchasing decisions by the use of digital platforms, and the construction and solidification of a brand's reputation. The results reveal that age, level of education and the quality of digital content in the purchasing decision process are key elements in the use of digital platforms and sharing experiences. Companies that embrace the importance of these elements can develop more effective marketing strategies, creating a positive online presence and cultivating lasting relationships with consumers. Consequently, they solidify and build a brand with a high reputation.

Constantly monitoring and adapting to emerging trends on digital platforms is crucial to success in today's dynamic and constantly evolving environment. Continuous adaptation and innovation in digital marketing strategies are essential to ensure the relevance and competitiveness of companies in the contemporary scenario.

The present investigation presents some limitations that could be overcome in future research. It is noteworthy that the study was not applied to any brand, it did not list the type of digital platforms used nor the level of frequency and purposes of their use.

Acknowledgment. This work was supported by national funds through FCT/MCTES (PID-DAC): UNIAG, UIDB/04752/2020 (DOI 10.54499/UIDB/04752/2020) and UIDP/04752/2020 (DOI 10.54499/UIDP/04752/2020).

References

1. Ryan, D.: Understanding digital marketing: A complete guide to engaging customers and implementing successful digital campaigns. Kogan Page Publishers (2020)
2. Hanlon, A.: Digital marketing: strategic planning & integration. Digital Marketing, 1–100 (2021)

3. Chaffey, D., Smith, P.R.: Digital marketing excellence: planning, optimizing and integrating online marketing. Taylor & Francis (2022)
4. Chen, J.C., Sénéchal, S.: The reciprocal relationship between search engine optimization (SEO) success and brand equity (BE): an analysis of SMEs. Eur. Bus. Rev. **35**(5), 860–873 (2023)
5. Meneses, G. D., Lasala, S.C.P., Amador-Marrero, M.: Search engine optimisation for social marketing. In: Effective Digital Marketing for Improving Society Behavior Toward DEI and SDGs, pp. 76–103. IGI Global (2024)
6. Terho, H., Mero, J., Siutla, L., Jaakkola, E.: Digital content marketing in business markets: Activities, consequences, and contingencies along the customer journey. Ind. Mark. Manage. **105**, 294–310 (2022)
7. Kaur, K., Kumar, P.: Social media: a blessing or a curse? voice of owners in the beauty and wellness industry. TQM J. **34**(5), 1039–1056 (2022)
8. DeLeon, J., Brown, L.W.: Understanding social media presence and financial success in digital competition. J. Strategy Manag. (2023)
9. Proserpio, D., Zervas, G.: Online reputation management: Estimating the impact of management responses on consumer reviews. Mark. Sci. **36**(5), 645–665 (2017)
10. Olaleye, S.A., Sanusi, I.T., Salo, J.: Sentiment analysis of social commerce: a harbinger of online reputation management. Int. J. Electron. Bus. **14**(2), 85–102 (2018)
11. Joglekar, J., Tan, C.S.: The impact of LinkedIn posts on employer brand perception and the mediating effects of employer attractiveness and corporate reputation. J. Adv. Manag. Res. **19**(4), 624–650 (2022)
12. van den Berg, A.C.: Participation in online platforms: examining variations in intention to participate across citizens from diverse sociodemographic groups. Perspect. Public Manag. Governance **4**(3), 259–276 (2019)
13. Secinaro, S., Brescia, V., Iannaci, D., Jonathan, G.M.: Does citizen involvement feed on digital platforms? Int. J. Public Adm. **45**(9), 708–725 (2022)
14. Kim, M.: A Study on the Effect of Participation Performance on Platform Media: Sociodemographic and Technology Acceptance Factors (2022)
15. den Berg, V., Annelieke, C., Giest, S.N., Groeneveld, S.M., Kraaij, W.: Inclusivity in online platforms: recruitment strategies for improving participation of diverse sociodemographic groups. Public Adm. Rev. **80**(6), 989–1000 (2020)
16. Ma, L., Zheng, Y.: Does e-government performance actually boost citizen use? Evidence from European countries. Public Manag. Rev. **20**(10), 1513–1532 (2018)
17. Thijssen, P., Van Dooren, W.: Going online. Does ICT enabled-participation engage the young in local governance? Local Government Stud. **42**(5). 842–62 (2016)
18. Nguyen, M.H., Hunsaker, A., Hargittai, E.: Older adults' online social engagement and social capital: the moderating role of Internet skills. Inf. Commun. Soc. **25**(7), 942–958 (2022)
19. Van den Berg, A.C., Giest, S.N., Groeneveld, S.M., Kraaij, W.: Inclusivity in online platforms: recruitment strategies for improving participation of diverse sociodemographic groups. Public Adm. Rev. **80**(6), 989–1000 (2020)
20. Kobul, M.K.: Socioeconomic status influences Turkish digital natives' internet use habitus. Behav. Inform. Technol. **42**(5), 624–642 (2023)
21. Dastane, O., Haba, H.F.: The landscape of digital natives research: a bibliometric and science mapping analysis. FIIB Bus. Rev. 23197145221137960 (2023)
22. Nguyen, T.X.H., et al.: Elderly people's adaptation to the evolving digital society: a case study in Vietnam. Soc. Sci. **11**(8), 324 (2022)
23. Cheung, C.M., Lee, M.K., Rabjohn, N.: The impact of electronic word-of-mouth: The adoption of online opinions in online customer communities. Internet Res. **18**(3), 229–247 (2008)

24. Ismagilova, E., Dwivedi, Y.K., Slade, E., Williams, M.D.: Electronic word of mouth (eWOM) in the marketing context: A state of the art analysis and future directions (2017)
25. Camilleri, M.A., Filieri, R.: Customer satisfaction and loyalty with online consumer reviews: Factors affecting revisit intentions. Int. J. Hosp. Manag. **114**, 103575 (2023)
26. Onofrei, G., Filieri, R., Kennedy, L.: Social media interactions, purchase intention, and behavioural engagement: the mediating role of source and content factors. J. Bus. Res. **142**, 100–112 (2022)
27. McWilliam, G.: Building stronger brands through online communities. MIT Sloan Manag. Rev. **41**(3), 43 (2000)
28. Yaylı, A., Bayram, M.: E-WOM: the effects of online consumer reviews on purchasing decisions. Inter. J. Internet Marketing Advert. **7**(1), 51–64 (2012)
29. Kudeshia, C., Kumar, A.: Social eWOM: does it affect the brand attitude and purchase intention of brands? Manag. Res. Rev. **40**(3), 310–330 (2017)
30. Voramontri, D., Klieb, L.: Impact of social media on consumer behaviour. Inter. J. Inform. Decision Sci. **11**(3), 209–233 (2019)
31. Kautish, P., Khare, A.: Antecedents of sustainable fashion apparel purchase behavior. J. Consum. Mark. **39**(5), 475–487 (2022)
32. Jin, X.L., Chen, X., Zhou, Z.: The impact of cover image authenticity and aesthetics on users' product-knowing and content-reading willingness in social shopping community. Int. J. Inf. Manage. **62**, 102428 (2022)
33. Sun, X., Zhang, Y., Feng, J.: Impact of online information on the pricing and profits of firms with different levels of brand reputation. Inform. Manag., 103882 (2023)
34. De Maeyer, P.: Impact of online consumer reviews on sales and price strategies: a review and directions for future research. J. Product & Brand Manag. **21**(2), 132–139 (2012)
35. Srivastava, M., Sivaramakrishnan, S.: The impact of eWOM on consumer brand engagement. Mark. Intell. Plan. **39**(3), 469–484 (2021)
36. Özsoy, D., Akbulut, E., Atılgan, S.S., Muschert, G.W.: Determinants of digital skills in Northeast Anatolia Turkey. J. Multicultural Discourses **15**(2), 148–164 (2020)
37. Liu, H., Shaalan, A., Jayawardhena, C.: The impact of electronic word-of-mouth (eWOM) on consumer behaviours. SAGE Handbook Digital Market. 136 (2022)
38. Bhaiswar, R., Meenakshi, N., Chawla, D.: Evolution of electronic word of mouth: A systematic literature review using bibliometric analysis of 20 years (2000–2020). FIIB Business Review **10**(3), 215–231 (2021)
39. Moradi, M., Zihagh, F.: A meta-analysis of the elaboration likelihood model in the electronic word of mouth literature. Int. J. Consum. Stud. **46**(5), 1900–1918 (2022)
40. Iqbal, J., Malik, M., Yousaf, S., Yaqub, R.M.S.: Brand reputation, brand experience, and electronic word of mouth toward smartphone: investigating the mediating role of brand love. J. Public Aff. **21**(3), e2455 (2021)

Activating a Brand Through Digital Marketing: The Case of 'Os Bonitos'

Sara Rocha[1] and Alexandra Leandro[1,2,3](✉)

[1] Coimbra Education School, Polytechnic Institute of Coimbra, Rua Dom João III - Solum, 3030-329 Coimbra, Portugal
{sfrocha,aleandro}@esec.pt
[2] CEOS.PP Coimbra, Polytechnic Institute of Coimbra, Bencanta, 3045-601 Coimbra, Portugal
[3] CECS UMinho, Campus de Gualtar, 4710-057 Braga, Portugal

Abstract. With the emergence of the internet and the growth of social networks, consumers began to use them to communicate with each other spreading ideas and sharing information. Therefore, this has impacted the communication between brands and their consumers, who recognized the opportunity to use them to promote their products, establish closer relationships with consumers, add value to the brand, and increase awareness. So, brands are using digital media as a marketing and communication strategy, to engage consumers and provide greater diversity of brand experiences. Thus, social networks are considered empowering of a brand's digital activation, as they promote a more human customer relationship, increase recognition, notoriety, and loyalty, in addition to being cost-effective, easy to use, and intuitive. This study aims to understand how the brand 'Os Bonitos' used digital media as a brand activation strategy. Specifically, it is intended to understand the objectives and practices of brand activation, as well as understand the relationship between the brand's content and follower interaction. To this end, a case study was conducted using a qualitative approach. Data collection was done through a semi-structured interview with the brand's founder and an analysis of the brand's social media, particularly its Instagram. Document analysis and content analysis of its posts were conducted. The results obtained from the semi-structured interview concluded that the brand activation of 'Os Bonitos' through digital media aligns with the stages and practices referenced in the literature review. On the other hand, with the results obtained from the analysis of the brand's Instagram, it was understood how the brand communicates with its consumers, namely, which post have the highest interaction and what is the correlation between the brand's content and the followers' interactions with it.

Keywords: Brand Activation · Social Networks · Digital Marketing · Communication · Consumer Experience

1 Introduction

Over the years, brand communication has evolved, with brand activation performing a deeper look at the brand's strategy and positioning [1]. Thus, brand activation is a communication strategy that aims to create contact and interaction between the brand

Á. Rocha et al. (Eds.): WorldCIST 2024, LNNS 990, pp. 58–67, 2024.
https://doi.org/10.1007/978-3-031-60328-0_6

and the consumer through experience, increasing satisfaction and creating and reinforcing the connection with the consumer [2–4]. As such, brand activation benefited from online communication, since its tool is interaction and experience with the consumer, these elements having been improved with the emergence of the internet [5]. With the exponential growth of social networks, consumers use social networks to disseminate ideas, obtain and share information [6]. Likewise, brands realized the opportunity to use them as a means of communication [7]. In this way, through social networks, brands can create valuable and original content, in a coherent and constant way. Furthermore, they can create communities of consumers, establish close relationships with them, as well as promote their products, add value to the brand and increase awareness [8]. Therefore, it is important for digital to be a fundamental ally in the process of activating a brand. As such, the brand's digital activation is one of the prominent scenarios that reaches a wider and more diverse target audience, focusing on its creative communication [9]. The general objective of this case is to understand how 'Os Bonitos' brand used digital marketing as a brand activation tool. 'Os Bonitos' is a brand of Portuguese ceramics and other gifts, founded in 2019 in Coimbra. Although the brand has a physical point of sale, it actively communicates through its social networks, which have been fundamental to its growth. A case study was conducted, and a qualitative methodology was used for its development. Collecting information is essential and, therefore, a semi-structured interview with the founder and responsible for brand communication, Filipa Santos, was fulfilled. In addition to this, the brand's social networks were also analyzed, namely its Instagram, through a document and content analyses.

2   Literature Review

2.1   Brand

Brands have become increasingly important, as they are considered an asset of organizations and has the vital role of differentiating themselves in the mind of the consumer, thus being one of the points of reference for consumers' choice decisions [10]. According to the American Marketing Association [11], a brand consists of a name, term, design, symbol, or any other characteristic that identifies the products or services of a seller or group of sellers and differentiates them from the competition. For Keller [12], whenever a marketer creates a new name, logo, or symbol for a new product, he creates a brand. This author refers to the brand as something that really created awareness, reputation, and prominence in the market, that is, something that resides in the minds of consumers. In this way, a brand is more than a product, as it can have dimensions that differentiate it in some way from other products designed to satisfy the same need [12]. These differences can be rational and tangible – related to the performance of the brand's product – or more symbolic, emotional, and intangible – related to what the brand represents [12].

2.1.1   Brand Management

According to Ruão [13], brand management involves analyzing the concepts of identity, communication, and associations with the brand. The first as the expressed meaning of the brand and its attitudes in the market, the second as the basic assumption of its existence and the third as the resulting perception in the minds of consumers [13]. For Aaker [14], brand management planning also involves monitoring the brand's identity, especially regarding differentiation. In conclusion, management can be activated in all relationships with consumers, from the first information unconsciously perceived by the consumer to sales and after-sales. Therefore, people interact all the time and create relationships with brands, and it is natural for consumers to become increasingly influenced by their management [10].

2.1.2 Brand Activation

Regarding the objective of communication, two types of communication can be considered, in global terms: commercial and corporate. Within the first we have product communication (product performances, objective characteristics) and brand communication (brand personality, the brand's imaginary territory defined by positioning) [15].

According to Morel et al. [1], brand activation is a deeper look at the brand strategy and its positioning, to find assets that have relevant consequences for the entire company. As such, a brand can be activated in different situations, summarized in four pillars: products and services, employees, identity, and communication [1]. Brand activation is defined as a marketing relationship created between the brand and consumers, with the aim of them better understanding it and considering it as an integral part of their lives [3]. It is the process of creatively activating customers in available communication sources, creating experiences with the brand, exposing its attributes [3] and emotionally connecting the customer to the brand [2]. It is also a communication strategy guided by contact and interaction between the brand and the consumer through experience, which increases their satisfaction with the brand and its products [2, 3]. The brand activation steps can be summarized in these five [3]: (1) Discovery: where marketers need to understand their brand, as well as their target audience's needs and wants and identify them into segments to create a process of better and more effective brand activation [3]. As such, the first step of a brand activation plan is to know your audience and their online and offline behavior, as what is transmitted in the digital context must be aligned with what is transmitted offline [16]; (2) Strategic development: it concerns the definition of the strategy, where the means of communication must be chosen and the resources that will be adopted must be identified [3]. Thus, media can be characterized by the brand's own channels, earned channels and paid channels. The first consist of communication channels or platforms owned by the brand and over which it has control, such as website, blogs, social networks, among others. Brand activation planning must be aligned in both online and offline forms of communication, to reach various points of contact with consumers [3, 17]; (3) Creative development: the focus in this phase is on the execution and creative side of brand activation. That is, several plans are developed and creative ways to execute the strategy are decided [3]. It is at this point that the communication content must be defined, which involves defining the creative strategies and executional elements of the advertisements that should ensure the desired responses from consumers [17]; (4) Implementation: according to pre-defined execution plans and creative forms, that is, the action is put into practice according to the plan that was previously developed [3]. Pratas and Brito [17] thus consider that the brand activation plan must be composed of various research, such as market studies, consumer trends, consumer behavior, potential customers, and permanent analysis of the competition; and finally, (5) Assessment: the last phase is evaluation. This determines whether the brand activation effort was effective or not, as well as whether the company's expectations were met [3]. According to the same authors [17], a campaign must go through a pre-evaluation, control of budget deviations, post-evaluations, and return-on-investment (ROI) analysis. This is the main metric for any marketing strategy and can be defined as a proportion between the investment in the campaign and the return or result obtained [17].

2.2 Digital Marketing

Digital marketing allows companies to reach many consumers and monitor their evolution, with the creation of content capturing fundamental attention, attracting, and retaining customers [18]. Faustino [19] defines digital marketing as the application of communication and marketing strategies to promote products or services through channels such as websites, blogs, social networks, mobile applications, and electronic devices. He also points out some advantages that digital marketing has over traditional marketing, such as target audience segments, real-time data analysis, lower and more assertive costs, interaction with the public and agility in implementing campaigns.

2.2.1 Social Media

According to Safko [20], the first part of the terminology "social media", that is, "social", relates to the instinctive need that humans need to connect with other humans. While the second part, "media", refers to the myriad of means one uses to realize these same connections with each other [20]. For Marques [21], social media, within the digital world, has grown very quickly, as it has benefited from great diversification, reaching diverse needs of different audiences. The same author considers that the advent of mobile and 4G was the perfect combination for exponential growth. Thus, social media or Web 2.0. is a wide range of Internet-based mobile services that allow users to participate in online exchanges, contribute user-created content or participate in online communities [22]. For Kotler and Keller [23], social media is a means where consumers share information through text, image, audio, and video, both with each other and with companies, and vice versa. In turn, Schivinski and Dabrowski [24] state that social media can be understood as a variety of digital sources of information that are created, initiated, circulated, and consumed by internet users as a way of educating each other. Others about products, brands, services, personalities, and problems.

2.3 Digital Brand Activation

Gendron [25] indicates that the effective activation of a brand inspires consumers to interact with it and, consequently, also leads consumers to interact and create relationships between themselves. As such, this cooperation creates value and awareness for the brand, boosting sales, as well as brand loyalty and loyalty. Brand activation can then include in your strategy: public relations; digital advertising; free product sampling; experiential marketing; and shopping marketing [25]. The experiences developed by brands as part of activation strategies are conducted in live or online events, which allow consumers to get a feel for the brand through engaging and interactive activities [26]. Still according to Saeed et al. [3], brand activation can also be developed through means such as television, the internet (such as social networks), newspapers or via mobile phone, and through these sources transmit the message and activate the brand. Through the digital medium there are countless opportunities for brand activation, as it allows the diversification of experiences, greater consumer involvement and generates brand value in consumers' daily lives. Digital platforms enable access to all digital activities at any time and place and, consequently, enable brands to interact with consumers through different forms of content in real time: photos, videos, texts, animations, among others [5].

3 Research Methodology

The case study method was used to understand and analyze the process of brand activation through digital media, as well as its way of being and acting with its consumers. Thus, the qualitative methodology will be used to develop this case study. The collection of relevant information is essential and, consequently, an interview was made to the brand's founder. The aim is also to collect relevant information from the digital media used by the brand. As such, a document analysis was developed of the content of their social media (Instagram), during a specific period.

The main question of this case study arises: in a world where companies and brands increasingly communicating through social networks and where people also filter what they want to see, how can a brand be activated, in this case the brand 'Os Bonitos', through digital media? That is, what digital communication strategies can be pinpointed to increase brand awareness among its consumers? On that account, the following specific objectives were defined: RO.1) know the objectives of activating the brand through digital media; RO.2) identify the activation practices of the brand through digital media; RO.3) identify which publications generate the most interaction with the brand's followers; RO.4) understand if there is any relationship between the content that the brand chooses to use and the interaction of followers in the brand's publications.

4 Analysis and Discussion of Results

The interview was conducted in person on May 4, 2023, and was recorded, with consent, for later transcription. To fulfill it, a script was prepared in accordance with the analysis dimensions addressed in the literature review, and which are also included in the analysis model previously presented. In other words, the analysis dimensions concern the brand activation stages identified by Saeed et al. [3]. Thus, the information obtained in the literature was essential for defining the topics to be addressed. With the interview categorization grid concluded, the next step was to understand what the interviewee thought about the role of brand activation through digital marketing. According to her, it is essential to define a strategy to activate the brand in the digital environment, which goes through several stages, which are in line with the brand activation stages described by Saeed et al. [3]. At each stage, various marketing tools are used to promote the brand's products and provide closer and more lasting contact with its consumers. In the following Table 1, significant answers are presented in relation to the categories established beforehand and to the specific research objective 1 (RO.1), as aforementioned.

Table 1. Semi-structured Interview Categorization Grid with answers, concerning the activation objectives of the brand through digital marketing

Analysis Dimension	Categories	Significant answers
Presence of brands in digital media	Disclosure	I consider it extremely important, because they are one of the main means of promoting the brand.
Brand Activation Role	Segmentation *Engagement*	It is faster access to reaching the target audience and creating a long-term relationship (…). We arrive faster than just promoting a page or website and people randomly arriving at our product. (…) through segmentation, personalized contact, and the connection we create with customers, we offer on social networks, (…), direct access to the page to come and discover our products, the space, and the website.
Brand Activation Objectives	Disclosure	We activated the brand in November 2019, so almost 4 years ago, when the brand went on sale (…). We opened first on social media, first on Instagram, then on Facebook and, later, the website.
	Loyalty Conversion	Mainly promoting the brand, that is, reaching people with the new brand 'Os Bonitos' (…). Make our products known and make them love them. It was our first way out onto the street. Subsequently, take the customer to try our products to create a closer and more lasting connection with the customer and lead them to purchase.

Other relevant information drawn from the interview with the founder are discussed below, in relation to Instagram's analysis and bibliographical context.

In general, the themes of content shared by the brand that received the most interaction with followers were publications related to Christmas and Easter, both of which fall within the main theme of the time framework. It was therefore noticed that the brand obtains greater interaction in publications on specific themes and commemorative dates, in line with the perspective of Afonso and Alvarez [16] who consider that these publications allow to diversify the content, increasing the reach and engagement of customers. Users, as the frequency of these publications means that consumers are in permanent contact with the brand.

To respond to the same research objective, which publication format and tone of publication generates greater interaction from the brand's followers was also assessed. It was found that reels, in both themes, are the type of publication that followers like most and that get the most interaction. In this way, these data are in line with the perspective of Marques [22] who considers that the creation of videos, in the main social media, is fundamental, since it is a format that provides a great reach and retains the attention of consumers. Likewise, Filipa Santos, in the interview, states that "the more reach our reels have, the more we understand what people like" (Filipa Santos), in other words, it is a format that the brand uses constantly and that captures attention of your followers.

Finally, to understand whether there are other indicators that influence or increase the interaction of followers with the brand's publications, the number of identifications in the publications and the number of hashtags were related to the categories that measure interaction. It was verified that there is no increase in interaction when external accounts are identified. Additionally, it was found that publications that use between 20 and 30 hashtags are those that generate the most interaction from followers. In short, through the categories "themes of shared content", "type of publication" and "tone of the publication", the brand defines what it intends to sell and which products, how to share and the best language to use. According to the brand's founder and Saeed et al. [3], in the strategic development phase of brand activation, it is essential to define the strategy and objectives, as well as the choice of means of communication, which meet the categories addressed. By relating the categories analyzed, the aim was to respond to the last research objective (RO.4). Based on their analysis, it was concluded that the content that received the most interaction, in general, is related to the presentation/demonstration of the product, its animation and creative photos. Through this content, it was verified that the brand transmits its message and puts its promise to the public into practice. That is, it makes your product known to the public, allowing them to see it from all angles and to understand and value the color and imagery behind it, as mentioned by the founder in the interview. It was realized that these contents shared by the brand, which relate to product and packaging, according to the interview, are "totally original, creative, that "come out of the box" and that, above all, attract the public's attention (…) where we show the reality of products and parts we sell" (Filipa Santos), also meeting the perspective of multiple authors [16, 19, 27] who argue that content should be the epicenter of the digital marketing strategy, involving the creation of interesting, creative, interactive and relevant content, in order to facilitate approach and access to the public, as well as influencing them to make the decision to buy. Finally, it was also verified that content that is published continuously generates greater interaction from followers, as well as content that is published less often, as it can generate novelty and expectation among followers.

In turn, there is an apparent relationship between interactive publications and the interaction of followers. Of the 247 publications, the brand shared only two interactive publications (giveaway and call-to-action), which received greater interaction, both in terms of likes, views, comments and saved than the average publication. These data are in line with the perspective of Faustino [19] who argues that the use of a call-to-action in publications, holding competitions and giveaways, creating Instagram 'stories' and creating videos more frequently, allows for an increase in interaction with publications.

Finally, it was possible to verify that publications with advertisements (66 of the 247), in general, obtain greater interaction from followers. Therefore, these data are in line with the perspective of the founder of the brand, as she considers it essential to advertise her products and her publications on social media. As such, these data are also in line with the perspective of Gouveia [28] who argues that designing ads on Facebook and Instagram are the most effective way to obtain greater reach and greater interaction on these platforms. In summary, it was possible to respond to all proposed research objectives. In Table 2, it is possible to find, in a summary form, the answers found for those objectives.

Table 2. Summary of answers to research objectives

---

1 – Know the activation objectives of the brand through digital media
The main objectives of brand activation are promotion, loyalty, and conversion

---

2 – Identify the activation practices of the brand through digital media
Discovery: first, a study was conducted on the product and the public and then, segmentation was carried out. **Strategic development**: objectives are defined (what to sell, what products, how to share and what they like most). Subsequently, the means of communication are chosen, which, in the case of the brand, are social networks and the website. Finally, a schedule and evaluation of the content must be carried out
Creative development: the message is defined, which involves making the product known, allowing the public to see it from all angles and to perceive and value the color and imagery behind them. The content is aligned with this message and, therefore, the brand communicates in a constant, interactive, and creative way, to capture the attention of its followers. To achieve this, your products are personalized, differentiating them from others. The use of psychological support is also considered an added value. **Implementation**: marketing channels and tools are used. Social networks (Facebook and Instagram) are the main center for exchanging content and interaction, allowing us to promote products, increase reach and create relationships with consumers. The website allows easy and quick access to products. Google Ads and Meta allow you to advertise on search engines and advertise products/content. E-mail marketing allows you to have more direct contact with the public and notify about new collections or promotions. Finally, personal tools to better edit content and improve user experience.
Assessment: is carried out through Instagram metrics and conversion into online and physical sales. A return-on-investment analysis is also conducted after the launch of advertisements or campaigns

---

3 – Identify which publications generate the most interaction with the brand's followers
The publications of temporal frame, namely Christmas and Easter, in video format (reels) and with mixed language were those that received the most interaction

---

4 – Understand if there is any relationship between the content that the brand chooses to use and the interaction of followers in the brand's publications
The content relating to the presentation/demonstration of the product, its creation, and creative photos, allowed for an increase in interaction. Likewise, publications that are interactive with the public and publications with advertisements also achieved greater interaction

---

5   Conclusions

Through the literature review, it was possible to understand that brand activation is a strategic communication action, which uses interaction and experience with the consumer as a tool [5]. To consolidate the literature review, a semi-structured interview was conducted with the founder and responsible for brand communication. This is aligned with the brand's positioning, which involves making the product known, allowing the public to see it from all angles and to understand and value the color and imagery behind them. The message is also in line with the content shared by the brand and, therefore, communicates in a constant, interactive, and creative way, to capture the attention of followers. In the implementation stage, various marketing channels and tools are used, including social networks, the website, Google Ads, Meta, E-goi and personal tools. Through analysis, it was found that social networks are the main center for exchanging content and interaction, allowing products to be promoted, increase reach, and create relationships with consumers. It was also found that Instagram is the most important social network for the brand, as it allows it to have greater reach. Finally, the evaluation stage is conducted using Instagram metrics, namely the number of likes, views, comments, and messages, as well as conversion into sales, both online and physical. A ROI analysis is also carried out after the launch of advertisements or campaigns. In addition to the semi-structured interview, a document analysis and content analysis of the brand's social networks, namely its Instagram account, were also conducted. Through this analysis it was possible to respond to RO.3. When analyzing the brands' publications, it was noticed that the themes preferred by the brand's followers are related to Christmas and Easter. It was also found that within these themes, publications in video format (reels) and using mixed language were those that generated the most interaction from followers. At last, when analyzing the brand's publications, it was noticed that there is a relationship between the content that the brand chooses to use and the interaction with it, since the content that was published frequently and that concerns the presentation/demonstration of the product, creation of even and creative photos, allowed an increase in interaction. However, content such as the creation of the product and its packaging, which were published less often, also received a lot of interaction. With this, it was concluded that content that is published continuously generates great interaction, as well as content that is published less often, as it generates novelty and expectation. Finally, it was also found that there is a relationship between publications that are interactive with the public and publications with advertisements since the use of these also allows for greater interaction on the part of followers.

Throughout this study, some limitations were encountered, such as the available literature on the specific topic of digital limitation. Not being able to listen to the brand's followers was also identified as a gap. Consequently, the authors suggest for future research, direct consultation methods with the brand's followers and monitor in-store and online sales following each type of post. However, the authors believe that this study offers significant contributions to the brand's communication and strategies. Finally, understanding the steps of digital activation allows brands to better identify and understand their target audience, as well as identify which digital channels are most effective in reaching them. It also helps to define goals and create content strategies, identify the necessary resources, better position the brand and establish metrics to evaluate the success of the strategies.

References

1. Morel, P., Preisler, P., Nystrom A.: Brand activation. Starsky Insight, 1–11 (2002)
2. Dissanayake, R., Gunawardane, N.: Brand activation: a review on conceptual and practical perspectives, vol. 14, no. 8, pp. 37–43 (2018)

3. Saeed, R., Zameer, H., Tufail, S., Ahmad, I.: Brand activation: a theoretical perspective. J. Mark. Consum. Res. **13**, 94–98 (2015)
4. Burnett, J., Hutton, R.B.: New consumers need new brands. J. Prod. Brand Manag. **16**(5), 342–347 (2007)
5. Dutra, R.: The Activation of the IKEA Brand in the digital environment. (Master's Thesis, University of Algarve). University of Algarve Repository (2021)
6. Dionísio, P., Rodrigues, J.-V., Faria, H., Canhoto, R., Nunes, R.C.: b-Mercator. Blended Marketing. Dom Quixote (2009)
7. Alves, G., Antunes, J.: New communication paradigm: social networks between brands and consumers. Viseu Polytechnic Institute (2015)
8. Rodrigues, S.A.: The impact of social networks on the relationship between brands and consumers. (Master's Dissertation, University Institute of Lisbon). Repository of the University Institute of Lisbon (2012)
9. Leal, B.D.A.: The Brand – Communication and Digital Activation. (Master's Dissertation, Porto Polytechnic Institute). Scientific Repository of the Porto Polytechnic Institute (2017)
10. Cordeiro, I.M.B.D.: Brand activation: an innovative approach to brand management? (Master's Dissertation, Universidade Católica Portuguesa). Institutional Repository of the Portuguese Catholic University (2015)
11. American Marketing Association. Marketing Definitions (2023). https://www.ama.org/the-definition-of-marketing-what-is-marketing/
12. Keller, K.: Strategic Brand Management: Building, Measuring, and Managing Brand Equity, 4th edn. Global Edition (2013)
13. Ruão, T.: Marcas e Identidades, 2ª Edição. Edições Húmus (2017)
14. Aaker, D.: Managing Brand Equity: Capitalizing on the Value of a Brand Name. The Free Press (1991)
15. Baynast, A., Lendrevie, J., Lévy, J., Dionísio, P., Rodrigues, V.: Mercator, Marketing in the Digital Age. Dom Quixote (2021)
16. Afonso, C., Alvarez, S.: Being Digital. How to create a striking online presence. House of Letters (2020)
17. Pratas, J.M. Brito, P.Q.: Distribution: Management of Points of Sale and Retail. Current Publisher (2019)
18. Silva, D., Pimenta, J., Pinheiro, M., Lopes, R.: Digital marketing: the power of social networks in selling educational products. J. Psychol. **15**(57), 579–582 (2021)
19. Faustino, P.: Digital Marketing in Practice, 6 edn. Highlighter (2019)
20. Safko, L., Brake, D.K.: The Social Media Bible: Tactics, Tools & Strategies for Business Success, 1st edn. Wiley, Hoboken (2009)
21. Marques, V.: Digital Marketing from A to Z. Digital 360 (2022)
22. Alves, C.M.S.A.: Marketing and Digital Communication Strategy on Social Networks for Hibiscus Restaurant. (Master's Dissertation, University of Minho). University of Minho Repository (2022)
23. Kotler, P., Keller, K.L.: Marketing Administration. Pearson Education (2007)
24. Schivinsk, B., Dabrowsk, D.: The effect of social media communication on consumer perceptions of brands. J. Market. Commun. 1–24 (2014)
25. Gendron, M.: From public relations to brand activation: integrating today's communications tools to move business forward. Glob. Bus. Organ. Excell. **36**(3), 6–13 (2017)
26. Justo, I.D.M.: Experiential Marketing in the SIC Brand Activation Strategy. (Master's Dissertation, Universidade Católica Portuguesa). Institutional Repository of the Portuguese Catholic University(2021)
27. Kotler, P., Kartajaya, H., Setiawan, I.: Marketing 4.0: Shift from Traditional to Digital. Current Publisher (2017)
28. Gouveia, M.: Digital Marketing – the complete guide. Reading Ideas (2022)

Social Marketing Importance
for the Sustainability of Third Sector
Organizations

Susana M. S. R. Fonseca[1]([✉]) [iD], Filipe A. P. Duarte[2] [iD], Ana Branca Carvalho[3] [iD],
Ana Guia[4] [iD], Maria José Madeira[2] [iD], and Geisa Machado[5]

[1] CI&DEI, Polytechnic Institute of Viseu, Viseu, Portugal
sfonseca@estgl.pt
[2] NECE-UBI (Universidade Da Beira Interior) Research Centre, Covilhã, Portugal
[3] CISeD, Polytechnic Institute of Viseu, Viseu, Portugal
[4] CERNAS, Polytechnic Institute of Viseu, Viseu, Portugal
[5] Polytechnic Institute of Viseu, Viseu, Portugal

Abstract. Currently, there is an increasing need for us to reinvent ourselves, to find solutions that allow us to respond to and address the needs of organizations, especially those in the third sector that do not have profit as their fundamental premise. However, this does not mean that they do not need to be sustainable. Therefore, we should consider managing these organizations in a more professional manner. Marketing, as a management tool, allows the development of strategies to achieve proposed objectives and improve the organization's image, attracting potential clients and thus promoting increased sales force or service diffusion. It is important to integrate this tool into the daily activities of organizations.

Social marketing aims to be a strategy that aligns the promotion of social causes with the values and principles advocated by organizations, benefiting both the organization and the cause. In addition to marketing, fundraising actions are necessary to promote the organization in society, taking it outside and thereby increasing donations.

This study seeks to understand what is meant by social marketing, fundraising, and to assess the receptivity within the population, as well as the applicability of the actions. It also aims to identify the profile of donors to social causes, verifying whether social marketing and fundraising have implications for donations and the image of third-sector organizations. To achieve this objective, a non-probabilistic convenience sample was used, and a survey was developed, resulting in 110 completed surveys within the time frame defined between February 1 and May 31, 2022.

Keywords: Social Marketing · Fundraising · Sustainability · Third Sector Organizations

Á. Rocha et al. (Eds.): WorldCIST 2024, LNNS 990, pp. 68–77, 2024.
https://doi.org/10.1007/978-3-031-60328-0_7

1   Introduction

Third Sector Organizations (TSOs) play a crucial role today in Portugal in providing care and intervening in social issues. The origin of these organizations dates to the 19th century, with a growing number up to the present day, experiencing exponential growth since the 1970s [1]. According to [2] and [3], the major challenge for TSOs is to develop a management structure adapted to their particularities. In other words, we must consider the reality of these institutions, viewing management as a social product rather than a market or state practice. For an organization to achieve positive results, it is necessary for the manager to use strategic means to plan and define goals, creating motivating circumstances to engage all resources in achieving objectives.

To ensure management more focused on the sustainability of TSOs, it is the manager's responsibility to increasingly consider marketing and fundraising and apply their principles to the organization. In the words of [4], in general terms, marketing is a tool that the manager can use to achieve the organization's objectives, focusing on customer satisfaction (value creation). On the other hand, we have fundraising, term referring to fundraising actions for organizations to develop, promote sustainability, and carry out their activities [5].

This article aims to reflect on the importance of using social marketing and fundraising for the sustainability of TSOs. Also intends to analyze the society's perception of these concepts and the added value in their use by these organizations. The article was structured into 5 parts, starting with the theoretical framework on TSO´s management, social marketing, and fundraising. Subsequently, the methodology is presented, followed by the presentation and discussion of the obtained results, concluding with the final considerations.

2   Theoretical Framework

The use of the term management dates to the 19th century. During this time, the need for companies to organize themselves to adopt the most effective way of managing their resources began to emerge [6]. This concept has been extensively studied by researchers and can take on different conceptions. According to [7], "management is a set of administrative practices implemented by the management of a company to achieve the objectives it has set" (p. 119). This definition implies that management is a set of methods that enable organizations to plan and make decisions to achieve predefined objectives.

For any organization to fulfill its function fully, the practice of management must be intrinsic to it. Each organization has its specificities, making the act of management a challenge that must be adapted to each one's contours. Taking example of TSOs, these organizations largely depend on volunteer work, donations, and societal involvement. Therefore, it is important for the manager to have a concern for the surrounding environment, always guided by ideals such as solidarity, public interest [8].

In the management of TSOs, it is imperative to understand that working in these organizations is often challenging, as it is indirectly linked to addressing current social issues. Thus, professionals must possess qualities and specificities to work in these organizations, where resource scarcity is often evident, relying on voluntary work to compensate for the lack of resources [9].

To attest to the difficulty of managing TSOs, [10] argue that the critical point in this management area is the premise that these organizations need donations to operate. To avoid detrimental management for the organization, there must be a focus on continuous improvement of management efficiency, requiring continuous planning of all actions to be developed [11]. Efficiency is associated with effectiveness, characterized by outlining action plans and consequently defining the mission, vision, and objectives of the organization [12]. Additionally, certain indicators must be considered when discussing TSOs management, such as the stakeholder indicator and the Materials Management indicator.

Regarding sustainability, it encompasses a focus on social change, environmental respect, and economic results that improve services and the effectiveness of social work [13]. The bases of sustainability, according to [14], are the environment, profit, and people, with [15] adding culture and space. For an organization to be fully sustainable, there must be harmony between economic-financial development, justice in society, and environmental preservation [16].

[17] outlines four basic characteristics for the sustainability of third-sector organizations: 1) making decisions that promote environmental protection/preservation; 2) seeking to develop the community where it is inserted; 3) focusing on strategic planning; 4) increasing sources of resources.

When it comes to the sustainability of TSOs, challenges arise due to changes in societal conditions, and organizations must adapt their sustainability practices accordingly [18]. Organizations need to seek sustainability by involving various financiers, including government, companies, foundations, partners, and generating their resources [19]. The sustainability of TSOs is seen as a strategic aspect, guiding the study of TSO management towards internal and external aspects [20]. Sustainability ensures a balance between economic-financial development, justice in society, and environmental protection [16].

In the current societal context, marketing plays a significant role, especially with the explosion of the Internet in the 1990s [21]. Marketing is crucial for the economic sustainability of organizations, and its scope has evolved over time. Initially focused on product segmentation and brand creation for customer loyalty, marketing has become a tool for achieving organizational objectives by creating customer value [4].

Marketing in organizations should consider both internal and external aspects. Internal marketing focuses on satisfying internal customers (employees) to motivate and stimulate them, ultimately achieving organizational objectives [10]. Understanding and addressing the needs and desires of internal customers is crucial before attending to external customers. Marketing is not a given in TSOs, and its practice is often neglected [23]. Hence, a clearer definition and implementation of marketing practices, including internal and external marketing, are essential for the economic sustainability of these organizations [24].

The concept of social marketing, focused on tying external clients to a cause or social plan, is challenging to define universally. For [25], social marketing arises from the need for organizations to connect external customers to a cause or social plan. Different perspectives define social marketing as strategically managing the process of social change, adopting new behaviors, attitudes, and practices, guided by ethical principles, human rights, and social equity [26]. We can consider that social marketing as the application of marketing principles to a social reality, while [27] defines it as a behavior change.

Social marketing is increasingly used by managers of social organizations to obtain resources for their organizations while fulfilling their missions [28]. Although often considered a given in TSOs, marketing, including social marketing, requires explicit attention and definition within these organizations [23].

In conclusion, the core of the social marketing concept revolves around changing behaviors and promoting social change. Organizations, especially TSOs, must leverage management tools, including social marketing, to contribute to the sustainability of their operations. These tools should be integral mechanisms that managers do not overlook.

3 Methodology

The present study was based on a qualitative and quantitative methodology. Regarding the research objective, it will be an explanatory investigation, and in terms of purpose, it will be a pure investigation. Analyzing the method, it is intended to use the case study is a method that focuses on describing, exploring, and understanding complex events influenced by various aspects. Therefore, due to these characteristics of the case study, we ultimately decided to use this study method to guide the research.

The methodology used in this article is based on a survey conducted with the population it is the community where an OTS is located, and the goal is to understand how fundraising can assist Social Marketing. Thus, this organization is an association of the faithful, constituted in canonical legal order, with the aim of meeting social needs and practicing Catholic worship acts, in harmony with its traditional spirit informed by the principles of Christian doctrine and morality, being recognized as a Private Institution of Social Solidarity.

To collect data, an open survey was conducted with the general community, where we sought to understand what is commonly understood by social marketing, fundraising, and to assess the receptivity of the population to these terms, as well as the applicability of actions. The survey was disseminated on social media for greater outreach. Respondents were individuals over 15 years old, and the same survey could be answered by anyone using social media in the period from February to May 2022. For this study, 110 surveys were collected.

4 Results and Discussion

The presentation and discussion of the results obtained begin with the characterization sociodemographic of the respondents. Out of the total respondents, 75% are female, and 25% are male. Regarding age groups, we can infer that most respondents were between 20 and 25 years old (29%). It was also noted that those who are older than 45 years are the ones with the fewest responses, possibly due to the survey being disseminated through email and social media.

In Fig. 1, when asked about their academic qualifications, it was verified that most respondents have a bachelor's degree, while a minority (around 7%) has lower academic degrees.

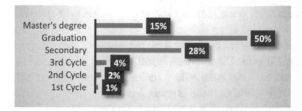

Fig. 1. Academic qualifications.

As mentioned in the theoretical framework, despite the concept of social marketing emerging only in the 1970s, in Fig. 2 most respondents are familiar with the concept (about 81%), while only 19% are not aware of it. According to [29], this concept was introduced by Kotler and Zaltman, who recognized that some principles of traditional marketing could be applied to changes in ideas or behaviors. Therefore, given that it is a relatively recent concept and there is still not much concern about it in the organizational world, it was expected that the majority would be unfamiliar with this concept.

Fig. 2. Knowledge of the Social Marketing Concept.

As previously mentioned, the aim of this study is to analyze the importance of social marketing and fundraising for the sustainability of OTSs. In this way, it was sought to understand the importance of these management tools for citizens. Table 1 shows that for most respondents (about 79%), they consider the application of marketing tools in OTSs to be important and very important. With the same objective, we asked about the importance of fundraising actions in OTSs, and the majority (54%) consider it very important for these organizations to carry out such actions, with only 5% considering this type of action not important. Based on [4], marketing is understood as a management tool that the manager cannot and should not neglect. By questioning the citizens, the majority consider it important for OTSs to apply marketing tools and principles in their management and daily activities. Also shows that most respondents also believe that organizations should invest in fundraising actions.

Continuing the analysis of Table 1, about 51% consider publicizing fundraising actions and the underlying cause to be very important. Also, for about 62% of respondents, knowing the purpose of the donations raised in the actions is very important. However, 5% do not consider it very important to have this transparency. For making donations, it is important to know the purpose of the donation so that the organization conveys a message of transparency and trust to the outside world. [30] supports this idea by stating that for fundraising to be truly valuable and important for OTSs, it is important that it is based on principles of transparency and coherence to foster donor loyalty and, on the other hand, attract new donors. Following this analysis, it was also asked about the importance of the presence of influential people in these actions. About 1% consider this factor not important at all, and only 36% consider it very important. Finally, regarding the importance of promoting actions on social media, we found that 50% of respondents consider it very important, which may be related to the fact that we live in a digital world where social media plays a fundamental role. [21] also notes that the exponential growth of marketing in the 90s was largely due to the explosion of the Internet. In this sense, it can correlate the importance that our respondents give to the advertising of fundraising actions, for example through social media, as a quick means of dissemination and access to many people.

Table 1. Degree of importance of social marketing/fundraising in Third Sector Organizations.

Results (Not Important to Very Important)					
Application of marketing tools	5%	8%	9%	34%	45%
Carrying out fundraising actions	5%	8%	8%	25%	54%
Advertising of actions to be developed	4%	8%	7%	30%	51%
Information about the purpose of donations	5%	6%	9%	18%	62%
Presence of influential people in fundraising actions	6%	8%	14%	35%	36%
Publicizing the action through social media	5%	7%	8%	30%	50%

Continuing the analysis by asking about the habit of making donations for fundraising actions, it observes (Fig. 3) that about 55% do not make this type of donation, while 45% contribute to fundraising actions for OTSs.

Fig. 3. Habit of Contributing with Donations to OTSs.

It is important to understand the reason why people do not contribute to fundraising actions. In this way, it is found that the majority who did not contribute is because they do not know the purpose of the fundraising action (about 55%), also citing reasons such as insufficient promotion, lack of transparency from OTSs, and not having resources, as constated in Fig. 4.

Next, it is sought to understand how long respondents have been supporting these organizations, and it found in Fig. 5 that about 70% have been supporting for more than 3 years, while about 15% have been supporting for less than 1 year. Also note that it is

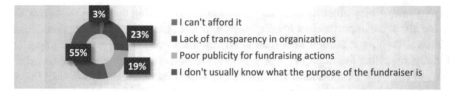

Fig. 4. Reasons for not contributing to fundraising.

during the Christmas season that they contribute the most (approximately 63%), which is likely since organizations promote fundraising actions more during this time.

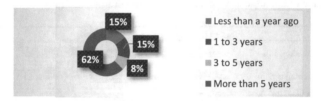

Fig. 5. For how long they have been supporting OTSs.

It is important to know the reasons that lead individuals to participate in fundraising, and in Fig. 6 verified that about 45% want to help others, especially those in greater need. They also cite empathy, solidarity, and the social cause itself as motivations.

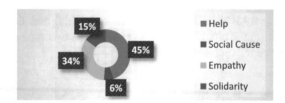

Fig. 6. Motivations/Reasons for Making Donations.

The next question aimed to understand the types of donations individuals are most likely to make to OTSs. The majority prefers to contribute with money (40%), while about 12% prefer volunteering for social causes. However, approximately 48% contribute with food and non-food items, including clothing and toys (Fig. 7). The main causes supported are those dedicated to children, with 50% of the support, and other causes such as animals, disability, elderly, cancer fight, and homelessness also receive support.

Finally, to gauge citizens' likelihood of contributing to social causes/fundraising actions in OTSs in the future, and in Table 2 the most important factor for people to contribute to the future is knowing the social cause (65%), closely followed by knowing the reason for fundraising (60%). What they give less importance to is related to the opinion of friends/family (15%) and factors related to disclosure and advertising. We conclude

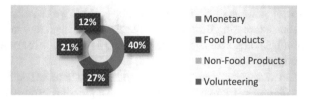

Fig. 7. Types of Donations.

from the data analysis that for making donations, knowledge of the cause/organization is more important than promotional/advertising actions.

Table 2. Probability of donating to OTSs.

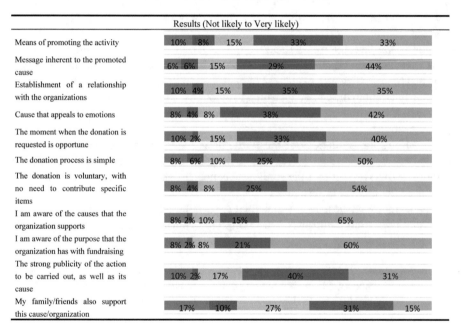

5 Conclusions

With this study, it observed that social marketing and fundraising contribute to the sustainability of OTSs. It realizes that through these actions, organizations can obtain resources to be more sustainable. The concept of social marketing is already starting to be disseminated and known, and most citizens state that OTSs should invest in these tools. However, the vast majority do not participate in these types of actions, citing reasons such as lack of financial capital, lack of transparency in organizations regarding the funds raised, as well as weak publicity of the actions. Thus, we can already see a warning to organizations addressing this issue, emphasizing the need to widely disseminate fundraising activities, so that this ceases to be a justifiable factor for non-participation in fundraising actions.

Another aspect that organizations should consider, which will eliminate another justifying reason for non-participation, is information about the purpose of the donations raised, as this increases the trust and transparency of the organization and promotes the obtaining of donations. It is important for the organization to inform the population about the purpose of donations raised with a particular activity, to increase the degree of trust and transparency, facilitating the obtaining of donations, as well as donor loyalty.

It also concludes that organizations seeking to promote fundraising actions should opt for the Christmas months, and organizations should not specify the type of donation, allowing the donation to be free and its process to be simple. Most respondents are familiar with Social Marketing and state that OTSs should invest in fundraising actions and the application of marketing tools in their daily management and organizational processes.

During the promotion of a fundraising activity, the strong advertising of it as well as its dissemination on social networks should be taken into consideration. Finally, it finds that there is a greater likelihood of contribution to fundraising actions if there is a message inherent to the cause, appealing to emotions, and if the purpose of the donation is known. It is also important when fundraising is done, as well as if it is free (without any imposition).

Regarding the limitations of the study, these focus on the data collection, as we should have had more survey responses to support the main conclusions of the study more robustly. The fact that the study was conducted during a pandemic context also influenced the number of responses.

As for future studies, there is potential for examining the comparison between for-profit and non-profit companies concerning the application of marketing tools in the management process.

Acknowledgments. This work is funded by National Funds through the FCT - Foundation for Science and Technology, I.P., within the scope of the project Refª UIDB/05507/2020. Furthermore, we would like to thank the Centre for Studies in Education and Innovation (CI&DEI) and the Polytechnic of Viseu for their support.

References

1. Quintão, C.: O terceiro sector e a sua renovação em Portugal: Uma abordagem preliminar. Instituto de Sociologia. Faculdade de Letras. Universidade do Porto (2011)
2. Tenório, F.: Gestão de ONGs: principais funções gerenciais. Editora da Fundação Getúlio Vargas (2006)
3. Teodósio, S.: Pensar o terceiro setor pelo avesso: dilemas e perspetivas da ação social organizada na gestão pública. In: encontro da associação nacional de pós-graduação e pesquisa em administração (2001)
4. Coelho, A., et al.: Introdução à gestão das organizações. Vida Económica (2011)

5. Cannon, C.: An Executive's Guide to Fundraising Operations: Principles, Tools and Trends. Hoboken (2011)
6. Grilo, R.: A teoria da gestão e a complexidade. [trabalho de conclusão da licenciatura em Gestão de Empresas]. Universidade de Évora (1996)
7. Chanlat, J.: Modos de gestão, saúde e segurança no trabalho. In Davel, E., Vasconcelos, J. (eds.) Recursos Humanos e subjetividade. Vozes (2002)
8. Kisil, M.: Organização Social e Desenvolvimento Sustentável: Projeto de Base Comunitária. In Ioschpe, B. (org.). Terceiro Setor: Desenvolvimento Social Sustentado. Paz e Terra (2007)
9. Santos, S.: Organização do Terceiro Setor. EdUnP (2007)
10. Júnior, A., Faria, M., Fontenele, R.: Gestão nas Organizações do Terceiro Setor: contribuição para um novo paradigma nos empreendimentos sociais. In: XXXIII Encontro da ANPAD. São Paulo/SP (2009)
11. Pereira, M.: Gestão para Organizações Não Governamentais. Tribo da Ilha (2013)
12. Maximiano, A.: Administração para empreendedores, 2 edn. Pearson (2011)
13. Mesquita, J.: A problemática da sustentabilidade económica. In Qualificação e sustentabilidade das organizações de economia social. EAPN Portugal (2014)
14. Guimarães, L.: Design e Sustentabilidade: Brasil produção e consumo. Design sociotécnico. Porto Alegre (2009)
15. Sachs, L.: Estratégias de transição para o século XXI: desenvolvimento e meio ambiente. Fundap (1993)
16. Sachs, I.: Caminhos para o desenvolvimento sustentável. Rio de Janeiro (2002)
17. Alves, J.: Sustentabilidade na Gestão de Organizações do Terceiro Setor: Um estudo dos Empreendimentos Sociais apoiados pela Ashoka. 242 f. In: Dissertação (Mestrado em Administração de Empresas) – Programa de Pós-Graduação em Administração. Universidade de Fortaleza (2008)
18. Armani, D.: O desenvolvimento institucional como condição de sustentabilidade das ONGs no Brasil In Ministério da Saúde. Secretaria de Políticas de Saúde (2001)
19. Cruz, C., Estraviz, M.: Captação de diferentes recursos para organizações da sociedade civil. Global, São Paulo (2000)
20. Carvalho, N.: Gestão e Sustentabilidade: Um estudo multicasos em ONGs ambientalistas em Minas Gerais. 157 f. Dissertação (Mestrado em Administração de Empresas) – Programa de Pós-Graduação em Administração. Universidade Federal de Minas Gerais: Belo Horizonte (2006)
21. Lendrevie, J., et al.: Mercator da língua portuguesa: Teoria e prática do marketing. Publicações Dom Quixote (2015)
22. Marques, C.: Projecto Manual de Acolhimento – Uma Abordagem de Marketing Interno. Relatório de Projecto Mestrado em Marketing, ISCTE Business School Instituto Universitário de Lisboa, Lisboa (2010)
23. Nomura, J., Souza, M.: Uma revisão Crítica do conceito de Marketing Social. Revista Gerenciais 3 (2004)
24. Dowbor, L.: O Mosaico partido: a economia além das equações. Vozes (2002)
25. Kotler, P.: Marketing para organizações que não visam o lucro. Editora Atlas (1978)
26. Schiavo, R., Fontes, B.: Conceito e Evolução do Marketing Social. II Curso de Capacitação em Marketing Social (1997)
27. Kotler, P.: Marketing social: estratégias para alterar o comportamento público. Campus (1992)
28. Soares, I.: O caso do plano de marketing da Estrutura Residencial para Pessoas Idosas da Santa Casa da Misericórdia de Espinho. (Dissertação de mestrado). ISSSP (2018)
29. Weinreich, N.: Hands-on Social Marketing. A Step-by-Step Guide to Designing Change for Good, 2ª edn. Sage Publications, Thousands Oaks (2011)
30. Jorge, H.: Estratégias de fundraising para a sustentabilidade financeira das organizações sem fins lucrativos: a experiência da Operação Nariz Vermelho. Escola de Ciências Sociais e Humanas do Instituto Universitário de Lisboa (2017)

The Impact of Process Automation on Employee Performance

Maria João Luz[1], Manuel José Serra da Fonseca[2,3]($\boxtimes$) (ID),
Jorge Esparteiro Garcia[2,4,5] (ID), and José Gabriel Andrade[6,7] (ID)

[1] ISAG – European Business School, Porto, Portugal
[2] Instituto Politécnico de Viana do Castelo, Viana do Castelo, Portugal
manuelfonseca@esce.ipvc.pt
[3] UNIAG - Unidade de Investigação Aplicada em Gestão, Viana do Castelo, Portugal
[4] ADiT-LAB, Instituto Politécnico de Viana do Castelo, Viana do Castelo, Portugal
[5] INESC TEC, Porto, Portugal
[6] Universidade do Minho, Campus de Gualtar, 4710-057 Braga, Portugal
[7] CECS - Centro de Estudos de Comunicação e Sociedade, Braga, Portugal

Abstract. Organizations aim to achieve operational efficiency capable of responding to high market competitiveness. The implementation of automation systems in organizational processes is a key factor in improving operational efficiency. This paper intends to contribute for a better understanding of the adoption of automation systems in organizations and analyze their impact on employee performance, considering the conditions under which they were implemented.

The methodology for this study was qualitative research, in which semi-structured exploratory interviews conducted with employees from the Accounts Receivable department of automotive sector companies were carried out. The main goal was to understand their perception of the use of automation systems in their work tasks. The results of this research led to the conclusion that automation systems, even when underutilized, are beneficial in reducing repetitive and manual tasks. Nevertheless, the way in which they are implemented has a direct impact on the motivation of employees to use them.

Keywords: Information technologies · Process automation · Performance · Accounts Receivable · Automotive industry

1 Introduction

Constant technological progress has been a crucial factor in today's world, as it has a significant influence on all human activities and consequently increases the degree of uncertainty and unpredictability of the future [1]. In the 1950s and 1960s, organizations began to adopt technological data processing systems to automate tasks such as stock control, accounting, and invoicing, among others [1]. Currently, in the corporate environment, the incorporation of information technologies into organizational processes is becoming increasingly mandatory, due to the large amount of data generated daily.

Organizations are gradually realizing that innovation and the ability to adapt are fundamental to remaining competitive in a constantly changing market. In line with technological advancement, another crucial factor that has been of great concern in the corporate environment is the valuation of human capital. Employees are increasingly considered an organization's most valuable resource, as they provide unique skills, knowledge and experiences that drive innovation, productivity, and company success [2]. In this context, this study was developed with the aim of understanding and analyzing the advantages and disadvantages of process automation software on the performance of employees in the departments responsible for managing accounts receivable (Account Receivables) in the automotive sector. More specifically, to identify the advantages and disadvantages of implementing organizational process automation technologies for companies, to determine the benefits and difficulties of implementing organizational process automation technologies for companies, to understand the behavior of employees when using automation processes, considering the conditions in which they were implemented, and to identify the added value of adopting automated systems in customer relations. This study focuses on the analysis and investigation of the subject adapted to the automotive sector, due to its size and representativeness in the national economy. More specifically, the research was carried out in the accounts receivable department, as this is where I currently work. Through the analysis of empirical and research work, relevant themes were determined to be addressed in the literature review, such as the characterization of the automotive industry at a national level and concepts and approaches to process management and the implementation of automation systems in organizations. In response to the general research objectives described above, a study was carried out using a qualitative methodological approach, to study and understand in detail the perception of employees of the use of automation systems in organizations. This paper is divided into 5 chapters. The first chapter, the Introduction and definition of the research problem, gives a brief background to the topic and the research objectives defined. The second chapter, the Literature Review, consists of an in-depth and detailed analysis of the literature, specifically on the relevant concepts of the topic. The third chapter, Methodology, describes the methodological approach chosen, the data instrument selected and the respective procedure and identifies the research objectives in detail. The fifth chapter, Analysis and Discussion of Results, consists of a characterization of the convenience sample and a descriptive analysis of the results obtained. The sixth chapter, Conclusions, presents the final conclusions of the study, identifies the limitations of the study, and outlines some recommendations for future research in this area.

2 Literature Review

The importance of the automotive industry in boosting and growing a country's economy is becoming increasingly apparent in society. As well as representing a huge economic weight, the automotive sector is central to trade, since the car is the most accessible and flexible means of transporting goods and has also become an important factor in increasing globalization [3].

2.1   Characterization of the Automotive Industry at National Level

Initially, the domestic car industry was poorly developed and qualified. It was only in the second half of the 20th century, through technological modernization, market competitiveness and the start of production at Autoeuropa, that it achieved notable progress [4]. The automotive sector in Portugal is represented by three business associations: Associação do Comércio Automóvel de Portugal (ACAP), Associação Nacional das Empresas do Comércio e da Reparação Automóvel (ANECRA) and Associação Nacional do Ramo Automóvel (ARAN) [5]. In Portugal, this sector is currently very important, both in terms of Gross Domestic Product and the national employment rate. According to the Associação do Comércio Automóvel de Portugal [6], there are 32,300 thousand companies in the automotive sector, which earn around 28 million euros a year, representing 5% of the national GDP. According to data collected by ACAP [6], more vehicles were produced between January and May 2023 compared to the same period in 2022, representing a growth of 29.4%. Light passenger vehicles recorded the highest number of vehicles produced and heavy vehicles recorded the greatest variation between 2022 and 2023 [6]. Portugal is known for hosting several car assembly plants for international brands such as Volkswagen, Peugeot, and Renault [6]. According to the National Automobile Association, the Volkswagen brand had the highest percentage change in sales between January and May 2023 compared to the same period in 2022, with growth of 140.70%, followed by the Dacia and Renault brands, with 126% and 73.10% respectively. On the other hand, Citroen, Toyota, and Seat recorded fewer sales in 2023. The national automotive sector is also made up of the automotive components industry, the development of energy and sustainable mobility solutions and the automotive cluster. According to data collected by the National Association of Automotive Trade and Repair Companies [7] the companies with the greatest weight in the automotive components industry at a national level are the Salvador Caetano Group, the Simoldes Group and Faurecia.

2.2   Organizational Processes

A process is a sequence of activities with the aim of obtaining a result. According to Vieira et al. [8], an organizational process is a set of interconnected and usually systematized activities that relate to the organizational structure. These processes can be formal or informal and can involve different areas of the company, such as production, sales, marketing, finance, human resources, among others. For Li and Fang [9], a process is made up of four basic elements: activities, the logical relationship between activities, the implementation of activities and barriers to activities. In operational terms, a process generates value from an input and produces an output [8]. According to Teixeira [10], an organizational process is any activity or set of activities that adds value to an input and, through the organization's resources, transforms it into an output for a specific customer. The aim is to make processes more efficient, flexible, and adaptable to changes in the market and customer needs [10].

2.3   Business Intelligence

According to Dias et al. [11] the initial concept of Business Intelligence (BI) is based on a set of methods used to facilitate business decision-making, based on computerized data and information systems. BI is understood as a technological and analytical process that extracts information from various sources and transforms it into knowledge to detect patterns, trends, and business opportunities [11]. The evolution of artificial intelligence and the emergence of the concept of big data has led to the implementation of Business Intelligence systems being increasingly studied and developed, especially in large companies [12]. A BI system is a set of online tools, such as data warehouses and online dashboards, which facilitate the analysis of a company's data, providing results and information in real time. Through BI, users can perform operations such as filtering, aggregating, and prioritizing financial, logistical, and commercial data, among others [13]. Although Business Process Management and Business Intelligence are distinct concepts, they are interconnected and have a complementary relationship. BPM and BI are part of a cycle of continuous improvement, in which processes are constantly analyzed, monitored, and improved. Analysis of the data extracted by BI makes it possible to identify faults and/or opportunities that allow BPM processes to be improved and optimized. In addition, as BI provides real-time reports, it allows for greater monitoring of organizational processes and analysis of the company's Key Performance Indicators (KPIs) [14]. Thus, due to the high level of accuracy of the data collected by BI, it is easier to make decisions within the scope of BPM.

2.4 Corporate Performance Management

The organizational community, from managers to financial directors and economists, recognizes that the management of an organization, in addition to defining a business strategy, must take performance management systems into account [15]. The traditional method of performance management is based on evaluating an organization's financial and production results, which is insufficient for business continuity. For Wade and Recardo [15], organizations that focus their performance management on the traditional method only evaluate past activities and results, leaving aside important factors for business growth, such as customer satisfaction, innovation, and human resource performance evaluation. Currently, the main keys to organizational performance management involve planning, calculation, and analysis. Planning requires internal and external information from various sources and helps managers define the organization's strategic goals. Calculation defines KPIs by analyzing and comparing the company's current performance with what is expected. The analysis involves the use of various techniques that allow managers to filter internal and external information and obtain relevant insights for business continuity [16]. Corporate Performance Management (CPM), also known as Enterprise Performance Management or Business Performance Management, is a set of methods that help organizations define, monitor, and manage corporate performance. CPM involves a holistic approach to managing the organization's performance, integrating various areas of the organization, such as marketing, finance, human resources, sales, among others, and aims to improve operational efficiency, drive business growth, and maximize the organization's profits [16].

3 Methodology

Exploratory interviews are a relevant tool in this study, as they allow ideas, concepts, and opinions to be shared freely through reflective dialog [17]. Unlike structured interviews, in which the questions are predetermined and strictly followed, exploratory interviews are characterized by a more open and flexible format, in which the researcher conducts the interview with a list of discussion topics, which are adapted throughout the conversation [17]. For the current study, individual semi-structured exploratory interviews were conducted, allowing the participant to feel more comfortable sharing genuine and relevant information, without the influence of other opinions and perspectives. In addition, it was possible to identify patterns and insights important to the study. The convenience sample for this study is made up of financial technicians from the accounts receivable department of 5 multinationals operating in the automotive sector, which, according to ARAN [5], have large sales volumes in Portugal and have financial services in Portugal. Given that the methodological approach chosen for this study was qualitative research through semi-structured exploratory interviews, an interview script was developed to serve as a starting point and as a guiding tool, ensuring that all relevant topics were covered. The interview script was constructed in such a way as to obtain the most relevant information in response to the proposed objectives, while at the same time, during the interviews, there was freedom for new ideas and topics to emerge from what was initially determined. The script consisted of 14 discussion topics, divided into 7 dimensions: 1. Assessment of the type of process automation system. 2. Motivations, advantages, and disadvantages. 3. Practice of automation systems 4. Impact of using automation systems in companies. 5. Training management for the use of automation systems. 6. Impact of repetitive tasks. 7. Impact on customer relations.

As already mentioned, the instrument used to collect data was an individual, face-to-face exploratory interview. To carry them out, a brief explanation of the topic and objectives of this study was given to each interviewee beforehand, so that they were sure of their decision to take part. Each interview was scheduled by email or telephone and conducted individually and face-to-face, allowing for direct and personal interaction. The interviews lasted between 20 and 45 min and were carried out in locations according to the availability of each interviewee, giving preference to noise-free locations. According to Peters and Halcomb [18], audio recording allows the raw interview data to be preserved, guaranteeing its integrity. It also facilitates detailed analysis of the participants' responses. However, it is necessary to respect the privacy and confidentiality of the information shared during the interview. As such, it is essential to have the consent of the participants, informing them of the purpose of the recording and how the data will be stored and protected [19]. Before each interview in this study, the participant was asked if the audio could be recorded for future analysis. Each interview was transcribed in full, and the qualitative analysis was carried out through open coding, using the Taguette software. Axial and selective coding was then carried out using Microsoft Excel and Microsoft Word, respectively.

3.1 Research Objectives

The implementation of process automation software has been a constantly growing reality in the corporate world. It was therefore considered important to study the employee's perspective on the impact of this software on their performance and job performance. The study's main objective was to study and analyze the advantages and disadvantages of process automation software on the performance of employees in the departments responsible for managing accounts receivable in the automotive sector. The specific objectives were defined in line with the main objective and to achieve detailed and explanatory results. Therefore, the specific objectives were as follows:

- Identify the advantages and disadvantages of implementing organizational process automation technologies for companies;
- To determine the benefits and difficulties of implementing organizational process automation technologies for companies;
- Understand the behavior of employees when using automation processes, considering the conditions in which they were implemented;
- Identify the added value of adopting automated systems in customer relations.

4 Analysis and Discussion of Results

4.1 Characterization of the Sample

Qualitative research aims to explore and understand phenomena in an in-depth and detailed way, for which the researcher can opt for a small sample, allowing them to gain a deeper understanding of the participants' perspectives, experiences, and information [19]. A small sample size was used for this study, allowing for a thorough analysis of the data. The convenience sample consisted of 5 participants from the same occupational group - financial technicians from the accounts receivable department in the automotive sector. The companies interviewed are multinationals operating mainly in the automotive components industry. GKN is a British company founded in 1759, specializing in automotive and aerospace engineering, with expertise in transmission systems, driveline, and electric mobility technologies. It currently has a Global Business Service (GBS) in the city of Porto, where it provides financial and accounting services to the company's various factories worldwide. The Stellantis Group is a multinational automotive conglomerate, formed by the merger of Fiat Chrysler Automobiles and Peugeot S.A., starting in January 2021. The merger has enabled the company to be a leader in the automotive industry, covering a wide range of products, from passenger cars to commercial vehicles and luxury vehicles. In addition, the Stellantis Group is focused on the transition to electric vehicles and the implementation of sustainable mobility solutions. It currently has a subsidiary based in the city of Maia, in Porto, PSA Services Portugal, which acts as a Shared Service Center. Faurecia is a French organization, founded in 1997, specializing in the development and production of vehicle interior technologies, and is one of the world's leading suppliers to the automotive components industry. In addition, Faurecia is actively involved in developing solutions for electric and hybrid vehicles, driving the transition to sustainable mobility. In Portugal, Faurecia has four production units, based in the north and center of the country, and a GBS, in Santa

Maria da Feira. NORS is a Portuguese company founded in 1946 and based in Porto. NORS operates in the motor vehicle parts and accessories distribution sector and has an extensive network of stores and distribution centers in Portugal, offering a wide range of products including mechanical parts, body parts, tires, electrical and electronic components, among other. Yazaki is a Japanese company, founded in 1941, specializing in the development and manufacture of electrical systems and components for vehicles. Currently, Yazaki is also involved in the development of technologies for electric and hybrid vehicles. In Portugal, Yazaki has a production unit located in Ovar.

4.2   Descriptive Analysis of the Dimensions

Dimension 1. Evaluation of the type of process automation system. The aim of the first dimension was to find out about the type of process automation system used in the interviewees' companies and to understand whether these systems are used for common or different activities. According to the information gathered by the professionals, the automation systems in the finance departments of the national automotive industry are mainly related to processing and sending invoices to customers. Currently, multinationals are incorporating the use of BI systems into their systems to facilitate the work of employees and, above all, to provide immediate data and information. For example, Microsoft Power BI, as discussed in the literature review, allows advanced analysis resources to be created in real time. **Dimension 2.** Motivations, advantages, and disadvantages. The aim of the second dimension was to analyze the positive and negative points related to the introduction of process automation technologies in organizations. From the employee's perspective, the aim was to understand the motivations that led the company to implement automation systems and to identify the main advantages and disadvantages. Implementing automation systems in time-consuming, manual processes allows companies to reduce the time spent on repetitive tasks, resulting in greater operational efficiency and increased productivity [20]. All the interviewees mentioned time as the main motivation that led companies to incorporate automation into organizational processes, as it enables them to reduce time spent on repetitive tasks and focus on activities with greater added value and that involve critical analysis by the employee. **Dimension 3.** Practice of automation systems. The third dimension aimed to study the use of automation systems by employees and to understand whether they improve or limit their performance. Automation systems are, of course, an aid in carrying out repetitive and methodical tasks. As such, the interviewees say that having tasks such as archiving documentation, extracting information from clients' B2B portals, allocating client payments in the system, etc., automated, facilitates workflow and allows them to focus on activities with greater added value and which require more critical thinking, resulting in improved operational efficiency. **Dimension 4.** Impact of the use of automation systems on companies. The purpose of the fourth dimension was to understand the interviewees' perspective on the costs of automation systems for the company and whether, despite the costs, they are an asset to the company. According to [20], implementing automation systems generally requires a significant initial investment. Costs can include the purchase of software and hardware, employee training and maintenance costs. Although the interviewees do not have information to prove the numerical costs of automation systems in the company, they believe that the high initial investment is reflected in gains in the medium and long term, given the increase in productivity that they make possible. **Dimension 5.** Training management for the use of automation systems. Kirange [20] states that active employee participation is one of the ten critical success factors when implementing BI systems. The company should encourage employees to report anything that is wrong or to suggest improvements, since they will be the main users of these systems. As such, the fifth dimension sought to understand the interviewees' perception of the influence of management on the use of automation systems and to analyze the importance of training in the use of these systems. The opinions of the interviewees were divergent, given that this is a point where there is a direct influence of organizational culture. Logically, if in an organization the boss leads through pressure and meticulous control, the employee will feel more coerced into using what is imposed on them and will not suggest improvements. **Dimension 6.** Impact of repetitive tasks. Organizations are increasingly looking to increase operational efficiency and productivity by incorporating the automation of repetitive and time-consuming tasks into their organizational processes. The aim of dimension 6 was to understand whether the interviewees still have repetitive tasks in their day-to-day lives and whether they think these could be replaced by automation systems. All the interviewees said that they still have repetitive tasks in their daily lives. Some are very methodical and could easily be solved with simple automation, others are more analytical in nature and could not be 100% automated, as this would generate more risks associated with automation. **Dimension 7.** Impact on customer relations. The accounts receivable department is responsible for managing and controlling a company's customer accounts, monitoring and recording customer payments and issuing invoices to customers. The whole process involves the customer and therefore contact with them is frequent [21]. As such, it was considered important to understand the impact of automation systems on customer relations, this being the seventh and final dimension of the study. The main objective was to understand whether the implementation of these systems is perceptible to the company's customers and what the main differences are between before and after implementation. All the interviewees recognized that, directly or indirectly, a company's automation systems, and especially their quality, have an impact on customer relations.

5 Conclusions, Limitations, and Future Work

This research aimed to contribute to a better understanding of how automation systems are perceived by employees, specifically by employees in the automotive industry's accounts receivable department, and how they impact their performance, considering the conditions under which they are implemented. In this context, a qualitative study was carried out with the aim of understanding employee behavior in the use of automated systems, determining the benefits for companies and understanding whether there is added value in terms of customer relations. It was possible to conclude that automation systems, even if not used to their best advantage, are an asset in terms of reducing repetitive and manual tasks, resulting in an increase in productivity and, therefore, an improvement in employee performance. Since automation allows time-consuming and methodical processes to be carried out, it enables employees to focus on activities with greater added value and which involve sense and critical analysis, consequently resulting in an improvement in workflow and employee motivation, as it allows them to demonstrate their skills and receive due recognition. In an organization where the manager imposes the use of these systems immediately, not allowing time for adaptation and not providing clear and objective training, the result will not be what is expected, there will be more errors, the advanced functionalities of the systems will not be understood or used correctly, resulting in the underutilization of available resources. In addition to the above, it was possible to conclude that the growing implementation of automation systems in organizations is reflected in employees' fear of losing their jobs, since they replace tasks performed by human capital and, consequently, reduce the number of headcounts needed to perform a function. In terms of customer relations and building customer loyalty, it is possible to conclude that automation systems have an impact, even indirectly, since they allow for more efficient communication and the handling of customer needs quickly and consistently. It was therefore possible to determine that the main differences in customer relations between before and after the implementation of automation systems are mainly in customer service since response times are lower and there is greater consistency in data. According to Kirange [20], the team should consist of technical and analyst staff, with the technical staff developing the system itself and the analyst staff adapting it to the needs of the organization.

The main limitations observed are related to the scope and subjectivity of the topic, as process automation is a broad and recent concept, of which there is still no in-depth and widespread knowledge. In addition, another limitation is related to the complex and multifaceted nature of performance, since employee performance can be influenced by a variety of factors, such as organizational culture, motivations, individual skills, among others. It is difficult to attribute performance improvement exclusively to automation systems, as there can be complex interactions between these systems and other factors. In future research on the subject, it is recommended that senior management be chosen as the research sample, to understand the impact of automation systems from a more strategic perspective. Senior managers have a strategic vision of the organization and are actively involved in planning and setting goals and objectives, so their access to privileged resources and information can enrich the research.

This research is relevant and pertinent to academia and the professional market since there is currently no critical and exploratory reflection on the subject of the Impact of Process Automation on Employee Performance.

Acknowledgments. The authors are grateful to the Foundation for Science and Technology (FCT, Portugal) for financial support by national funds FCT/MCTES to UNIAG (UIDB/04752/2020 and UIDP/04752/2020).

References

1. Moraes, G.D.D.A., Terence, A.C.F., Escrivão Filho, E.: A tecnologia da informação como suporte à gestão estratégica da informação na pequena empresa. JISTEM-J. Inf. Syst. Technol. Manag. **1**, 27–43 (2004)
2. Silva, C.F.: Capital intelectual: gestão do capital humano nos escritórios de contabilidade. Revista Uniaraguaia **15**(3), 202–218 (2020)

3. Silva, D.A.L.C.: Implementação de um projeto mecânico de um pórtico automatizado (Doctoral dissertation). Instituto Superior de Engenharia do Porto, Porto (2022)
4. Fernandes, J.L.F.: Competitividade da Indústria Automóvel Portuguesa (Doctoral dissertation). Instituto Politecnico do Porto, Porto (2017)
5. Sousa, M.M.D.: Dashboard para análise do setor automóvel em Portugal: os anos 2019 e 2020 e consequências da pandemia (Doctoral dissertation). Instituto Superior de Economia e Gestão, Lisboa (2021)
6. Associação do Comércio Automóvel de Portugal. Estatísticas (2023). https://www.acap.pt/pt/estatisticas
7. Associação Nacional das Empresas do Comércio e da Reparação Automóvel. Mercado (2023). https://www.anecrarevista.pt/category/news/mercado/
8. Vieira, Y.S., Andrade, E., Seixas, J., Roberto, J., Junior, J.: Gestão de processos: métodos para a melhoria contínua. Revista Científica Multidisciplinar Núcleo do Conhecimento 11(9), 05–13 (2022)
9. Li, F., Fang, G.: Process-aware accounting information system based on business process management. Wirel. Commun. Mobile Comput. 2022 (2022)
10. Teixeira, P.A.D.S.: Gestão por processos numa instituição do ensino superior (Doctoral dissertation). Faculdade de Ciências e Tecnologia, Lisboa (2013)
11. Dias, B., Gonçalves, C., Silva, M.: Business Intelligence como driver da Análise de Indicadores Académicos. Cadernos Investigação do Mestrado em Negócio Eletrónico 2 (2022)
12. Božič, K., Dimovski, V.: Business intelligence and analytics for value creation: the role of absorptive capacity. Int. J. Inf. Manag. 46, 93–103 (2019)
13. Ain, N., Vaia, G., DeLone, W.H., Waheed, M.: Two decades of research on business intelligence system adoption, utilization, and success–a systematic literature review. Decis. Supp. Syst. 125, 113113 (2019)
14. Jeston, J., Nelis, J.: Business Process Management. Routledge, Abingdon (2014)
15. Wade, D., Recardo, R.J.: Corporate Performance Management: How to Build a Better Organization Through Measurement-Driven Strategic Alignment. Routledge, Abingdon (2001)
16. Richards, G., Yeoh, W., Chong, A.Y.L., Popovič, A.: Business intelligence effectiveness and corporate performance management: an empirical analysis. J. Comput. Inf. Syst. 59(2), 188–196 (2019)
17. Campenhoudt, L.V., Quivy, R.: Manual de investigação em ciências sociais. Gradiva Publicações (2008)
18. Peters, K., Halcomb, E.: Interviews in qualitative research. Nurse Res. 22(4), 6 (2015)
19. Brooks, J., Horrocks, C., King, N.: Interviews in qualitative research. Interv. Qual. Res. 1–360 (2018)
20. Kirange, S.: Role of business intelligence in decision-making for SMEs. Khoj J. Indian Manag. Res. Pract. 41–47 (2016)
21. Singh, R.P., Singh, R., Mishra, P.: Does managing customer accounts receivable impact customer relationships, and sales performance? an empirical investigation. J. Retail. Consum. Serv. 60, 102460 (2021)

Effect of Social Media on Workplace Procrastination Among Employees in Bosnia and Herzegovina

Suada Pestek[ID], Almir Pestek[(✉)][ID], and Amra Kozo[ID]

School of Economics and Business, University of Sarajevo, Trg Oslobodjenja-A.Izetbegovic 1, 71000 Sarajevo, Bosnia and Herzegovina
{almir.pestek,amra.kozo}@efsa.unsa.ba

Abstract. The paper deals with workplace procrastination, which results from using the Internet and social media during working hours. Practice shows that the appearance of procrastination causes numerous negative consequences, both for employees and organizations. Although procrastination is exceptionally well represented in literature and researched from different angles, this phenomenon has not been given sufficient attention in Bosnia and Herzegovina (B&H), and therefore the number of papers on this topic is negligible. In order to study the effects of workplace procrastination caused by the use of social media, a field survey was conducted on a sample of 701 employees in B&H. Our results align with the findings of other studies worldwide and offer new insights into the topic.

Keywords: procrastination · social media · Bosnia and Herzegovina

1 Introduction

The availability of a broad spectrum of content on online platforms offers many possibilities for making human life easier, more comfortable, and more pleasant. Platforms are designed to allow users to socialize and communicate with people they already know offline or to meet new people and network despite physical separation, different experiences, or language. The combination of entertainment and social benefits has a strong psychological effect, and it is therefore not surprising that the use of social media also becomes a habit and often an addiction [1]. The design of platforms and tools encourages social interaction and the level of use (for instance, by creating the "ideal self"). On the other hand, the use of the Internet and social media during working hours for reasons unrelated to the job is becoming common. Frequent and uncontrolled use of the Internet, particularly social media, during working hours leads to problem behavior – workplace procrastination.

Procrastination can be observed as the voluntary delay despite the expectation that one will be worse off as a result of that delay, or delay or postponing a task or an activity that would ideally be completed in the present [2] and which can hurt productivity and wellbeing requirements [3]. The causes of procrastination in the workplace can differently direct employees' energy toward short-term and enjoyable goals, often at the expense of performing crucial long-term tasks. Organizational procrastination losses can include decreased overall efficiency, lower security, poorer performance of (in-house) computer networks, costs related to removing viruses and spyware software, etc.

Thus, the main aim of this paper is to explore the phenomenon of workplace procrastination due to the effect of social media, i.e., whether the use of social media in the workplace affects workplace procrastination. First, a theoretical background on the phenomenon of procrastination is provided, followed by data and method, as well as the study's findings. Finally, results are discussed, and the conclusion is provided.

2   The Phenomenon of Procrastination and Social Media

Procrastination is an unnecessary delay that produces a feeling of personal distress [4]. It is a cognitive avoidance of tasks and engagement in activities implicitly aimed at diverting attention from things that should be done [5]. Workplace procrastination can be defined as putting off work-related activities during working hours due to engagement in various non-work related activities [6, 7] and refers to sub-optimum and self-undermining behavior that negatively affects both the performance and productivity of the organization and the productivity and wellbeing of employees themselves [8, 9].

The Internet and virtual environments have the potential to provide short-term comfort, excitement and/or distraction [10] for employees, and therefore, conscious or unconscious use of the Internet is often put ahead of essential work tasks, while online media are often chosen instead of performing more or less critical tasks [11–14]. Inappropriate use of the Internet is thus the most frequent way for employees to waste considerable (and measurable) time at work [8, 15]. In literature and practice, online procrastination is also called 'cyberloafing' or 'cyberslacking.'

Research shows that the appearance and level of workplace procrastination depend on difficulties in determining priorities and self-confidence [2], duration of employment and position in the organization [16–20], financial situation [21], job satisfaction [20–23], salary level [17], education and qualifications [16, 20], gender [17–24], as well as on self-control and conscientiousness [2, 7, 25–27]. Conscientiousness and emotional stability are strong predictors of performance [18, 24, 28–30], and the same is true of self-control [31–35]. Conscientious people attempt to fulfill their duties, are typically focused on performing tasks, and are less likely to procrastinate. Emotionally stable people have smaller needs to waste time and energy regulating their emotions, have greater capacity to allocate resources to tasks at hand, and are less likely to lose focus. A high level of self-control is related to positive outcomes, and a low level to negative ones. It can be claimed that procrastination is a self-regulatory failure [17]. It happens because people with low self-control are very impulsive, quickly react to environmental stimuli, and search for momentary pleasure while lacking persistence at work.

Use of the Internet in the workplace for personal needs decreases in organizations that restrict employees' use of computers and technology in the workplace through organization policy, technological deterrence, or both [19, 20, 24]; these factors affect employees' procrastination. For this reason, it is necessary to design inherently motivating tasks instead of over-relying on employees' self-motivation.

3   Data and Method

For the needs of the paper, an online survey was conducted in 2022, which used a convenient sample and a structured questionnaire developed based on the presented papers [9, 25, 36–38]. The questionnaires were distributed to respondents in B&H via the GoogleDocs platform and paid Facebook advertisements. The survey was filled in by 776 respondents (out of 2,693 who participated in the survey – response rate 28.8%); 701 questionnaires were valid and were used for the analysis.

The socio-demographic characteristics of the sample are presented below:

- Female respondents prevail in the sample (77.9%), which is one of the limitations of the research.
- Most respondents are aged 26–35 (38.2%) and 36–45 (32.7%). Respondents aged 18–25 account for 15.3%, and those older than 46 for 13.8%.
- Respondents with completed university education prevail in the sample (70.8%), while 23% of the respondents have a master's or a doctoral degree.
- The respondents' self-reported personal net monthly income mostly ranges between 250 and 500 EUR (33.1%) and between 500 and 750 EUR (31,8%). The average salary in B&H in 2022 was around 550 EUR.
- 62.6% of the respondents are married or live in a common-law relationship, and 50.9% of the respondents have children.
- 18.3% of the respondents hold managing positions.
- 37.8% of the respondents have up to five years of working experience, 20.8% between five and ten, 26.1% from 11 to 20, and 15.3% have over 20 years of working experience.
- 7.3% of the respondents use the Internet for less than one hour per day in total (workplace and free time), 38.1% use it for between one and three hours, 27.1% use it for four to five hours, 11.6% for six to eight hours, while 16% use it for more than eight hours.
- 37.9% of the respondents use the Internet in the workplace for less than one hour per day, 28.7% from one to three hours, 12.3% from six to eight hours, 12% from four to five hours, 4.4% for more than eight hours, while 4.7% do not use the Internet in the workplace at all.
- 54.1% of the respondents use social networks from one to three hours per day in total (workplace and free time), 18.7% for up to one hour, 15.8% for four to five hours, 6.8% for six to eight hours and 4.6% for over eight hours.
- For accessing the Internet or social networks, the respondents use a mobile phone (53.5%), a desktop (33.1%), or a laptop (13%).
- A total of 73.3% of respondents are allowed to use social networks during their working hours in the organizations they work for. The most often used networks are Facebook (27.7%), Viber (25.3%), and YouTube (21%).

4   Research Findings

The level and intensity of procrastination due to social media were viewed as the response to the question: "How much time a day do you spend on social networks in the workplace for your personal needs?". Respondents who do not use social networks in the workplace for their personal needs or those who use them for less than one hour per day prevail in the sample (57.1%). Additionally, 16.4% of the respondents use them for one to three hours, while 5.8% use them for over four hours. A total of 20.7% of respondents do not use social networks in the workplace for their personal needs at all. The average duration of using social networks in the workplace for personal needs amounts to 2.78 h (standard deviation 2.18, coefficient of skewness 1.40, and coefficient of kurtosis 1.30).

The key findings of the research on procrastination tendency according to different socio-demographic characteristics reveal the following:

- Gender does not affect the level of procrastination, although, on average, men have a somewhat higher level of procrastination (2.94 compared to 2.74 h for women). The nonparametric Mann-Whitney (MW) test for two independent samples shows that the difference is not statistically significant (U = 400726, z = -0.783, p = 0.433 > 0.05).
- Age affects the level of procrastination. Employees aged from 18 to 25 (3.71 h), from 26 to 35 (3.09 h), and 56 and older (3.94 h) have a significantly higher level of procrastination than persons aged from 36 to 45 (2.16 h) and from 46 to 55 (2.19). The nonparametric Kruskal-Wallis (KW) test for more than two independent samples shows that there is a statistically significant difference between the groups (H = 55.234, p = 0.000 < 0.05).
- Marital status affects the level of procrastination. Employees who live alone have a significantly higher level of procrastination (3.19 compared to 2.54 h). Nonparametric KW test for more than two independent samples reveals a statistically significant difference (H = 15.369, p = 0.000 < 0.05).
- Whether a respondent has children or not affects the level of procrastination. On average, employees who do not have children have a somewhat higher level of pro- crastination than those who do (3.13 compared to 2.44 h). The nonparametric MW test for two independent samples shows that there is a statistically significant difference (U = 49544, z = -4.854, p = 0.000 < 0.05).
- Education level affects the level of procrastination. Employees who have completed secondary education have a significantly higher level of procrastination than those who have completed higher education (3.34 compared to 2.54 h). Nonparametric KW test for more than two independent samples shows a statistically significant difference (H = 16.768, p = 0.000 < 0.05).
- Monthly income affects the level of procrastination. Employees with incomes lower than 500 EUR or higher than 1,250 EUR have a significantly higher level of procras- tination (3.64 and 3.13, respectively, compared to 2.27 h for incomes from 500 to 1,250 EUR). Nonparametric KW test for more than two independent samples shows a statistically significant difference (H = 28.676, p = 0.000 < 0.05).
- Working experience affects the level of procrastination. Employees with shorter work- ing experience have a significantly higher level of procrastination (3.41 h compared to 2.52 for working experience from five to ten years, 2.30 for working experience from 11 to 20 years, and 2.38 for working experience longer than 20 years). Nonparamet- ric KW test for more than two independent samples shows a statistically significant difference (H = 88.614, p = 0.000 < 0.05).
- Position in the organization does not affect the level of procrastination. On aver- age, employees have a somewhat higher level of procrastination than managers (2.80

compared to 2.66); however, the difference is not statistically significant. The non-parametric MW test for two independent samples shows no statistically significant difference (U = 35425, z = -0.66, p = 0.509 > 0.05).

If the respondents' statements about the total time spent using the Internet and the level of procrastination expressed by the time the respondents spend daily on social networks in the workplace for personal needs are cross-tabulated, it can be concluded that the more total time a respondent uses the Internet daily, the intensity of procrastination is greater. In the nonparametric chi-square test of the interdependence of qualitative variables, the chi-square value amounts to 116,445 (with p = 0.000 < 0.05), which means that there is an interdependence between the respondents' responses about the total time spent using the Internet and the level of procrastination. Besides, the more time a respondent uses the Internet in the workplace, the greater is the intensity of procrastination. In the nonparametric chi-square test of interdependence between qualitative variables, the chi-square value amounts to 418.9 (with p = 0.000 < 0.05), which means that there is an interdependence between the respondents' responses about the time of using the Internet in the workplace and the level of procrastination. There is a direct correlation between the scope and frequency of using social media in the workplace and the appearance of procrastination. The more time a respondent uses social networks, the greater is the intensity of procrastination. In the nonparametric chi-square test of interdependence between qualitative variables, the chi-square value amounts to 306.143 (with p = 0.000 < 0.05), which means there is an interdependence between the respondents' responses about the time spent using social networks and the level of procrastination.

The respondents rated 11 statements that pertain to the scale of their own procrastination [9], with three added statements marked with *. All the statements were measured on a 1 to 5 scale. Cronbach alpha coefficient amounts to 0.935(>0.7), which means that these 14 variables can be aggregated into one variable or construct which expresses procrastination. As the average of responses or ratings of agreement with statements in the questionnaire that pertain to the procrastination construct, a variable named "view of one's own procrastination" was calculated (Table 1).

The average rating for the construct "procrastination" is 1.94, with a standard deviation (SD) of 0.82. The respondents gave the lowest average rating in the construct "procrastination" for the statement "Use of social networks in the workplace is more important for me than performing work-related tasks", while the highest average rating was given for the statement "There is nothing wrong in my using social networks for personal needs during working hours." Normality was not satisfied either for initial statements in the questionnaire (p values of the Kolmogorov-Smirnov test are lower than 0.05), nor for the derived construct.

We will compare views of one's own procrastination by different groups:

- On average, men have a somewhat higher view of their own procrastination (mean 2.04 and SD 0.860) than women (mean 1.91 and SD 0.807), except for the statement "I start doing something but I end up on social networks", although there is a statistically significant difference for only two out of the 14 statements (MW test: "I wait for the last minute of the deadline to complete something", z = -2.503, p = 0.012 < 0.05 and "I often use social networks as an excuse to not complete work-related tasks", z

Table 1. View of one's own procrastination.

Statement	Descriptive statistics			Kolmogorov-Smirnov test for normality	
	N	Mean	Standard deviation	Statistics	P
Use of social networks in the workplace is more important for me than performing work-related tasks *	687	1.55	0.926	0.380	0.000
Because of using social networks I postpone accomplishing tasks *	690	1.59	0.914	0.358	0.000
There is nothing wrong in my using social networks for personal needs during working hours *	687	3.02	1.316	0.182	0.000
Because of the need to use some of social networks I unnecessarily postpone accomplishing important tasks	659	1.94	1.112	0.262	0.000
I spend time on social networks to avoid tasks I do not like	660	1.90	1.114	0.283	0.000
I wait for the last minute of the deadline to complete something	660	1.90	1.168	0.287	0.000
I postpone making difficult decisions	656	2.13	1.238	0.241	0.000
I often use social networks as an excuse to not complete work-related tasks	658	1.64	0.942	0.337	0.000
I am hopeless in wasting time on social networks	658	1.81	1.041	0.293	0.000
I know I waste time on social networks but it seems to me that I cannot help it	662	2.01	1.151	0.259	0.000

(*continued*)

Table 1. (*continued*)

Statement	Descriptive statistics			Kolmogorov-Smirnov test for normality	
	N	Mean	Standard deviation	Statistics	P
When something is very difficult, I believe that postponing will help	658	1.86	1.111	0.296	0.000
I start doing something but I end up on social networks	658	1.94	1.124	0.268	0.000
I fret for not beginning to work on time, but I keep postponing work	660	1.87	1.078	0.287	0.000
I often linger too much on social networks although I know how important it is to start working	655	1.86	1.089	0.293	0.000
View of my own own procrastination	690	1.94	0.820	0.125	0.000

Source: Author's research

$= -2.140$, $p = 0.032 < 0.05$). Thus, like in the case of the intensity of procrastination, gender does not affect the view of one's own procrastination.

- Persons aged up to 35 and 56 and older typically have somewhat higher views of their own procrastination (mean for 18–25 2.09 and SD 0.943; mean for 26–35 1.97 and SD 0.841; mean for 56–65 2.16 and SD 0.621) than persons aged 36 to 55 (mean for 36–45 1.84 and SD 0.711; mean for 46–55 1.91 and SD 0.859), though the difference is statistically significant only for two out of the 14 statements (KW test: "I start doing something but I end up on social networks", $H = 13.272$, $p = 0.010 < 0.05$ and "There is nothing wrong in my using social networks for personal needs during working hours", $H = 23.985$, $p = 0.000 < 0.05$). Thus, age does not significantly affect the view of one's own procrastination.
- Persons who live alone mostly have a higher view of their own procrastination (mean 2.07 and SD 0.880) compared to persons who share the household (mean 1.87 and SD 0.777), and the differences are significant for six out of the 14 statements related to the view of one's own procrastination (KW test: "There is nothing wrong in my using social networks for personal needs during working hours", $H = 11.709$; $p = 0.003 < 0.05$; "I am hopeless in wasting time on social networks", $H = 9.789$, $p = 0.007 < 0.05$; "I postpone making difficult decision", $H = 13.167$; $p = 0.001 < 0.05$; "I start doing something but I end up on social networks", $H = 12.186$; $p = 0.002 < 0.05$; "I fret for not beginning to work on time, but I keep postponing work", $H = 10.979$; $p = 0.004 < 0.05$; "I often linger too much on social networks although I

know how important it is to start working", H = 10.144; p = 0.006 < 0.05). Thus, marital status affects the view of one's own procrastination.

- For all the statements related to the view of one's own procrastination (except for the statement "Because of the need to use some of social networks I unnecessarily postpone accomplishing important tasks"), as well as for the construct view of one's own procrastination, respondents who do not have children have a higher view of their own procrastination (mean 2.02 and SD 9.869) than those who have children (mean 1.87 and SD 0.764). The difference is statistically significant for five statements related to the view of one's own procrastination and for the construct view of one's own procrastination (MW test: "There is nothing wrong in my using social networks for personal needs during working hours", z = -3.934, p = 0.000 < 0.05; "I postpone making difficult decisions", z = -2.292, p = 0.022 < 0.05; "I start doing something but I end up on social networks", z = -2.676, p = 0.007 < 0.05; "I fret for not beginning to work on time, but I keep postponing work", z = −2.455, p = 0.014 < 0.05; "I often linger too much on social networks although I know how important it is to start working", z = -2.123, p = 0.034 < 0.05). Thus, whether a respondent has children or not affects the view of his own procrastination.
- Differences in the view of one's own procrastination do not have the same direction for different statements, nor are statistically significant when respondents with different levels of education are compared (KW test, p < 0.05). Thus, the level of education affects the view of own procrastination.
- Differences in the view of own procrastination do not have the same direction for different statements between groups with different monthly income. Since the differences are not statistically significant (KW test, p < 0.05), it can be concluded that monthly income does not affect the view of own procrastination.
- Differences in the view of one's own procrastination do not have the same direction for different statements among groups with different working experience. If it is complemented with the result that the differences are mostly not statistically significant (KW test, p < 0.05), it can be concluded that working experience does not affect the view of one's own procrastination.
- On average, employees have a somewhat lower view of their own procrastination (mean 1.92 and SD 0.806) than managers (mean 2.03 and SD 0.875), which is contrary to the conclusion in the case of the intensity of procrastination; however, the difference is not statistically significant (MW test, p < 0.05). Thus, position in the organization does not affect the view of one's own procrastination.

5 Discussion

Literature [39] suggests how studies on procrastination so far have mainly focused on students' samples. Hence, in this study, workplace procrastination caused by the use of social media has been explored in a sample of employees in B&H. Obtained results showed that although, on average, men have a somewhat higher level of procrastination, this difference is not statistically significant, which is similar to some of the previous research, i.e., [35]. Next, employees aged up to 35 and 56 and older have a significantly higher level of procrastination than those aged 36 to 56. Interestingly, some recent research, i.e., [40], shows a negative relationship between age and decisional procrastination, meaning that decisional procrastination decreases as employees get older. Also, employees who live alone or do not have children have a higher level of procrastination. The results of this study showed that employees with completed secondary education have a significantly higher level of procrastination than employees with a higher level of education. On the contrary, some previous studies, i.e., [20], showed the opposite, implying that employees with higher education levels more frequently use the Internet for personal purposes during work.

Obtained findings showed interesting results regarding personal income since employees with income lower than 500 EUR or higher than 1,250 EUR have a significantly higher level of procrastination. A similar result was confirmed in previous studies, i.e., [17] and [21], which discuss how procrastination is higher among lower-income employees. Some of the previous research, i.e., [17], argues how the length of work experience is related to procrastination, and our study confirms that employees with a shorter working experience have a significantly higher level of procrastination. Even though some of the previous research, i.e., [16], showed that professional workers report higher levels of procrastination compared to non-skilled workers, in our research, when we compared procrastination among employees and managers, obtained results showed that although, on average, employees have a somewhat higher level of procrastination than managers, this difference is not statistically significant.

Finally, the more time a respondent uses social networks, the greater is the intensity of procrastination; in other words, there is a direct relationship between the scope and frequency of using social media in the workplace and the appearance of procrastination.

6 Conclusion

This research has identified several significant relationships between the socio-demographic characteristics of employees and the worldwide spread phenomenon of workplace procrastination. A self-reported assessment might be recognized as a potential limitation of the conducted study since it may influence obtained responses. Therefore, further studies besides self-reported assessments should call for investigating employees' procrastination from the perspectives of their coworkers and managers, including a more comprehensive organizational perspective. In this manner, future studies might offer nuanced and deeper understandings of workplace procrastination and ensure an inclusive perspective.

Given the lack of papers dealing with this topic in B&H, particularly within organizations of business sectors, we hope that this research will open space for additional research into workplace procrastination and help managers design the workplace and create and improve the system of motivation, control, and self-control in the workplace.

References

1. Kuss, D.J., Griffiths, M. D., Karila, L., Billieux, J.: Internet addiction: a systematic review of epidemiological research for the last decade. Current Pharmaceutical Design, **20**(25) (2014)
2. Steel, P.: The nature of procrastination: A meta-analytic and theoretical review of quintessential self- regulatory failure. Psychological Bulletin, **133**(1) (2007)
3. Alblwi, A., Stefanidis, A., Phalp, K., Ali, R.: Procrastination on Social Networks: Types and Triggers. Proceedings of the 6th International Conference on Behavioral, Economic and Socio-Cultural Computing (2019)
4. Solomon, L.J., Rothblum, E.D.: Academic Procrastination: Frequency and Cognitive-Behavioral Correlates. Journal of Counseling Psychology, **31**(4) (1984)
5. Beard, K.W., Wolf, E.M.: Modification in the proposed diagnostic criteria for Internet addiction. Cyberpsychol. Behav. **4**(3) (2001)
6. Metin, BM, Taris, T.W., Peeters, M.C.W.: Measuring procrastination at work and its associated workplace aspects. Personality and Individual Differences, vol. **101** (2014)
7. Van Eerde, W.: A meta-analytically derived nomological network of procrastination. Personality and Individual Differences (2003)
8. Zoghbi-Manrique-de-Lara, P., Olivares-Mesa, A.: Bringing cyber loafers back on the right track. Industrial Management & Data Systems, **110**(7) (2010)
9. Meier, A., Reinecke, L., Meltzer, C.E.: "Facebocrastination"? Predictors of using Facebook for procrastination and its effects on students' wellbeing. Comput. Hum. Behav. **64** (2016)
10. Widyanto, L., Griffiths, M.: Internet addiction: does it really exist? In: Gackenbach, J. (ed.) Psychology and the Internet: Intrapersonal, Interpersonal and Transpersonal Applications, 2nd ed. Academic Press, New York (2006)
11. Lavoie, J., Pychyl, T.A.: Cyberslacking and the procrastination superhighway: a web-based survey of online procrastination, attitudes, and emotion. Soc. Sci. Comput. Rev. **19**(4) (2001)
12. Myrick, J.G.: Emotion regulation, procrastination, and watching cat videos online: Who watches Internet cats, why, and to what effect? Comput. Hum. Behav. **52** (2015)
13. Reinecke, L., Hofmann, W.: Slacking off or winding down? an experience sampling study on the drivers and consequences of media use for recovery versus procrastination. Hum. Commun. Res. **42**(3) (2016)
14. Hofmann, W., Reinecke, L., Meier, A.: Of sweet temptations and bitter aftertaste: Self-control as a moderator of the effects of media use on wellbeing. In: Reinecke, L., Oliver, M.B. (eds.) The Routledge handbook of media use and wellbeing: International perspectives on theory and research on positive media effects. Routledge, New York (2017)
15. Griffiths, M.: Internet abuse and Internet addiction in the workplace. Journal of Workplace Learning, vol. **7** (2010)
16. Hammer, C.A., Ferrari, J.R.: Differential incidence of procrastination between blue-and whitecollar workers. Current Psychol. **21**(4) (2002)
17. Nguyen, B., Steel, P., Ferrari, J. R.: Procrastination's impact in the workplace and the workplace's impact on procrastination. Int. J. Sel. Assessment **21**(4) (2013)
18. Woods, F.: A Study into the Relationship Between Cyberloafing, Procrastination and Conscientiousness in the Workplace. Dublin Business School (2014)
19. Ugrin, J.C., Pearson, J.M., Odom, M.D.: Cyber-slacking: self-control, prior behavior and the impact of deterrence measures. Rev. Bus. Inf. Syst. **12**(1) (2008)
20. Garrett, R.K., Danziger, J.N.: On cyberslacking: workplace status and personal Internet use at work. CyberPsychol. Behav. **11**(3) (2008)
21. Steel, P.: The Procrastination Equation: How to Stop Putting Things Off and Start Getting Stuff Done. Random House, Toronto (2011)
22. Stanton, J.M.: Company profile of the frequent internet user. Commun. ACM. **45**(1) (2002)

23. Vitak, J., Crouse, J., LaRose, R.: Personal Internet use at work: Understanding cyberslacking. Comput. Hum. Behav. **27**(5) (2011)
24. Ozler, N.D.E., Polat, G.: Cyberloafing phenomenon in organizations: determinants and impacts. Int. J. eBusiness eGovernment Stud. **4**(2) (2012)
25. Tuckman, B.W.: The development and concurrent validity of the procrastination scale. Educational and Psychological Measurement (1991)
26. Lay, C.H., Brokenshire, R.: Conscientiousness, procrastination, and person-task characteristics in job searching by unemployed adults. Curr. Psychol. **16**(1) (1997)
27. Ocansey, G., Addo, C., Onyeaka, H. K., Andoh-Arthur, J., Oppong Asante, K.: The influence of personality types on academic procrastination among undergraduate students. Int. J. School Educ. Psychol. **10**(3) (2022)
28. Ferrari, J.R., Pychyl, T.A.: "If I wait, My Partner Will Do It:" The role of conscientiousness as a mediator in the relation of academic procrastination and perceived social loafing. North Am. J. Psychol. **14**(1) (2012)
29. Kim, K., Triana, M., Chung, K., Oh, N.: When do employees cyberloaf? an interactionist perspective examining personality, justice, and empowerment. Hum. Resource Manage. **55**(6) (2015)
30. Chak, K., Leung, L.: Shyness and locus of control as predictors of internet addiction and internet use. Cyberpsychol. Behav. **7**(5) (2004)
31. Blunt, A.K., Pychyl, T.A.: Task aversiveness and procrastination: a multi-dimensional approach to task aversiveness across stages of personal projects. Personality Individual Differences **28**(1) (2000)
32. Ferrari, J.R.: Compulsive procrastination: Some self-reported characteristics. Psychol. Rep. **68**(2) (1991)
33. Schnauber-Stockmann, A., Meier, A., Reinecke, L.: Procrastination out of habit? the role of impulsive versus reflective media selection in procrastinatory media use. Media Psychol. **21** (2018)
34. Reinecke, L., Vorderer, P., Knop-Huelss, K.: Entertainment 2.0? the role of intrinsic and extrinsic need satisfaction for the enjoyment of facebook use. J. Commun. **64**(3) (2014)
35. Ferrari, J. R., O'Callaghan, J., Newbegin, I.: Prevalence of procrastination in the United States, United Kingdom, and Australia: arousal and avoidance delays among adults. North Am. J. Psychol. **7**(1) (2005)
36. Lay, CH: At last, my research article on procrastination. J. Res. Personality **20**(4) (1986)
37. Maloney, P.W., Grawitch, M.J., Barber, LK: The multi-factor structure of the Brief Self-Control Scale: Discriminant validity of restraint and impulsivity. J. Res. Personality **46**(1) (2012)
38. Meerkerk, G-J., Vermulst, A.A., van den Eijnden, R.J.J.M., Garretsen, H.: The Compulsive Internet Use Scale: Some psychometric properties. Cyberpsychol. Behav. **12**(1) (2009)
39. Yan, B., Zhang, X.: What research has been conducted on procrastination? evidence from a systematical bibliometric analysis. Front. Psychol. **13**, 809044 (2022)
40. Kocak, O., Varan, H. H., Dashtbali, M., Bennett, R., Barner, Y.: How does social media addiction affect decisional procrastination? mediation role of work-family life balance. Europ. J. Environ. Public Health **7**(1) (2023)

Challenges of Using E-commerce in Bosnia and Herzegovina from the Perspective of Online Store Owners

Almir Pestek[✉] [iD] and Nadija Hadzijamakovic [iD]

School of Economics and Business Sarajevo, Trg oslobodjenja – A.Izetbegovic 1, 71000 Sarajevo, Bosnia and Herzegovina
{almir.pestek,nadija.hadzijamakovic}@efsa.unsa.ba

Abstract. Over the past two decades, the e-commerce sector experienced substantial growth as customers recognized the numerous benefits of online shopping. While providing plenty of benefits for customers, e-commerce growth affected businesses as well. Online stores offer myriad possibilities for businesses to improve their sales, while they also represent an excellent opportunity for small and medium-sized enterprises (SMEs) to grow their businesses. Several drivers of the overall online store success are found in literature: quality of online shopping experience, security, website quality, ease of use, value in terms of the product choice and prices, website reputation and the process of trust building with customers. In addition to buyers' trust, the sellers' trust plays an important role in the continuity of their use of online marketplaces. Since the existing literature predominantly deals with the perspective of customers, the focus of this study is on the presentation of challenges concerning business and success factors for online stores. The paper presents the findings conducted in 2023 on a sample of 162 online store owners from Bosnia and Herzegovina (BiH). The findings confirm some of the findings from literature, but also provide a specific overview of online sellers' perceptions across different categories.

Keywords: e-commerce · online store · Bosnia and Herzegovina

1 Introduction

Over the past two decades, the e-commerce sector experienced substantial growth as customers recognized the numerous benefits of online shopping, including the accessibility and convenience of online stores, time saved during shopping, enhanced information search, the possibility to read and write reviews, a wider selection of products offered and other [1, 8, 15]. Recent years accelerated the growth even more due to the COVID-19 pandemic, which besides the health and economic measures caused an increase in online shopping [2, 17]. While providing plenty of benefits for customers, e-commerce growth affected businesses, by creating new jobs across various sectors and new online market opportunities [7, 8, 10, 16]. While the growth in e-commerce impacted different industries, it had a particular influence on the retail sector, where the establishments

Á. Rocha et al. (Eds.): WorldCIST 2024, LNNS 990, pp. 99–109, 2024.
https://doi.org/10.1007/978-3-031-60328-0_10

and significant updates in commercial online stores replaced or complemented physical stores [9, 12, 23]. This created a shift in buyers' behavior and their perceptions of retail space, while generating new customer demands, such as multiple payment and delivery options, return policies and other buying alternatives [13, 16, 23].

Online stores offer myriad possibilities for businesses to improve their sales including lower operational costs and efficiency compared to traditional retail stores, customer data collection opportunities, enhanced connection with global markets, platform convenience for multiple products display, improved visual representation of products and other [8, 10, 16]. They also represent an excellent opportunity for small and medium-sized enterprises (SMEs) to grow their businesses [10]. However, some studies suggest there is a significant difference between the adoption of e-commerce among small and medium-sized enterprises (SMEs) and large enterprises (LEs). Some of the reasons include the lack of resources among SMEs to implement sophisticated digital technologies or the resistance to change in organizational culture [4, 6, 10, 19]. While business-to-business (B2B) enterprises seem to be motivated into e-commerce adoption via pressures coming from participation and competition on global markets, business-to-customer (B2C) enterprises are mostly influenced by consumer attitudes and variables specific to local contexts [10]. Nevertheless, some general concerns of businesses that operate in the online marketplace include customer and financial data security, retention of customer loyalty, the gap between customer expectations of the physical product compared to online representation of the product, pressures on the delivery and logistics systems and legislative issues [8, 25]. Factors influencing customer loyalty in physical and online stores are quite distinct and should be approached differently [22]. Online marketplace gives customers the opportunity to compare offers and prices across multiple stores in a few clicks, which creates an additional challenge for e-sellers to retain customers' loyalty. Some of the factors that influence customer loyalty to online stores include business credibility, website experience and reputation [5, 22, 24].

Several drivers of the overall online store success are found in literature: quality of online shopping experience, security, website quality, ease of use, value in terms of the product choice and prices, website reputation and the process of trust building [5, 18, 24].

In addition to buyers' trust, the sellers' trust plays an important role in the continuity of their use of online marketplaces [5]. Logistics and order management, product management, product uploading, and provision of training are some of the issues that affect sellers' trust in online systems and service quality [5, 21]. Furthermore, the key drivers affecting the adoption and success of e-commerce websites in B2C markets seem to be consumers' validation of quality content and concerns over security and privacy [5, 6]. For both factors, the advancements in social commerce have brought significant benefits for online store owners. The process of making buying decisions is intertwined with numerous opportunities given by social commerce, such as the availability of objective information about products offered, interaction with other users, better understanding of customers' needs etc. [1, 15]. According to the 2021 European E-commerce Report, Facebook, Instagram, and YouTube are the most frequently used social media channels among European online stores [17].

The rapid development of e-commerce also elevated the trend of direct delivery to customers, with more and more customers paying attention to the speed of delivery, which created the need for businesses to follow up with multiple delivery options [11]. According to data from the 2021 European E-commerce Report, the European online store delivery services experienced an increase in the number of options over the last decade. Furthermore, alongside all the positive opportunities that the growth of online marketplace brought to online store owners, users' trust and privacy/security concerns remain one of the most important challenges, while e-customers' lack trust is especially visible when it comes to online payments [17, 18].

E-customers feel secure when they are assured that their personal data is protected by the e-seller and that online stores may build customers' trust by having a clear display of privacy policy agreements and cookies policy [20]. Some authors [14] argue that encryption in today's e-commerce activities is mandatory to provide customers with a sense of security and confidentiality. Online word of mouth is also emphasized as the important role in building customers' trust [3].

Since the existing literature predominantly deals with the perspective of customers, the main goal of this study is to investigate and present the challenges concerning business and success factors for online stores.

2 Methodology and Sample Structure

The research was conducted in December 2023 in cooperation with the E-commerce Association in Bosnia and Herzegovina. The survey was developed by the authors based on relevant studies, and previously conducted research presented in the theoretical part. The survey was distributed online to online stores in BiH, and 162 valid responses have been collected.

Basic characteristics of the sample are presented below:

- The online store business domains are presented in Table 1. The dominant products sold by online stores are clothing (10.0%), consumer electronics (8.6%), computer equipment (7.2%), household items (6.9%) and cosmetics and body care (6 .4%),
- 29.0% of respondents have been in business for less than 12 months, while 22.8% from 1 to 2 years, 24.1% from 2 to 5 years and 24.1% for more than 5 years,
- 29.6% of respondents employ one person, and 10.5% have no employees except the owner,
- 88.9% of respondents are companies from BiH, while 7.4% are international companies,
- 69.1% of respondents sell products and services only in BiH, 27.8% in BiH and abroad, and 3.1% only abroad,
- 62.3% of respondents sell products and services to both physical and legal persons, 34.0% only to physical persons and 3.7% only to legal persons,
- 59.9% of respondents own a physical store as well, 24.1% do not have a physical store, 9.9% do not have a store and do not plan to have one, while the remaining 6.1% do not have a store and do not plan to open one,
- 38.9% of respondents use Wordpress as a platform, 21% use a custom solution, and the rest use different platforms (Shopify, Magento, Prestashop, Instagram, Facebook, etc.),

– 77% of respondents sell physical (tangible) products, 10.3% digital products and services (courses, audio books, software, etc.), 8% services (insurance, etc.) and 4.6% experiences (tickets, travel, etc.),
– 16.7% of respondents offer up to 10 items (without variations – color, size, etc.), 18.5% from 11 to 100 items, 27.2% from 101 to 1,000 items, 27.1% from 1,001 to 10,000 items, while 7.4% from 10,001 to 100,000 items,
– 29.4% of respondents deliver products via express mail (state/public operators), 35.1% via express mail (private operators), and 17.1% have their own delivery. 8.2% of respondents offer store pick-up. Only 0.8% of respondents deliver products by regular or registered mail,
– 27.4% of respondents have the card payment option, 20.6% bank transfer, 29.5% cash on delivery payment, 7.3% PayPal or similar services, and 14.3% payment at the office,
– 22.9% of respondents have a loyalty program, 40.1% of respondents do not have a loyalty program, 29.6% do not have one but intend to introduce it, 7.4% do not have one and do not want to introduce one,
– 41.4% of respondents use SSL/TTS encryption as a measure to ensure transaction security and data protection, 22.8% use a PCI DSS certified Payment Service Provider, 13.6% use other technologies, and 11.7% hire experts for safety.

Table 1. Business domains of the online stores

Business domain	Number of responses	%
Clothes	36	10.0
Consumer electronics	31	8.6
Computer equipment	26	7.2
Household items	25	6.9
Cosmetics and body care	23	6.4
Shoes	17	4.7
Watches	16	4.4
Products & services related to wellness	16	4.4
Machines & tools	16	4.4
Jewelry	15	4.2
Furniture & décor	14	3.9
Toys	13	3.6
Other	13	3.6
Food & drink	11	3.1
Software	11	3.1
Sports & fitness equipment	9	2.5

(*continued*)

Table 1. (*continued*)

Business domain	Number of responses	%
Books	9	2.5
Equipment for hobbies	9	2.5
Tickets	4	1.1
Clothes, products and equipment for babies and children	4	1.1
Office material	4	1.1
Promotional material, gifts, souvenirs, and decorations	4	1.1
Rent-a-car	3	0.8
Food supplements	3	0.8
Car parts and tires	3	0.8
Agricultural materials and equipment	3	0.8
Music	2	0.6
Travel	2	0.6
Insurance	2	0.6
Food & equipment for pets	2	0.6
Cleaning detergents	2	0.6
Accessories & haberdashery	2	0.6
Lighting, electrical, audio, water, and thermal equipment	2	0.6
Other	8	2.4
Total	360	100.0

3   Presentation of Empirical Research Findings

The challenges faced by the owners of online stores are shown in Table 2. It is evident that the dominant problems related to external factors on which businesses have no or limited influence are the following: consumer habits in BiH and tendencies towards electronic commerce (22%), disloyal competition on the domestic market related to online sales of goods and services from unregistered businesses (17.8%), limited/small market size in BiH (12.8%) and legislation in BiH (7.8%). It is interesting to note that the respondents do not recognize competition from abroad as a significant challenge, which is in line with the findings of a recent study on customer attitudes towards e-commerce [18] according to which 78% of online buyers in BiH make purchases in BiH (63.6% websites and 15.1% social media), and only 22% abroad (19.7% websites and 1.6% social media).

The findings on the main obstacles to the development of online stores and e-commerce in BiH are listed in Table 3. Following the findings presented in Table 2, it is evident that the owners of online stores in BiH face difficulties in: attracting and retaining customers (15.4%), lack of knowledge about e-commerce (13.7%), and different internal technical problems (11.9%). Other obstacles concern legislation, financial

Table 2. Challenges for online stores and e-commerce in BiH

Challenges	Number of responses	%
Customers' habits/affinity for e-commerce	99	22.0
Unfair competition on the domestic market (black market)	80	17.8
Market size	58	12.9
Legislation	35	7.8
System maintenance costs	29	6.4
Ensuring the quality of services for customers	28	6.2
Choice of courier service	25	5.6
The challenge of delivering products and services to different parts of the country	19	4.2
Communication and coordination with institutions	19	4.2
Servicing of complaints	16	3.6
The challenge of ensuring the security of transactions	14	3.1
Customer data protection	14	3.1
Competition from abroad	9	2.0
Delivery costs - public operators	3	0.7
Other	2	0.4
Total	450	100.0

resources, marketing, relationships with suppliers and customers, labor shortage, payment systems and software solutions. On the other hand, factors such as the customer support, IT equipment and hosting seem to cause the least problems for the respondents.

The main factors for the success of online stores in BiH are presented in Table 4. Among others, the following factors stand out: quality and availability of products and services (10.7%), prices of products and services (9.9%), speed and efficiency of delivery (9.2%). Other dominant factors include promotion and branding, user interface, transaction security and data protection, return policies, customer service and sales promotion activities. Advertising in traditional media seem to be the least important factor.

The respondents recognize the availability of labor as one of the problems, and 45.7% of the respondents have already faced this problem. Furthermore, 21.4% of respondents believe that there is enough qualified labor at the disposal, 31.5% that there is not enough qualified labor, and 48.1% of respondents believe that there are workers who are partially qualified. The way in which respondents overcome problems related to availability of labor is through the engagement of external collaborators (40.9%), by providing training for workers (30.7%), increasing salaries and benefits (14.8%) and increasing the flexibility of working hours (13.6%).

Table 3. The main obstacles to the development of online stores and e-commerce in BiH

Obstacles	Number of responses	%
Difficulties in attracting and retaining customers	62	15.4
Lack of knowledge about e-commerce	55	13.7
Technical problems (problems with website development or integration with other systems)	48	11.9
Lack of financial resources to start online sales	36	9.0
Difficulties with suppliers	35	8.7
Difficulties in marketing products or services	28	7.0
Labor shortage	28	7.0
Legislation	27	6.7
Payment system	27	6.7
Software solutions	27	6.7
Difficulties in the logistics and product delivery process	13	3.2
Customer support	8	2.0
Hosting	5	1.2
IT equipment	3	0.7
Total	402	100.0

Table 4. Factors of success for online stores in BiH

Factors of success	Number of responses	%
Quality and availability of products and services	117	10.7
Prices of products and services	108	9.9
Speed and efficiency of delivery	100	9.2
Promotion on social media	96	8.8
Free shipping above a certain purchase amount	83	7.6
Ease of site navigation	79	7.2
Transaction security and data protection	76	7.0
The right to return purchased goods/services within a defined period	74	6.8
Page loading speed	68	6.2
Visual and content branding	63	5.8
Approval of holiday discounts	48	4.4

(continued)

Table 4. (*continued*)

Factors of success	Number of responses	%
After-sales activities	41	3.8
Approval of discounts on purchases above a certain amount	36	3.3
Gift vouchers /certain % discount on the next purchase	36	3.3
Newsletter	25	2.3
Approval of discounts on card payments	21	1.9
Promotion through traditional channels	19	1.7
Total	1090	100.0

The key qualifications that respondents expect from employees are presented in Table 5. Experience with online marketing tools, as well as working experience with e-commerce, sales and website management are extremely important qualifications that are required from employees. It is important to note that qualifications related to computer and data processing software, data analysis and statistics are also among the requirements. A university degree is the least important qualification requirement.

Table 5. The key qualifications that respondents expect from employees

Key qualifications	Number of responses	%
Experience with online marketing tools	105	19.3
Working experience in e-commerce	89	16.4
Experience with website management	80	14.7
Experience in sales	73	13.4
Computer and data processing software related knowledge	62	11.4
Experience with data analysis and statistics	58	10.7
English language knowledge	48	8.8
Completed additional education programs	14	2.6
Completed University degree	10	1.8
Other	4	0.7
Total	543	100.0

4 Conclusion

The findings of this paper are in line with the findings of research on e-commerce users in BiH [18] according to which customers value the trustworthiness of the seller and positive experience in terms of the promised versus delivered value among all. However, in certain aspects, the lack of trust remains an issue for e-customers, especially when it comes to online payments [17, 18]. Accordingly, the findings in this paper revealed that some of the major factors responsible for the success of online stores in BiH include the quality and availability of products and services, convenient prices, but also the speed and efficiency of delivery. Furthermore, the importance of promotion on social media should not be neglected either. This is also in accordance with the findings from literature [1, 15, 18] which suggest that social media is an important marketplace for online shoppers, while it brings numerous benefits for e-sellers. In addition, the research findings suggest that the majority of respondents operate their online stores domestically, and that they do not recognize the competition from abroad as a significant challenge. On the other hand, the limited/small market size in BiH is reported to be one of the major challenges for the success of online stores, while the biggest challenge seems to be the consumer habits and tendencies towards e-commerce. This suggests that the perception of online sellers in BiH is largely shaped by consumer behavior, especially when it comes to the perception of competition. Recent findings [10, 18] support this notion, suggesting that e-customers in BiH are reluctant towards e-shopping from e-sellers abroad, which clearly affects the perception of competition by online sellers in BiH as well. Although the theoretical framework for e-commerce is abundant and suggests a significant research potential for this topic, research in the context of BiH is still scarce. The expectation is that this study will motivate other researchers to carry out deeper analyses of the e-commerce market in BiH. As a result of the rapid growth of e-commerce globally, factors such as competition in online shopping or consumer attitudes are changing rapidly, and practices are quickly replaced, so a longitudinal study may provide more conclusive findings. Hence, the authors will reconduct this study with the E-commerce Association in Bosnia and Herzegovina in February 2024, aiming to obtain more detailed insights and make grounds for a deeper understanding of the results. Additionally, the recommendation for future studies is to extend this study by conducting qualitative research which would provide a more holistic review of the results and an in-depth analysis of the constructs of interest. Also, since the available data are limited to BiH, it would be interesting to see similar studies conducted in other countries of the region in order to improve the universality of this research.

References

1. Attar, R.W., Shanmugam, M., Hajli, N.: Investigating the antecedents of e-commerce satisfaction in social commerce context. British Food J. **123**(3), 849–868 (2021)
2. Beckers, J., Weekx, S., Beutels, P., Verhetsel, A.: COVID-19 and retail: the catalyst for e-commerce in Belgium? J. Retail. Consum. Serv.Consum. Serv. **62**, 102645 (2021)
3. Chetioui, Y., Lebdaoui, H., Chetioui, H.: Factors influencing consumer attitudes toward online shopping: the mediating effect of trust. EuroMed J. Bus. J. Bus. **16**(4), 544–563 (2021)

4. Činjarević, M., Pijalović, V., Peštek, A., Lazović-Pita, L., Karić, L.: Heading out SMEs to the e-commerce highway: drivers of the e-commerce perceived usefulness among SMEs in Bosnia and Herzegovina. Manage. J. Contemporary Manage. Issues 26(1), 3–20 (2021)
5. Cui, Y., Mou, J., Cohen, J., Liu, Y.: Understanding information system success model and valence framework in sellers' acceptance of cross-border e-commerce: a sequential multi-method approach. Electron. Commer. Res.Commer. Res. 19, 885–914 (2019)
6. Gibbs, J., Kraemer, K.L., Dedrick, J.: Environment and policy factors shaping global e-commerce diffusion: a cross-country comparison. Inf. Soc. 19(1), 5–18 (2003)
7. Gregory, G.D., Ngo, L.V., Karavdic, M.: Developing e-commerce marketing capabilities and efficiencies for enhanced performance in business-to-business export ventures. Ind. Mark. Manage. 78, 146–157 (2019)
8. Gupta, S., Kushwaha, P.S., Badhera, U., Chatterjee, P., Gonzalez, E.D.S.: Identification of Benefits, Challenges, and Pathways in E-commerce Industries: An integrated two-phase decision-making model. Sustainable Operations and Computers (2023)
9. Hagberg, J., Sundstrom, M., Egels-Zandén, N.: The digitalization of retailing: an exploratory framework. Int. J. Retail Distrib. Manage. 44(7), 694–712 (2016)
10. Hamad, H., Elbeltagi, I., El-Gohary, H.: An empirical investigation of business-to-business e-commerce adoption and its impact on SMEs competitive advantage: the case of Egyptian manufacturing SMEs. Strateg. Chang.. Chang. 27(3), 209–229 (2018)
11. Ignat, B., Chankov, S.: Do e-commerce customers change their preferred last-mile delivery based on its sustainability impact? Int. J. Logistics Manage. 31(3), 521–548 (2020)
12. Jahanshahi, A.A., Zhang, S.X., Brem, A.: E-commerce for SMEs: empirical insights from three countries. J. Small Bus. Enterp. Dev.Enterp. Dev. 20(4), 849–865 (2013)
13. Kim, T.Y., Dekker, R., Heij, C.: Cross-border electronic commerce: distance effects and express delivery in European Union markets. Int. J. Electron. Commer.Commer. 21(2), 184–218 (2017)
14. La Lau, R., La Lau, R.: Traffic encryption: SSL/TLS. Practical Internet Server Configuration: Learn to Build a Fully Functional and Well-Secured Enterprise Class Internet Server, pp. 227–237 (2021)
15. Lin, X., Wang, X., Hajli, N.: Building e-commerce satisfaction and boosting sales: the role of social commerce trust and its antecedents. Int. J. Electron. Commer.Commer. 23(3), 328–363 (2019)
16. Liu, X., Tong, D., Huang, J., Zheng, W., Kong, M., Zhou, G.: What matters in the e-commerce era? modelling and mapping shop rents in Guangzhou. China. Land Use Policy 123, 106430 (2022)
17. Lone, S., Harboul, N., Weltevreden, J.W.J.: 2021 European E-commerce Report (2021). https://rb.gy/de9vy1. Aaccessed 05 Nov 2023
18. Peštek, A., Hadžijamaković, N.: Customer attitudes toward E-commerce: case of Bosnia and Herzegovina. In: 8th International Scientific Conference CRODMA 2023 Proceedings, pp. 79–90. University of Zagreb, Faculty of Organization and Informatics, Varaždin (2023)
19. Peter, M.K., Dalla Vecchia, M.: The digital marketing toolkit: a literature review for the identification of digital marketing channels and platforms. New trends in business information systems and technology: Digital innovation and digital business transformation, pp. 251–265 (2021)
20. Strzelecki, A., Rizun, M.: Consumers' security and trust for online shopping after GDPR: examples from Poland and Ukraine. Digital Policy, Regulation and Governance 22(4), 289–305 (2020)
21. Sun, H.: Sellers' trust and continued use of online marketplaces. J. Assoc. Inf. Syst. 11(4), 182–221 (2010)
22. Swaminathan, S., Anderson, R., Song, L.: Building loyalty in e-commerce: impact of business and customer characteristics. J. Mark. Channels 25(1–2), 22–35 (2018)

23. Tolstoy, D., Nordman, E.R., Hånell, S.M., Özbek, N.: The development of international e-commerce in retail SMEs: an effectuation perspective. J. World Bus. **56**(3), 101165 (2021)
24. Wang, Y.S.: Assessing e-commerce systems success: a respecifcation and validation of the DeLone and McLean model of IS success. Inf. Syst. J. **18**(5), 529–557 (2008)
25. Zhang, H., Jia, F., You, J.X.: Striking a balance between supply chain resilience and supply chain vulnerability in the cross-border e-commerce supply chain. Int J Log Res Appl **26**(3), 320–344 (2023)

Analyzing São Paulo's Place Branding Positioning in Promotional Videos (2017–2019)

José Gabriel Andrade[1,2] , Adriano Sampaio[3,4] , Jorge Esparteiro Garcia[5,7,8(✉)] ,
Álvaro Cairrão[5] , and Manuel José Serra da Fonseca[5,6]

[1] Universidade do Minho, Campus de Gualtar, 4710-057 Braga, Portugal
[2] CECS - Centro de Estudos de Comunicação e Sociedade, Braga, Portugal
[3] Universidade Federal da Bahia, Barão de Jeremoabo, Salvador 1649-023, Brazil
[4] Grupo de Pesquisa Logos – Comunicação Estratégica, Marca e Cultura, Salvador, Brazil
[5] Instituto Politécnico de Viana do Castelo, Viana do Castelo, Portugal
jorgegarcia@esce.ipvc.pt
[6] UNIAG - Unidade de Investigação Aplicada em Gestão, Viana do Castelo, Portugal
[7] ADiT-LAB, Instituto Politécnico de Viana do Castelo, Viana do Castelo, Portugal
[8] INESC TEC, Porto, Portugal

Abstract. This research aims to analyze the positioning theory and discourse within São Paulo's Place Branding from 2014 to 2019, investigating the symbolic representations employed by the São Paulo Tourism Bureau to emphasize its branding endeavors. The methodology employed a framework based on Semprini's [10] Project/Manifestation approach and Discourse Analysis. The impetus behind this study arises from the substantial investments made by cities to craft comprehensive disclosure strategies and establish place branding for their respective regions. We observed aspects of Communication and Digital Marketing in the three promotional videos produced by SPTuris in 2014, 2017, and 2019, which underwent meticulous analysis. Our findings unveiled a consistent thematic discourse despite shifts in political administration. The 2014 video accentuated multiculturalism and cosmopolitanism, while the 2017 edition highlighted experiential marketing, business, consumption, and cosmopolitan elements. Remarkably, the 2019 presentation featured images emphasizing receptivity. Themes such as Culture, Arts, and Gastronomy were recurrent across all videos. The scrutinized discourse reaffirms São Paulo's capital as a trendsetter within Brazil.

Keywords: Promotional Videos · São Paulo · Place Branding · Tourism · Culture

1   Introduction

In 'Invisible Cities,' Italo Calvino (1990) presents the narratives recounted by the Venetian traveler Marco Polo (1254–1324) about the cities conquered by the Mongol emperor Kublai Khan, whom he served for many years. Polo's stories not only replace the eyes of the ruler but also his understanding of the surroundings and the territory he inhabits. This novel describes an experience that we acknowledge in our present-day, yet it is mediated by the strategic communication associated with territories.

Á. Rocha et al. (Eds.): WorldCIST 2024, LNNS 990, pp. 110–116, 2024.
https://doi.org/10.1007/978-3-031-60328-0_11

A recent subject in communication studies concerns the notion of 'place brand-ing.' Studies by Kotler [7], and Dinnie [2] heavily influence this field with a marketing approach. For instance, Kotler [7] suggests that territories should concentrate efforts to position their brands, aiming to gain visibility and competitiveness in the tourism mar-ket. According to Semprini [9] three concepts guide the discursive positioning of brands: credibility, legitimacy, and seduction. Each brand seeks to build its positioning within a universe of meanings, a possible world that must be manifested through self-promotion strategies. Furthermore, it must be coherent so that the public identifies familiar identity traits in these manifestations (e.g., advertising campaigns) [9].

The relationship between brand and audience is constructed through expectations surrounding the brand and the discursive unity translated in its positioning. Semprini [10] also proposes that brands organize themselves based on the textual, possible, and real worlds. These are founded, respectively, on their symbolic or sensitive expressions; on a narrative about the brand; and on the values that constitute the context in which the brand operates. With this premise, each brand constructs a possible world in each of its manifestations. Building upon these premises, we propose the analysis of two promotional videos from São Paulo: 'São Paulo, a multicultural city,' 'Viva! São Paulo,' and 'How can you not love São Paulo?' produced by São Paulo Turismo (SPTuris), the official events and tourism company of the city. The videos were produced in 2014, 2017, and 2019, with durations of four minutes, one and a half minutes, and half a minute, respectively, and are available on YouTube.

With this analytical exercise, we aim to identify the discursive positioning of the São Paulo brand during the analyzed period and the signs activated by SPTuris to highlight the brand's project. As preliminary results, we note that in the first, more documentary-styled video, São Paulo is discursively positioned as multicultural and cosmopolitan, with a notable emphasis on its cultural dimension. In the second, promotional in nature, the city's tourism body focuses on thematic axes such as entertainment, culture, business and major events, consumption, and gastronomy. In the third promotional video, the human aspect of the city is emphasized, with images evoking love, friendship, and receptivity.

2 Methods

Based on this contextual data, we proceed to analyze the institutional videos of the city of São Paulo, promoted by SPTuris. The analysis of the videos is organized into three stages: The first aims to delineate the textual/manifest world [3] within the production. It will be characterized by the discourse analysis of the videos. The second stage outlines the dimension of the possible world [4, 6, 10] and will be aided using the Greimasian semiotic square to observe the promises made by the place brand in question and the relationships between the terms in the elementary structure of signification. Finally, the third stage maps the social practices [3] that support the pillars advocated through these productions. It seeks to identify the relationships of the videos with the contextual data they suggest, briefly introduced in this article.

3 Promotional Video: São Paulo – Embracing Multiculturalism

The first video, entitled "São Paulo: A Multicultural City" (Fig. 1) lasts for four minutes and adopts a structure akin to a short documentary. In this production, the city is portrayed through various dimensions, such as architecture and tangible heritage, arts and culture, gastronomy, multiculturalism, and cosmopolitanism. To convey these dimensions, there are six characters/speakers and a narrator who, through redundancy, reaffirms the selected essential information. The video commences with the caption 'largest city in South America' while showcasing aerial shots of buildings, bridges, and thoroughfares. Subsequently, the text 'home to over eleven million inhabitants' accompanies footage of a crowd bustling through the streets or heading towards the metro.

Each of the six 'characters' narrating São Paulo has their statements validated by their profession displayed on-screen. They respectively represent engineering, music, gastronomy, theater, and education/research. Through their dialogues, they unveil the city's tangible and architectural heritage, the diverse encounters the city facilitates, the quality of services offered, exemplified through gastronomy, the richness and vibrancy of the arts, and the inclusive nature of a city that embraces different social groups and cultures. It's noteworthy that this latter characteristic is emphasized by an educational and research agent, a teacher, prompting us to consider qualifications and professional training as a significant dimension in defining the municipality. Throughout the video, scenes of avenues, buildings, cultural landmarks, and the greenery of parks abound, yet it's the city's multiculturalism and cosmopolitan character that take the spotlight. The first 'character' testifying about the city is engineer Toni Francesco, who introduces viewers to the first skyscraper in Latin America. His foreign accent can be perceived as an indicator that São Paulo comprises people from diverse places, a notion reinforced at other junctures in the video, solidifying the city's cosmopolitanism. Danilo Brito, a São Paulo musician, speaks about music in the city's Old Center and emphasizes that 'São Paulo is a very romantic city and has space for everyone here.' Reinforcing this idea, a successful Northeasterner in São Paulo, chef Francisco Gameleira, the third character, asserts, '[…] I was born in a small town in Rio Grande do Norte and ended up here […] In gastronomy, São Paulo is on par with Europe, the United States, with everything.' Italian Giovanni Bruno, the renowned owner of the Il Sogno di Anarello trattoria, declares, 'We had the great privilege of transforming São Paulo into the world because the world is in São Paulo and in Brazil.' The dialogues of these four characters, witnessing and experiencing São Paulo, advocate that the city not only welcomes immigrants but also offers opportunities to all, providing the chance to experience local grandeur. This is made possible because the city amalgamates experiences, cultures, and arts from around the world. Thus, through their experiences and stories, these four witnesses substantiate the discursive construction of São Paulo as a cosmopolitan city.

The sequence then presents images of a stadium under construction, along with football scenes, as the narrator informs that São Paulo will host the opening7 of the FIFA World Cup. Soon after, the Argentine director naturalized Brazilian, Hector Babenco, touches upon São Paulo's extensive production of symbolic goods and the richness of its artistic life. He describes it as "[…] a rare combination where everything is possible. And I can tell you that at the cultural consumption level, you won't see in any of those cities what you can see in São Paulo any week of the year." The narrator reinforces this

Fig. 1. Frames from the video 'São Paulo: A Multicultural City. Source: SPTuris [11]

claim, attributing to the city the characteristic of being "[…] one of the greatest cultural hubs in the world." The final testimony regarding São Paulo in the video comes from Luis Krausz, a professor of Hebrew and Jewish literature at USP, who states: "I see São Paulo as a city of coexistence, a city of dialogue, a city of exchanges, right? And a city where, fortunately, different groups do not live segregated from each other as happens in other cities around the world even today. In this sense, I believe São Paulo is also an example for the world".

In the testimonies of Babenco and Krausz, the cultural element emerges as notewor-thy, although stemming from different perspectives. The former is closer to the idea of symbolic goods and artistic consumption, while the latter approaches it from a more sociological perspective. Despite this, both value and highlight São Paulo's cultural richness and its capacity to produce symbolic goods, whether in the form of a theatrical performance or the production of knowledge. This, according to the proposed narrative, is directly associated with encounters between different social groups, nationalities, and territorialities. The city's association with diversity and non-segregation is reinforced by the credibility discursively built through the words of a university professor—particu-larly from a renowned public institution like USP. Simultaneously, Krausz's testimony reaffirms São Paulo as a hub for knowledge production.

4   Video: Embrace São Paulo

From the second one-minute video, we spotlight the introduction, akin to the first video, featuring the city's skyscrapers and bustling traffic. The visuals of "São Paulo: Live it All" (Fig. 2) reinforce the statements of a narrator who enthusiastically proclaims: Celebrating the diversity of culture. Celebrating the history within museums. Embracing the grand spectacles. Feeling the energy of the parks. Seizing the opportunities of fairs and events. Exploring the shopping choices. Indulging in delightful gastronomy. Embracing the lively nightlife. São Paulo: embrace it all.

The production has a more promotional character compared to the first video, which was more documentary in nature, featuring quicker shots, direct language, and less didacticism. The video's discourse invites the viewer, using the imperative 'live,' to explore São Paulo's richness across its diverse attractions and city dimensions. Moreover, it delves into a secondary meaning of 'live,' inviting a celebration of diversity. The use

Fig. 2. Frames from the São Paulo: Live It All. Source: SPTuris [11]

of the imperative and repetition reinforces the invitation, creating a sense of complicity while also offering advice to the video's imagined audience [1].

At the video's onset, in prompting the viewer to 'experience the diversity of culture,' scenes of a museum and graffiti-covered walls are displayed. These scenes present two perspectives that enable an exploration of the term 'culture's' polysemy, portraying a canonical view - art as an artistic expression - and an expanded, sociological perspective - the living culture that is constantly made and remade in daily codes on the streets, in the networks that people create and live in, attributing meaning to their ordinary and extraordinary experiences [5]. These two notions of culture clash in ideologies and values while synthesizing mixtures, emphasizing the cosmopolitanism attributed to São Paulo. Building upon this reading and now approaching the second interpretive level, we seek the narrative underlying São Paulo's brand promise. The city is depicted as a multicultural and cosmopolitan metropolis. This forms the video's initial and concluding message, exploring dimensions such as art, culture, consumption, business, environment, technology, and entertainment. As cosmopolitanism stands as São Paulo's primary characteristic, the opposite of a cosmopolitan city would be provincial. In contrast to the provincial, there exists the non-provincial, which seems to distinguish São Paulo from other capitals in Brazil. As indicated in the videos, São Paulo is home to diverse cultures, peoples, and nations, making it a metropolis, not a province. Within this interplay between the non-provincial and the non-cosmopolitan, emerges a context where São Paulo embodies the non-provincial. This second video also draws attention to new elements absent in the first, such as environmental sustainability dimensions (featuring parks) and consumption (highlighting commercial appeal)."

5   Video: How Not to Love São Paulo

Entitled "How Not to Love São Paulo?", the 30-s video was produced by the São Paulo City Hall on the occasion of the city's anniversary (Fig. 3). The following text is narrated: "How not to love São Paulo? The city of all peoples, all colors, all climates. The city that welcomes, that teaches, that pulsates. With its own language, its own habits, its own life. How not to love São Paulo? The city that rises early and stays awake until late. That entertains, employs, transforms. Those who live in São Paulo know what it is to love this city, so full of love, diversity, and plurality. Happy anniversary, São Paulo".

The format of this third video resembles the second, also short, fast-paced, and promotional. Its language is direct and informal. All three share common settings, such as bustling streets, graffiti, and museums. Additionally, they characterize the city as cosmopolitan and diverse.

Fig. 3. Frames from the video 'How Not to Love São Paulo?' Source: SPTuris [11].

6   Conclusions

In this study, we aimed to identify the discursive positioning of the São Paulo brand between 2014 and 2019, alongside the visual cues utilized by SPTuris to illustrate the essence of the brand. Employing Semprini's [10] project/manifestation model and discourse analysis as our methodology, we analyzed three nationally promoted videos of the São Paulo brand released in 2014, 2017, and 2019.

Despite shifts in municipal administration, the strategic guidelines of the São Paulo brand remained consistent, bolstering its establishment. Elements such as culture, arts, business, and consumption were highlighted in the videos, contributing to the portrayal of São Paulo as a cosmopolitan city, reinforcing its portrayal as a cultural hub and trendsetter in Brazil. The videos emphasized complementary contrasts, showcasing nature in parks versus urbanization in towering buildings, the utilitarian and practical aspects of shopping versus the creativity found in fairs and arts, and the duality between the local (paulistano) and the cosmopolitan.

This approach presents São Paulo as an inclusive city, a space for encounters, and a cosmopolitan metropolis. Through the video analyses, we highlighted the concept of cosmopolitanism. The phrase from the founder of the Italian Cantina, as mentioned in the videos, reinforces the idea that this metropolis encompasses the world, valuing local and regional cultures. Contrasting cosmopolitanism, the videos explore São Paulo as a city of coexistence, emphasizing the absence of segregation between groups, as mentioned by the professor from USP. This perspective strengthens the city as a unique, multicultural, and inclusive place.

The discursive positioning of the São Paulo brand, as constructed in the videos, stands out in its relationship between cosmopolitanism, cultural diversity, and the concept of a welcoming city for all. However, we observed that promotional strategies often simplify the complexity and genuine identity of a city in tourism campaigns. We propose that more authentic promotional strategies could be achieved through dialogical approaches, integrating active community engagement and participation in municipal tourism and cultural plans.

References

1. Dayan, D.: Televisão o quase público. José Carlos Abrandes e Daniel Dayan (org.), Televisão: das Audiências aos Públicos. Livros Horizonte e CIMJ, Lisboa (2006)
2. Dinnie, K.: Nation Branding: Concepts, Issues, Practices. Elsevier, Burlington (2008)
3. Fairclough, N.: Discurso e mudança social. Editora UNB, Brasília (2001)
4. Floch, J.-M.: Le changement de formule du quotidien approche dune double exigence: la modernité du discours et la fidélité du lectorat. In: Les Medias: expériences, recherches, actuelles, applications. IREP, Paris (1985)
5. Geertz, C.: A interpretação das culturas. LTC, Rio de Janeiro (2000)
6. Greimas, A.: Del sentido II. Gredos, Madrid (1989)
7. Kotler, P.: Marketing de lugares: como conquistar crescimento de longo prazo na América Latina e no Caribe. Prentice Hall, São Paulo (2006)
8. Semprini, A.: El marketing de la marca: una aproximación semiótica. Paidós, Barcelona (1992)
9. Semprini, A.: Analyser la communication: comment analyser les images, les médias, la publicité. Editions L'Harmattan, França (1996)
10. Semprini, A.: A marca pós-moderna: poder e fragilidade da marca na sociedade contemporânea. Estação das Letras, São Paulo (2006)
11. SPTuris. São Paulo Turismo (2023). https://spturis.com

The Influence of TikTok in Portuguese Millennials' Footwear Consumer Behaviour

Alexandre Duarte[1](✉) 🆔 and Luís Albuquerque[2] 🆔

[1] ICNOVA/UNL, Av. Berna, 26 C, 1069-061 Lisbon, Portugal
alexandreduarte@fcsh.unl.pt
[2] FCH/UCP, Palma de Cima, 1649-023 Lisbon, Portugal

Abstract. This study focusses on the influence of TikTok on Portuguese millennials' footwear consumer behaviour. Recent data shows that there are over 3.5 millions of users in Portugal (almost half of the country's total social media users), but little research has been done so far to understand how the platform impacts users and their consumer behaviour, specifically regarding the footwear industry.

This investigation emphasizes how different strategies in Marketing (content marketing, influencer marketing, word-of-mouth, virality) are perceived by the consumers and to what extent it influences their purchase intentions.

A mixed methodology was implemented, including a quantitatively analysis of an online survey with 142 valid participants, and then a qualitative analysis of two focus groups, to complement the information.

The results concluded that TikTok is very influential with Portuguese millennials' footwear consumption.

Keywords: Content Marketing · Consumer Behaviour · electronic Word-of-Mouth · Influencer Marketing · Millennials · Social Media Marketing · TikTok

1 Introduction

The following study emerges from the need of understanding how TikTok, one of the largest social media networks of today's digital landscape, directly influences consumer behaviour, especially in Portuguese millennial consumers on footwear industry.

The ease associated with how consumers discover, explore, evaluate, and purchase products nowadays is directly linked to the technological evolution of social media, e-commerce, digital platforms, and social commerce tools [3]. According to [25] there were over 4.5 billion social media users worldwide, which represents about 59% of the planet's population. Social media has replaced the influential role traditional media have in younger generations' social behaviour [9], and in their buying patterns. These new platforms are great tools for companies because they are highly engaged, interactive, massified, reactive, user-friendly, and cost-efficient [23].

This study is, therefore, relevant for brands to understand what is driving influence in millennials' consumer behaviour, as they'll be soon the next big consumption group. Their motivations, their consumer behaviour, how they relate to TikTok and how this platform influences their consumption behaviour will be analysed in the next chapters.

Á. Rocha et al. (Eds.): WorldCIST 2024, LNNS 990, pp. 117–126, 2024.
https://doi.org/10.1007/978-3-031-60328-0_12

2 Theoretical Framework

2.1 The Rise and Power of TikTok

TikTok is a short-form video social media platform that has gained immense popularity worldwide since its launch in 2018. Founded by Chinese tech company ByteDance, TikTok has been proliferating since it merged with musical.ly, and it has since become a global phenomenon like no other platform has [6], with over 1.9 billion users worldwide, ranking as the 10th most valuable brand in the world in 2023, with a market value of 65 billion dollars, just behind Google and surpassing Facebook [26].

TikTok has redefined the way people create, consume, and interact with media, as the company does not see itself as a social media platform but rather as an entertainment outlet [30] that requires the user's full-attention span compared to other competing platforms like Instagram or Twitter that have a more passive component to them, and the user may access them while multitasking with other activities.

In Portugal, there are over 3.5 million TikTok users that spend an average of 50 min per day on the platform.

Beyond its social impact, TikTok has also become a powerful marketing tool for businesses and influencers. Brands have recognized the platform's potential to reach large and engaged audiences, and many have used TikTok to promote their products and services [19]. According to a study by [26], TikTok is the 5th most-used social media among marketers worldwide with 30% of them affirming that they use TikTok to promote the companies they worked for.

2.2 TikTok's Influence on Consumer Behavior

As a social platform that has allowed users to seek their motivations for fame, self-expression, and feel recognized [4], TikTok has impacted culture in an explosive way through virality and trends. From old popular music being back to the top of worldwide charts to the book industry breaking sales records over a popular TikTok hashtag, to the normalisation of luxury brands to day-to-day style, TikTok has impact to its users, as 71% of users believe that TikTok is a driver for what is trending worldwide [29].

By the source of its content, TikTok has become a driver of individual behaviour and not just mutually exclusive to consumption, whether by its DIY videos (or do-it-yourself videos, traditionally teaching people how to do certain things without a third party involved), fashion hacks, or beauty tips, TikTok is valued by its users as a medium that makes them go out and explore new products, ideas, and trends to better their lives, as the information being perceived by users comes across as more customised and authentic then of other platforms.

2.3 Millennials as Consumers

Millennials are a complex generation to define. It is so complex that researchers call them all sorts of other things: Generation Y, Nexters, Echo Boomers, Generation XX, and Generation 2000, among several others [21] and there's no consensus in defining their exact age gap [22]. In this study, we will consider Pew Research Centre's article that defines the millennials as the ones born between 1981 and 1996 (they would be aged 27 to 42 in 2023).

In 2021, according to the World Economic Forum, millennials surpassed Generation X (or baby boomers) as the generation with the most influence globally. There were over 1.7 billion millennials, meaning that 1 in 4 people belong to this cohort. This generation already dominates the marketplace, with an estimated 75% of millennials representing the entire global workforce by 2025.

This generation is driven by a type of consumption that adds value by fulfilling their expectations, the allure of trying new things (Forbes, 2018), the gratification of an experience in their consumption, and they're overall more willing to spend money on quality and durability of purchased goods [20], being more sophisticated and selective than previous generations [15].

Also, sustainability and sustainable shopping is a big concern for millennials, as they are concerned with several aspects like family planning, the food they are consuming, the clothes they are wearing, and what their offices' recycling policy are. Nevertheless, their purchasing power ultimately undertones how sustainable they can be.

3 Methodology

3.1 Research Questions and Objectives

The main question that guided this investigation is: *"To what extent does TikTok influence Millennials' purchase intentions in the Portuguese footwear industry?"*. Nevertheless, other objectives were settled:

- Determine the influence of TikTok on Portuguese Millennials' footwear purchasing behaviour.
- Examine the correlation between TikTok's features (including variables previously mentioned in the hypothesis such as advertising, content, influencer marketing, and user-generated content) and the actual purchase intention of Portuguese Millennials' consumption of footwear.
- Determine which of the previously mentioned variables have the most impact on the Portuguese millennial consumer.
- Assess the impact of post-TikTok usage on Portuguese Millennials' awareness of footwear brands and/or products.

3.2 Methods, and Strategies

This study used a mixed methodology, as it is considered that the use of qualitative and quantitative research methods contributes to more solid and rigorous research [18]. As for the quantitative method, an online survey was conducted between September 15th and October 26th, 2023, and made available on Google Forms. The questionnaire was previously pilot tested by 10 participants to ensure the validity of the constructs and align possible discrepancies in the questions.

The first questions were eliminatory. The respondents were asked about age and usage of TikTok, and only millennials and TikTok users were allowed to proceed.

The research followed a sequential explanatory strategy where the qualitative analysis was based on the preliminary results produced by the quantitative analysis [13]. The hypotheses sought to this research were:

- H1. TikTok influences Portuguese Millennials' footwear consumer behaviour.
- H2. Influencers' TikTok presence influences Portuguese Millennials' Footwear consumer behaviour.
- H3. Content on TikTok influences in Portuguese Millennials' Footwear consumer behaviour.
- H4. Out of all social media platforms, TikTok is the one that is influencing positively the most consumption of footwear for Portuguese Millennials
- H5. Users' content and perceived word-of-mouth on TikTok influence in Portuguese Millennials' footwear consumer behaviour.
- H6. TikTok is influencing Portuguese millennials' footwear purchasing behaviour.
- H7. TikTok is useful in consumer purchasing decisions.

3.3 Survey

An online survey was applied as it is an easy tool to reach the community, to share through various groups of people and it doesn't involve any costs to the researcher [5, 25]. Social media platforms of the authors were used to spread the survey, being publicised on several platforms such as Instagram, LinkedIn, Facebook, WhatsApp groups, and through personal messages.

The sample was selected through a non-probabilistic snowball technique by convenience, as the participants are selected to whom the researcher has access [11] and it is not representative once it doesn't represent the totality of millennial TikTok users and millennial footwear consumers in Portugal.

3.4 Focus Group

With its origins in sociology [10], the focus group is considered a rich qualitative research tool that comes from an informal and carefully planned debate between previously screened individuals. [17] defines this interaction to get clearer perceptions about a defined area of interest, proving to be a considerable source of a wide variety of ideas [14].

The first set of questions focusses on the motivations behind TikTok's use and was based on [7] study. The second set of questions aims to understand the participants' perception and overall level of influence from TikTok in their day-to-day activities, focusing on their footwear consumption habits. Finally, participants are asked to explain what influences them the most on TikTok.

The sample selected was also choose by a non-probabilistic snowball sampling by convenience. Participants were found by directly asking in the researchers' personal networks (Instagram and LinkedIn) if they were (1) millennials, (2) had TikTok, and (3) were willing to join an online focus group. In total, 11 people responded to the invitation, but three of them backed down, on the previous days of conducting the focus group. So, two groups of four millennials (in each group) were interviewed throughout October 2023. Ages ranged from 27 to 29 years old. Each session lasted around 25 min and were recorded via Google Meet.

4 Results

4.1 Results from the Survey

Out of the total of 563 collected participants in the online survey, only 142 answers were validated, considering the six criteria that each answer had to meet to be considered:

- Be a millennial, according to Pew Research Centre's official definition.
- Be Portuguese
- Have a TikTok account (at least have an account and the app installed on their mobile phones)
- Have had interest in purchasing anything from TikTok's influence.
- Have had interest in purchasing footwear from TikTok's influence.
- Have had purchased footwear from TikTok's influence.

From the 142 validated answers, 44,4% of them (63) were female, 54,2% were male (77), and the following 1,4% (2) didn't want to disclaim their gender (0,7%) or identified with another gender (0,7%). 91 of them (64,1%) state that they use TikTok at least once a day, and 43 participants (30,3%) use it at least two/three times a week.

Regarding motivation, Duarte and Dias' study [8] that applied the theory of uses and gratifications in TikTok's use by Gen-Z were used. 81% of the respondents affirmed that they use TikTok for entertainment, while 77% claim use it for information about products and brands and, finally, 75,4% said they were on TikTok to see what's trending. The bottom two reasons that fell below the average include create content (39,4%) and connecting with people (31%), which is aligned with research that showed that the level of personal interaction between users (including messaging) on TikTok is much smaller compared with other social media platforms [30].

5 Discussion

5.1 Addressing the Research Questions

The main research question was "To what extent does TikTok influence Millennials' purchase intentions in the Portuguese footwear industry?". To answer that, the following hypothesis were considered (H1. TikTok influences Portuguese Millennials' footwear consumer behaviour; H4. Of all social media platforms, TikTok is the one that positively influences the most footwear consumption for Portuguese Millennials; H6. TikTok is influencing Portuguese millennials' footwear purchasing behaviour; and H7. TikTok is helpful in consumer purchasing decisions).

The participants of this study reported that they visit TikTok to find information about products and brands (65,2%) and see trends (68,6%), aligned with the report in [27] that stated that Portuguese millennials liked to follow brands online and like to engage with their content. The participants are also very engaged, with 64,1% saying they visit every day TikTok, but that is no surprise considering that in 2018, 67% of Portuguese millennials claimed to prefer video content so that TikTok can fit right like a glove with those desires [27].

From the focus group, participants deemed TikTok is very influential and helps to inspire to find new products, new recipes, new people, new points of view and creativity.

The secondary research question was "What kind of features from TikTok influence Portuguese millennials the most for footwear consumption?" To answer that, the following hypothesis were considered (H2. Influencers' TikTok presence influences Portuguese Millennials' Footwear consumer behaviour, H3. Content on TikTok influences in Portuguese Millennials' Footwear consumer behaviour, H5. Users' content and perceived word of mouth on TikTok influence Portuguese Millennials' footwear consumer behaviour). We proceeded to analyse some variables that are strongly present in Tik-Tok, including content, word-of-mouth, and influencer marketing, all with high levels of influence, highlighting at first the content in TikTok (in the univariate analysis), which seemed incredibly influential (80,87%) but then when interpreted the data using the correlation of Spearman, between all variables, we came to find that influencer marketing is also enormously influential in consumer behaviour (0,842 in a scale −1 to 1).

Content poses a critical factor for TikTok, as the platform is known for being incredibly successful in entertaining entire communities with an everlasting stream of impactful, meaningful, original, and user-generated content [29]. TikTok is unique for creating a curated online space of fluid content that can reach the entire world [2] proving to be an incredible opportunity for brands who want to create genuine relationships with their customers. Even though millennials who participated in this survey are not necessarily in the app to create content (39.4% out of N = 142), they like to see content from the brand (65.2%) and see trends (68.6%). When analysing the univariate variables, one of the more substantial factors (out of the 36 items) is 'Brands should use TikTok to enhance shopping, and the only negative one is "I would like footwear advertised to me by TikTok'", meaning that the participants want to more organic involvement from brands.

When considering data from the question on the survey (that appeared in Section V: TikTok's influence on consumer behaviour - From a scale of 1 (strongly disagree) to 7 (strongly agree), please rate the following scenarios in how they apply to you), the goal was a first attempt to funnel the reasons why users purchased footwear products on TikTok. What came from that were the variables ('I saw an influencer reviewing and or/wearing the product,' 'I actively searched for the product,' 'I saw content from the brand,' 'The product was advertised to me on TikTok,' 'I saw another user review of the product (whether content or comments in the comment section),' and 'I saw the product trending/going viral'). Most of them are positive (or above the neutral level, i.e., 4), except for advertisement, which is not a factor for the study participants.

TikTok, as a search engine, can present some perks, as highlighted in the study, with 75% of the sample (N = 142) stating that they use TikTok to see what's trending and 70% find information about products and brands. Regarding the focus group, all agree that the word-of-mouth component of TikTok is one of the most influential factors of TikTok with Francisco stating: "People are more authentic on TikTok [...] It's a platform that truly allows people to be themselves, and that's because video is video. It has movement, it has a tone of voice, it has a story and an experience, and it's more palpable than the photo, which, you know, doesn't have a background," and João said "Especially if I am looking for information about a subject online, I am very dependent on my peers' feedback and online reviews. I check all possible sources".

Other users' feedback is fundamental to millennials and consumers alike as they search for information before actively shopping (an increase versus other generations) and influence/pursuing recommendations from word of mouth, including friends, influencers, online reviews, and social media, while also sharing a concern to sharing the feedback right back. Peer-recommended websites influence Portuguese millennials' visits online [27].

Influencer marketing is an influential factor in TikTok. In every analysis, influencer marketing ranked as the most influential factor (in the correlation analysis) or one of the two influential factors (in the scores analysis). That correlates with findings in literature review [12].

A decisive variable in question that this study was unable to analyse was virality, which TikTok has gained a reputation for (the ability to constantly trend everything from music to fashion, to slang, to recipes is constantly appearing in media), as 71% of users believe that TikTok is a driver for what is trending worldwide [31]. Viral behaviours are directly linked to increased intent purchase behaviour [1] Users agree that is an influence, as we see in our sample answers, with a 6.1 mean out of 7, the strongest out of all six variables. In the focus group, participants deemed virality incredibly important because it makes them explore and find new products and trends. However, it can be too overwhelming (thus losing its impact).

6 Conclusions

This study contributes to increase the knowledge of TikTok on millennials' footwear consumer behaviour by analysing content, influence power, virality, and word-of-mouth of this platform.

Social media has been a powerful tool in influencing individuals' consumer behaviour. It has completely changed the way consumers engage by putting the focus on them and has eliminated forever the one-way communication that brands were accustomed to interacting with consumers. TikTok is one of the most recent social media platforms that has completely disrupted the social media scene, by becoming the fastest-growing brand (in terms of users) and its impact is well-documented worldwide, becoming a potent marketing tool for businesses and influencers.

Millennials are the first digital natives, so they know what it is like to grow digital and have technological skills embedded in them. They are the buyers of today, as they are the biggest consumers in the world, and they share a bunch of characteristics as consumers that include being price sensitive but indulgent in impulsive buying; they shop online, yet they want to find the next big thing that can define them; they prefer experiences and traveling, but most live pay check to pay check; they want their purchases to mean something, to add value to the price they are willing to pay; and they use the internet like no other generation, especially for their purchases. They're also very dependent on other people's feedback, including peers, social groups, and influencers. Portuguese millennials, on the contrary, are not still the biggest drivers of consumption in Portugal (they fall behind Generation X). They like to interact with brands on social media, and one in three millennials has bought something directly influenced by social media. They are particularly interested in apparel and footwear [16].

This research shown the Portuguese millennials surveyed in this study are present on TikTok, that they are influenced by the platform (with a large sum of them considering TikTok as the most influential platform in today's landscape) and influenced explicitly by influencers and other users' word-of-mouth/content. The focus group validated most of the findings and quantified examples of how and what millennials are purchasing footwear.

7   Managerial Implications, Main Limitations, and Further Possibilities

By testing whether factors such as influencer marketing, the type of content, and electronic word-of-mouth that other studies have shown to be decisive in the consumer decision-making process, can be applied to TikTok, this study helps managers to make founded decisions, in particular by a specific generation, millennials, by delving deeper into their purchasing behaviour, more specifically in the footwear industry.

The main limitation of this study was the relatively short number of valid participants in the quantitative survey. Regarding the focus groups, maybe it would be interesting to add more diversity, such as nationality, cultural backgrounds, or income, for example, as the only criteria used here was the age.

Finally, it would be interesting to see TikTok being applied to a consumer behaviour model (e.g., AIDA, or the Engel, Kollat & Blackwell model) to develop both the research on how social media, and see if those models apply to the findings, since both fields (social media and consumer behaviour) are known for its volatility, as evidenced by the literature review.

Funding Agency.   This study was funded by national funds through the FCT - Fundação para a Ciência e a Tecnologia within the project UIDB/05021/2020.

References

1. Alhabash, S., Almutairi, N., Lou, C., Kim, W.: Pathways to virality: psychophysiological responses preceding likes, shares, comments, and status updates on Facebook. Media Psychol. **22**(2), 196–216 (2019)
2. Bower, A.: Tapping Into TikTok as a Resource for Entertainment and Engagement. https://www.adweek.com/commerce/tapping-into-tiktok-as-a-resource-for-entertainment-and-engagement/. Accessed 8 Nov 2021
3. Bronner, F., de Hoog, R.: Social media and consumer choice. Int. J. Mark. Res. **56**(1), 51–57 (2014)
4. Chu, S.C., Deng, T., Mundel, J.: The impact of personalization on viral behavior intentions on TikTok: the role of perceived creativity, authenticity, and need for uniqueness. J. Mark. Commun. **30**, 1–20 (2022)
5. Creswell, J.W., Creswell, J.D.: Research Design: Qualitative, Quantitative, and Mixed Methods Approaches. Sage Publications, New York (2017)

6. Dias, P., Duarte, A.: How Portuguese adolescents relate to influencers and brands on TikTok. J. Digit. Soc. Media Mark. **10**(1), 82–95 (2022)
7. Dias, P., Duarte, A.: TikTok practices among teenagers in Portugal: a uses & gratifications approach. Journalism Media **3**, 615–632 (2022). https://doi.org/10.3390/journalmedia304 0041
8. Duarte, A., Dias, P.: TikTok: usos e motivações entre adolescentes em Portugal. Chasqui. Revista Latinoamericana de Comunicación **1**(147), 81–103 (2021). https://doi.org/10.16921/chasqui.v1i147.4419
9. Duffett, R.G.: Influence of social media marketing communications on young consumers' attitudes. Young Consum. **18**(1), 19–39 (2017)
10. Freitas, H., Oliveira, M., Jenkins, M., Popjoy, O.: The Focus Group, a qualitative research method. J. Educ. **1**(1), 1–22 (1998)
11. Gunter, B.: Media Research Methods: Measuring Audiences, Reactions, and Impact. Media Research Methods, pp. 1–320 (1999)
12. Gwi (n.d.): Millennials report: What matters to millennials in 2023? - GWI. https://www.gwi.com/reports/millennials
13. Ivankova, N.V., Creswell, J.W., Stick, S.L.: Using mixed-methods sequential explanatory design: from theory to practice. Field Methods **18**(1), 3–20 (2006)
14. Jaccard, J., Jacoby, J.: Theory Construction and Model-Building Skills: A Practical Guide for Social Scientists. Guilford Publications, New York (2019)
15. Jackson, V., Stoel, L., Brantley, A.: Mall attributes and shopping value: differences by gender and generational cohort. J. Retail. Consum. Serv. **18**(1), 1–9 (2011)
16. Klarna: Main social platforms used to purchase a product in selected countries worldwide in 2022 [Graph]. In Statista. https://www.statista.com/statistics/1341374/top-social-commerce-platforms-by-country/. Accessed 24 Oct 2022
17. Krueger, R.A.: Focus Groups: A Practical Guide for Applied Research. Sage Publications, New York (2014)
18. Meirinhos, M.F.A.: Desenvolvimento profissional docente em ambientes colaborativos de aprendizagem a distância: estudo de caso no âmbito da formação contínua (Doctoral dissertation, Instituto Politécnico de Bragança (Portugal) (2006)
19. McKinsey: Understanding and shaping consumer behavior in the next normal (2022). https://www.mckinsey.com/capabilities/growth-marketing-and-sales/our-insights/understanding-and-shaping-consumer-behavior-in-the-next-normal
20. Meeting millennials where they shop: Shaping the future of shopping malls. McKinsey & Company, 20 January 2017. https://www.mckinsey.com/industries/real-estate/our-insights/meeting-millennials-where-they-shop-shaping-the-future-of-shopping-malls
21. Moreno, F.M., Lafuente, J.G., Carreón, F.Á., Moreno, S.M.: The characterization of the millennials and their buying behaviour. Int. J. Mark. Stud. **9**(5), 135–144 (2017)
22. Pew Research Center: Where Millennials end and Generation Z begins|Pew Research Center, 22 May 2023. https://www.pewresearch.org/short-reads/2019/01/17/where-millennials-end-and-generation-z-begins/
23. Power, A.: What is social media? Br. J. Midwifery **22**(12), 896–897 (2014)
24. Statista: Number of social media users in Portugal from 2019 to 2028 (in millions) [Graph]. In Statista. https://www.statista.com/statistics/569032/predicted-number-of-social-network-users-in-portugal/. Accessed 11 Sept 2023
25. Statista: Number of social media users worldwide from 2017 to 2027 (in billions) [Graph]. In Statista. https://www.statista.com/statistics/278414/number-of-worldwide-social-network-users/. Accessed 28 Sept 2023
26. Statista: Number of TikTok users worldwide from 2018 to 2027 (in millions) [Graph]. In Statista. https://www.statista.com/forecasts/1142687/tiktok-users-worldwide. Accessed 5 June 2023

27. SOL: Millennials. Portugueses são os que mais seguem lojas nas redes sociais (2018). https://sol.sapo.pt/2018/04/30/millennials-portugueses-sao-os-que-mais-seguem-lojas-nas-redes-sociais/
28. TikTok. New studies quantify TikTok's growing impact on culture and music. Newsroom|TikTok, 21 July 2021. https://newsroom.tiktok.com/en-us/new-studies-quantify-tiktoks-growing-impact-on-culture-and-music
29. TikTok For Business: How the Entertainment Industry is Driving Culture on TikTok. TikTok for Business (2022). https://www.tiktok.com/business/en-US/blog/entertainment-culture-catalysts-tiktok?redirected=1
30. TikTok isn't a social media platform, according to TikTok. Fast Company. https://www.fastcompany.com/90746981/tiktok-social-video. Accessed 31 July 2023
31. TikTok Insights: Research Tool|Learning and resources (n.d.). TikTok for Business, 22 April 2022. https://www.tiktok.com/business/en/insights

4th Workshop on Open Learning and Inclusive Education Through Information and Communication Technology

Promoting Inclusion in the Brazilian Educational Scenario: Actions for Teacher Training

Cibelle A. H. Amato[1] , Cibele C. da S. Spigel[1] , Gerson O. E. Muitana[1] ,
Andressa G. Saad[1] , Maria Angelica de P. Couto[1] , and Valéria F. Martins[2]([envelope])

[1] Programa de Pós-Graduação em Ciências do Desenvolvimento Humano,
Universidade Presbiteriana Mackenzie, São Paulo, Brazil
{cibelle.amato,cibele.spigel,mariaangelica.couto}@mackenzie.br
[2] Programa de Pós-Graduação em Computação Aplicada, Universidade Presbiteriana
Mackenzie, São Paulo, Brazil
valeria.farinazzo@mackenzie.br

Abstract. Although there are public policies in Brazil to include students with disabilities at all educational levels, they are not always effective in effectively reducing barriers to access to education for these students. These barriers are due to infrastructure in schools, prejudice, lack of adequate skills to accommodate these students, even the lack of preparation of teachers to meet the individual educational demands of students. This work presents a series of actions to contribute to teacher training in several educational levels in the Brazilian context.

Keywords: Teacher Training · Inclusion · Disabilities · ICT

1 Introduction

The establishment of inclusive education in an educational institution is not just about its implementation; It is imperative to ensure not only student access, but also their permanence at school. To achieve this objective, it is essential to consider the individual educational demands of students, in order to guarantee an environment conducive to learning. Only in this way is it possible to ensure quality education that serves all students effectively.

The 2018 UN report on disability and development addresses the inequality that exists between people with and without disabilities in relation to the poverty rate, access to education and health. According to these data, there are at least 1 billion people with some disability worldwide, representing 12% of the population (ONU, 2018).

The WHO highlights that the population aged 60 and over is growing rapidly, projecting an increase of 223% between 1970 and 2025, reaching approximately 694 million elderly people. By 2025, a total of around 1.2 billion people in this age group are expected to reach two billion by 2050. Notably, 80% of this increase will occur in developing countries. This phenomenon reveals a crucial demographic shift, demanding greater attention to issues related to aging across the world (ONU BR, 2020).

Á. Rocha et al. (Eds.): WorldCIST 2024, LNNS 990, pp. 129–138, 2024.
https://doi.org/10.1007/978-3-031-60328-0_13

To promote education for all, accessibility and digital inclusion must be implemented by improving each person's skills and access to IT resources within the school context (Spigel et al., 2022) (Martins et al., 2019). The use of information and communication technologies (ICT) is seen as a crucial tool for education to reach everyone, regardless of their singularities.

The objective of this paper is to present the results of five actions to promote teacher training for inclusion at different educational levels and with different inclusion approaches in the Brazilian context: inclusion of elderly people; in higher education, addressing the creation of accessible material; for visually impaired people at different educational levels; for students with autism, for elementary school and, neurodevelopmental disorders for early childhood.

2   Theoretical Foundations

2.1   Inclusion in the Brazilian Context

The concept of inclusion has undergone changes in the Brazilian education system and is often confused with the concept of integration, as can be seen in public policies on inclusive education. In the Brazilian context, the Federal Constitution of 1988 stands out for guaranteeing education as a fundamental right and the Brazilian Law for the Inclusion of Persons with Disabilities, published in 2015, known as the Statute of Persons with Disabilities (Sylvestre, Martinho, and Amato, 2023).

Inclusion encompasses the acceptance of diversity in the school environment and goes beyond simply considering students' individual educational difficulties. It requires a broader understanding, involving the need to modify teaching methods, train teachers and incorporate information and communication technologies. This covers all aspects of the teaching-learning relationship and social interaction, with the aim of providing all students with the opportunity to develop, as far as possible, their academic, personal, and social skills (Tezani, 2004) (Crochik, 2011).

The challenge of ensuring access to quality and equitable inclusive education for all is still global, being considered one of the 17 Sustainable Development Goals (SDGs) that make up the global agenda for 2030 (ONU BR, 2015). In relation to the Elderly Person, there is no doubt that their legal recognition as a subject of rights comes with the Federal Constitution of 1988 (Brasil, 1988) and later with the National Policy on the Elderly (Law No. 8,842/1994) and the Statute of the Elderly (Law No. 10,741/2003) (Brasil, 1994) (Estatuto do idoso, 2003).

Public inclusive education policies play a fundamental role in seeking to provide high-quality education for all individuals, in addition to guiding the inclusive approach adopted by schools. However, it is important to highlight that it is not only the legislation in force in the country that will have the power to overcome the physical and attitudinal barriers present in educational institutions.

2.2   Disabilities

In the Brazilian reality, people with disabilities and elderly people are understood as vulnerable groups, included among "minorities". These people are still invisible to the

State in terms of prioritization in the execution of Public Policies. According to the School Census (Brasil, 2023) the number of enrollments of students with special needs grew 33.2% across the country, from 2014 to 2018. In the same period, the percentage of those included in classes also increased from 87.1% to 92.1% common. According to Census data, the public network records the highest number of students in regular classrooms. In 2018, 97.3% of students with special educational needs were enrolled in these classes. In the private network, this percentage was 51.8%.

Regarding the elderly, there was an increase in the number of enrollments in higher education of 46.3%. It was the age group that showed the greatest growth since 2013. The data indicates that Distance Education (EAD) is experiencing faster growth compared to face-to-face teaching among the elderly. In a period of one year (from 2016 to 2017), distance enrollments increased by 19%, while in-person teaching grew by just 3%. When focusing exclusively on people over 65, we observed a 5% increase in in-person enrollments and a notable 27% increase in distance learning (Brasil, 2023).

For these students with specific educational needs, the digital environment is not considered a barrier but rather the possibility of expanding and democratizing the teaching-learning environment. However, adapting to the digital environment still represents a challenge for educational institutions and teachers who need to prioritize the implementation of different strategies to facilitate access and permanence of students with disabilities and the elderly in the digital environment, in addition to recognizing the importance of qualifying all people involved in the education process.

2.3 Teacher Training About Inclusion

The teacher is a key player in any teaching-learning process. In the context of identifying developmental changes or with elderly people, it is no different. He is usually the first to notice signs that may suggest some developmental change. Teachers have been an alternative reference due to their natural role and the attention they have given to observing children's strengths and difficulties for many hours throughout the day. Therefore, they have a great chance of detecting these types of disorders among their students and referring them for appropriate evaluation in order to obtain early intervention services. However, one of the biggest obstacles in this process is the lack of skills and knowledge of these professionals to identify (Taresh, et al., 2020).

Teachers who work with the elderly, whether in Third Age Universities or in other specific projects, face the challenge of discovering an effective way of teaching this audience. It is necessary to investigate whether there are existing and duly proven methodologies on how to teach the elderly population more effectively. Research shows that teachers have a very low level of knowledge regarding identification, assessment, early detection and little competence in teaching children with special needs. (Bukvic, 2014) (Wiliyanto, 2017). According to De Leeuw, Happé and Hoekstra (2020), it is essential that, in addition to training teachers, digital tools are also developed and adapted for people with special educational needs.

Therefore, teachers play an important role in the early identification of children with any suspicion of TND or in leading the learning process with elderly people. However, what is observed in practice is that little attention is given to research that evaluates knowledge, provides training and develops a computerized evaluation system to help these professionals.

2.4 Related Works

Wiliyanto (2017) carried out a survey with the aim of evaluating the level of understanding of teachers in relation to concepts, identification, needs and challenges in implementing the identification of children with special educational needs. The study indicated that the main needs of teachers were the absence of a training program, lack of development and standardization of digital identification instruments. Brook, Watemberg and Geva (2000) conducted a survey with the aim of investigating teachers' knowledge and attitudes towards ADHD (Attention Deficit Hyperactivity Disorder) and learning difficulties. The results revealed that teachers who did not receive specific training to deal with the disorder had mistaken knowledge about the main causes and risk factors associated with the condition. They wrongly claimed that ADHD was merely the result of the influence of parental attitudes.

3 Methodology

To meet the objectives presented in this paper, the steps of the methodology used to implement actions to promote the inclusion and accessibility of students with disabilities are presented, mainly regarding the teachers' training at different educational levels. These actions are aimed at providing knowledge and practice for different disabilities. The training courses are part of the research of students at the master and doctoral levels at a private Brazilian university in a postgraduate program in the Interdisciplinary area. The main steps in this process are highlighted below:

- Bibliographic review of the topics involved in the projects, such as Brazilian legislation on inclusion and accessibility disability studies, also obtained from subjects in the Human Developmental Sciences Graduate Program.
- Construction of the course strategy, such as the definition of the target audience (for example, the educational level at which the teacher acts), the definition of the medium used to deliver the course (face-to-face, hybrid or online), the definition of the technological structure that will support the course, number of hours, availability of support material, among other things.
- Construction of data collection tools, such as collecting the target audience's profile, perceptions on inclusion, subject domain, perception about institutional support, and so on. Furthermore, it is necessary to create a collection tool that will occur after the course with data related to learning.
- Construction of the course content: using the knowledge acquired under the supervision of researchers in an appropriate learning environment.
- Creation of eBooks: several publications also supported students (educators) by providing complementary material in eBooks.
- Validation of course/e-book content: eBook content was validated by experts called judges.

- Preparation of the course as a university extension course so that it could be made available to audiences external to the university, whether in a local, regional or national context, through invitations on social networks.
- Application of actions (courses): the actions are shaped through extension courses aiming to reach an audience external to the university, with the possibility of being taken nationwide.

4 Results About Teacher Training in the Brazilian Context

Among the various works on the topic developed by the group, in a postgraduate context, five of them can be highlighted. All work was developed in Brazil.

4.1 Actions to Promote the Inclusion of Elderly People in the Educational Context

The work by Marino et al. (2021) deals with the difficulties in teaching the elderly. Aging imposes limits on the body and mind. Even in the face of these limitations - difficulty memorizing, anomie, decline in hearing acuity and visual acuity, elderly people are returning to school, taking short courses and attending university again.

These students are facing difficulties in relation to the materials that need to be adapted to their specific needs. Teachers, then, are faced with a new challenge: receiving these students over 60 and including them effectively.

Thus, the main objective of this work was to verify how easy it is for teachers to create or transform teaching material into accessible digital teaching material, aimed at these students. To this end, an ecosystem called SELI (Smart Ecosystem for Learning and Inclusion) was used (Martins et al., 2020) The participation of these professionals took place in two courses – One for creating inclusive teaching material and the other for adapting teaching material, both using this ecosystem. The project took place remotely in response to the Covid-19 pandemic.

As a result of this work, a mapping was carried out between the difficulties that are generally encountered in the elderly with the media resources that can mitigate such difficulties, such as, for example, inserting written feedback/captions/images for the problem of sound acuity. The use of the ecosystem to create accessible digital teaching material can be seen in Fig. 1. When choosing an image, the teacher can select the option of a descriptive text as shown in this figure, which is one of the possibilities for generating accessibility to an image. The model was validated by experts in Inclusive Education in a workshop during a Brazilian conference on information technology in education.

Fig. 1. Image editor screen and inclusion of descriptive text for the image (Source: author)

4.2   Actions to Promote Knowledge About Accessibility in the Context of Higher Education

The work of Spiegel et al. (2022) addresses the growing percentage of students with disabilities who have increasingly reached higher education. However, in Brazil, unlike basic education teachers, higher education teachers are not always subjected in their training, both initial and continuing, to the assumptions of inclusive education. This lack of training often leaves them with a feeling of incapacity when faced with a student with a disability in the classroom, from how to act to the pedagogical part, triggering an increase in dropout rates, as the student is faced with the lack of accessibility, with materials not adapted to their limitations, and the teacher unprepared to deal with this audience.

For higher education teachers, technical competence is required, that is, mastery in their area of competence, but not specific teacher training, focusing on didactic and pedagogical issues, however, even in these, they are not inclusive practices. Therefore, this study investigates the perception of higher education teachers about inclusive education and how the adaptation of their materials, through Information and Communication Technologies (ICT), corroborated with the Universal Design for Learning (UDL) guidelines can contribute to your pedagogical practice with a student with a disability.

As a result, an eBook was created (Fig. 2), with general guidelines and accessibility recommendations, and two courses on inclusive education were taught: the first – Inclusive attitudes in the classroom: possible paths – addresses the types of disability, the seven dimensions of accessibility and the DUA, and the second – Inclusive attitudes in the classroom: making my material accessible – provides guidance on adapting material using ICT, plus accessibility recommendations. The course had the participation of teachers from all regions of Brazil and was taught remotely. Currently, the eBook is used throughout the university, which is the focus of this study. The e-Book can be accessed at the link: https://issuu.com/cibele.spigel/docs/atitudes_inclusivas_ebook.

Fig. 2. Ebook image

4.3 Actions to Promote the Inclusion of Students with Visual Impairments

Couto (2023) focuses on actions to promote the inclusion of people with visual impairment (VI) in teaching. This deficiency brings important limitations to access to knowledge. Thus, the use of technological resources and specific tools guarantees the equalization of opportunities in the teaching-learning process for sighted students.

In order for these opportunities for access and knowledge to be guaranteed, teachers must be prepared to use these differentiated resources in the classroom. In this way, the educational environment becomes truly inclusive, with teachers prepared and confident in their pedagogical actions and satisfied students, therefore, the necessary tools for better academic development. The course was aimed at teachers of different educational levels.

The results of this research were an eBook (Fig. 3) was created for educators, with general guidelines and accessibility recommendations for VI, and a course was taught with two modules, covering everything from initial concepts about accessibility to optical and non-optical resources for inclusion, audio description, among other resources.

Fig. 3. Ebook Image

4.4 Actions to Promote the Inclusion of Students with ASD

The study by Saad (2023) aimed to research the process of building an informative digital trail to support teachers and school staff in Elementary School, in the process of school inclusion of children with Autism Spectrum Disorders (ASD). The contents were developed with the aim of providing quality information, based on scientific evidence, in accessible language, promoting involvement, motivation and engagement of users, specifically teachers and school staff professionals. The Trail was composed of six modules with the following themes: 1: The importance of Inclusive Education; 2: Neurodevelopmental Disorders; 3: Autism Spectrum Disorder (ASD); 4: The role of the school community in the inclusion of students with ASD; 5: Promoting the Learning of students with ASD; 6: Sharing experiences and multiplying knowledge. 37 Elementary School I and II educators from across the country participated in the research (they completed the entire trail).

As a result, the trail was built on the Google Classroom platform, in addition to an Ebook, shown in Figs. 4 and 5.

Fig. 4. Image of the Track on the Google Classroom Platform

Fig. 5. Ebook Image

4.5 Actions to Check for Signs of Neurodevelopmental Disorders in Early Childhood Children

This research, which is ongoing, aims to verify, even in early childhood, deficits in the development of many basic skills that may suggest some suspicion of neurodevelopmental disorders (NDD). Therefore, identifying changes as quickly as possible, especially during preschool, is extremely necessary, and the teacher is essential in this process.

Therefore, training this professional with content related to NDD and providing technological tools for identification becomes the objective of this ongoing project. Therefore, it is necessary for the teacher to obtain knowledge, children can receive a diagnosis and early interventions. This study aims to train teachers in NDD subjects and develop a web-based computerized system for early identification of preschool children at risk for NDD diagnosis.

5   Conclusions

The intrinsic nature of inclusion demands effective action, transformation, and appreciation of diversity. This implies preparation for the full exercise of citizenship and the redefinition of the social function of the school, which plays a crucial role in providing education, recognized as a public good, the right of all and the responsibility of the State and the family.

To achieve inclusion, public policies must direct their efforts to provide means that enable schools to offer inclusive education. In this context, the development of public policies that encourage collaboration between the public sector and the private sector emerges as a viable alternative. This partnership can significantly contribute to overcoming the deficiencies highlighted by specialized literature, providing additional resources and expertise to promote a more open, diverse, and accessible educational environment for all students.

In the same way, stimulating the expansion of accessibility and digital inclusion benefits the participation of all people. It is essential to consider those with special needs that are sometimes invisible. This includes elderly people, those with low education, those with intellectual and mental difficulties or limitations, physical, sensory, motor disabilities and/or reduced mobility, in addition to those with temporary limitations, among other examples. By promoting digital inclusion comprehensively, we are not only creating a more accessible environment, but also ensuring that technology is truly inclusive, meeting the diverse needs and characteristics of the population in a broad and equitable way.

As future work, the use of the described actions can be provided, in a context of extension actions, strengthening the relationship between the university and schools of different educational levels. Thus, more educators from these schools could take advantage of the opportunities already mapped in this study.

Acknowledgment. This work was carried out with the support of the Coordination for the Improvement of Higher Education Personnel - Brazil (CAPES) - Excellence Program - Proex 1133/2019.

References

Brasil. Constituição. Constituição da República Federativa do Brasil, de 5 de outubro de 1988
BRASIL, Ministério da Previdência e Assistência Social Lei n. 8.842. Política Nacional do Idoso.
Brasília: DF, 4 de janeiro de 1994
BRASIL. Instituto Nacional de Estudos e Pesquisas Educacionais Anísio Teixeira (Inep). Censo
da Educação Básica 2022: notas estatísticas. Inep, Brasília, DF (2023)
Brook, U., Watemberg, N., Geva, D.: Attitude and knowledge of attention deficit hyperactivity
disorder and learning disability among high school teachers. Patient Educ. Couns. **40**(3), 247–
252 (2000)
Bukvić, Z.: Teachers competency for inclusive education. Eur. J. Soc. Behav. Sci. **11**(4), 407–412
(2014)
De P. Couto, M.A.: Recursos em Sala de Aula para Alunos com Deficiência Visual - Orientações
para Professores de Educação Básica e Nível Superior (2023). ISBN: 978-65-264-0430-0
Crochick, J., et al.: Análise de atitudes de professoras do ensino fundamental no que se refere
à educação inclusiva. Educação e Pesquisa, São Paulo, vol. 37, no. 3, pp. 565–582, set./dez
(2011)
de Leeuw, A., Happé, F., Hoekstra, R.A.: A conceptual framework for understanding the cultural
and contextual factors on autism across the globe. Autism Res. **13**(7), 1029–1050 (2020)
Estatuto do idoso: lei federal n° 10.741, de 01 de outubro de 2003. Secretaria Especial dos Direitos
Humanos, Brasília, DF (2004)
Marino, M.C.H., Amato, C.A.H., Silveira, I.F., Eliseo, M.A., Martins, V.F.: Supporting teachers
who work with the elderly through SELI. In: Şimşek, B., Akyar, Ö.Y., Oyelere, S.S., Demirhan,
G. (eds.) Reflections on Inclusion and ICT in the context of Smart Ecosystem for Learning and
Inclusion Project, 1st edn., vol. 1, pp. 27–49. Hacettepe Universitesi Yayinlari, Ankara (2021)
Martins, V.F., et al.: A smart ecosystem for learning and inclusion: an architectural overview. In:
Gervasi, O., et al. (eds.) ICCSA 2020. LNCS, vol. 12249, pp. 601–616. Springer, Cham (2020).
https://doi.org/10.1007/978-3-030-58799-4_44
Martins, V.F., Amato, C.A.D.L.H., Ribeiro, G.R., Eliseo, M.A.: Desenvolvimento de aplicações
acessíveis no contexto de sala de aula da disciplina de Interação Humano-computador. Revista
Iberica de Sistemas e Tecnologias de Informacao **E17**, 729–741 (2019)
ONU, Disability and Development Report. ONU, New York (2018)
ONU BR – Nações Unidas no Basil. A Agenda 2030 (2015)
ONU BR. Assembleia Geral da ONU declara 2021–2030 como Década do Envelhecimento
Saudável (2020)
Saad, A.G.D.F.: Construção de uma trilha informativa de suporte à inclusão de alunos com TEA,
para professores do ensino fundamental: indicadores de efetividade, Tese de doutorado (2023)
Spigel, C.C., Amato, C., Eliseo, M.A., Silveira, I.F., Martins, V.F.: Experience report on the
integration of instructional design and accessibility in the construction of didactic material. In:
INTED2022 Proceedings, pp. 4837–4846. IATED (2022)
Sylvestre, A.A.G., Martinho, S.M.S., de la H. Amato, C.A.: Política de Inclusão no Ensino e na
Saúde. Cadernos De Pós-Graduação Em Distúrbios Do Desenvolvimento **22**(2), 16–29 (2023)
Taresh, S., et al.: Pre-school teachers' knowledge, belief, identification skills, and self-efficacy in
identifying autism spectrum disorder (ASD): a conceptual framework to identify children with
ASD. Brain Sci. **10**(3), 165 (2020)
Tezani, T.C.R.: Os caminhos para a construção da escola inclusiva: a relação entre a gestão escolar
e o processo de inclusão. Dissertação (Mestrado em Educação) – Universidade Federal de São
Carlos, São Paulo (2004)
Wiliyanto, D.A.: Analysis of teachers' understanding level, needs, and difficulties in identifying
children with special needs in inclusive school in Surakarta. J. Educ. Learn. (EduLearn) **11**(4),
352–357 (2017)

1st Workshop on Environmental Data Analytics

Impact of Preprocessing Using Substitution on the Performance of Selected NER Models - Methodology

Miroslav Potočár[(✉)] and Michal Kvet

University of Žilina, Žilina, Slovakia
{Miroslav.Potocar,Michal.Kvet}@fri.uniza.sk

Abstract. This paper investigates the effect of preprocessing, specifically word substitution by pseudo words, on the performance of selected named entity recognition (NER) models. The study focuses on explaining the methodology used during the experimental process. The paper comprehensively describes the dataset used, the process of word substitution with pseudo words, the process of model training, the process of executing the test scenario, the performance evaluation criteria and the limitations of the experiment. This paper contributes to the evolving area of Natural Language Processing by providing a comprehensive examination of the impact of preprocessing using substitution strategy on the performance of selected NER models.

Keywords: named entity recognition · preprocessing · substitution · pseudo words

1 Introduction

Named Entity Recognition (NER) plays a key role in solving various Natural Language Processing tasks. It allows the extraction of entities, such as persons, organizations, and places, from unstructured text. New NER models are regularly emerging, which are achieving increasingly better results on specific domains. However, little attention has been paid to text preprocessing, which may be a critical factor in the overall performance of the models. This paper investigates the impact of a particular preprocessing technique - pseudo word substitution - on the performance of selected NER models, namely hidden Markov model (HMM), conditional random fields (CRF), gated recurrent unit (GRU), bidirectional long short-term memory network (BiLSTM) and our Naïve model. Substitution involves replacing a particular word in a sequence with a pseudo word that to some extent reflects one of the features of that word. Such a preprocessing method is intended to improve the model's ability to generalize.

In addition to the approach used and the associated model, the preprocessing of the input data can have a significant impact on the performance of the system [5]. As noted by Hickman et al. [2] certain text preprocessing procedures can

A. Rocha et al. (Eds.): WorldCIST 2024, LNNS 990, pp. 141–150, 2024.
https://doi.org/10.1007/978-3-031-60328-0_14

help improve the accuracy of subsequent text analysis. Standard preprocessing procedures include stopword removal, lowercase conversion, and stemming. Contractions expansion (converting abbreviations and abbreviated words to their full form) is commonly used in text analysis [3]. Similar preprocessing procedures can be used for the NER task [5]. Despite the potential impact of preprocessing on the resulting performance on various NLP tasks, there has been little attention given to this topic.

This study is designed to explain the methodology used to investigate the impact of substitution on the performance of selected NER models. In the Sect. 2, we will explain the concept of word substitution by pseudo words and also present the main ideas behind the origin of the idea of replacing words by pseudo words representing certain features of the original word. Section 3 focuses on the methodology itself. Here, we describe in detail the selected dataset, the process of replacing words with pseudo words along with the tested scenarios, the process of model training, the process of executing the test scenario, the observed metrics along with the method of performance evaluation, and finally, we conclude with the limitations of the experiment.

By clarifying the methodological background of our experiment, we set the foundation for a deeper understanding of how preprocessing may lead to changes in the performance and robustness of the NER model.

2   Concept of Pseudo Word Substitution

The idea of using word substitution with pseudo words to investigate its impact on the performance of different models arose when reviewing the work of Bikel et al. [1] where pseudo words were used as one of feature that were used to solve a NER task. In our work, we do not use these words as additional features, but use them directly to replace words in sequences.

As stated in the original work [1], the intuition behind the use of these words is clear:

- In Roman languages, a capital letter at the beginning of a word is often good evidence that it is a name. Therefore, it makes sense that if we come across an unknown word that begins with a capital letter, we replace that word with a pseudo word that represents that information.
- If we consider each word consisting of numeric characters as a unique number, we would need an infinitely large vocabulary. Certain forms of numeric characters tend to represent the same information. For example, a four-digit number often represents the year, numbers separated by slashes often represent the date, numbers containing a comma usually represent monetary amounts, and numbers with a period may represent percentages.

We have taken the categories and order of features from the original work. To these features we have assigned a custom pseudo word (tag) to be used in the substitution. We have also defined the rules that must be fulfilled for a word to be replaced by a pseudo word. Almost all of the rules have the form of a regular expression. The only exception is the pseudo word representing the first word in a sentence. Here we needed an index of that word within the sentence when evaluating the condition. The individual pseudo words (tags), the conditions, their order along with an example and intuition can be seen in Table 1 which is a modification of the table from the original work. The meaning of the individual regex symbols can be found at this page. We suggest that pseudo words may help to reduce the vocabulary and increase the model's ability to generalize, making the model better at dealing with unknown words.

Table 1. Word features, pseudo word tag, conditions, examples and intuition behind them

Word feature	Tag	Condition	Example	Intuition
twoDigitNum	[TDN]	^\ d{2}$	90	Two-digit year
fourDigitNum	[FDN]	^\d{4}$	2023	Four-digit year
digitAndAlpha	[CDA]	^(?:.*[a-zA-Z].*\ d.*)| (?:.*\ d.*[a-zA-Z].*)$	A8956-67	Product code
digitAndDash	[CDD]	^(?:[\ d\-]*\ d-[\ d\-]*)$	09-96	Date
digitAndSlash	[CDS]	^(?:[\ d/]*\ d/[\ d\/]*)$	11/9/89	Date
digitAndComma	[CDC]	^(?:-?[\ d]*\ d,[\ d\,]*.?\ d*)$	23,000.00	Monetary amount
digitAndPeriod	[CDP]	^(?:-?\ d+.\ d+)$	1.00	Monetary amount, percentage
otherNum	[ON]	^-?\ d+$	456789	Other number
allCaps	[AC]	^[A-Z]+$	OSN	Organization
capPeriod	[CP]	^([A-Z]([a-z]{0,2}| [a-z][A-Z])\.\ s*)+$	OSN	Organization
firstWord	[FW]	*word index = 0*	*first word of sentence*	No useful capitalization information
initCap	[IC]	^[A-Z][a-zA-Z]*$	Sally	Capitalized word
lowerCase	[LC]	^[a-z]+$	can	Uncapitalized word
other	[OW]	^[\ s\ S]*]+$	,	Punctuation marks, all other words

3 Methodology

3.1 Data

In our research we have used the *CoNLLpp* dataset [6], which is a corrected version of the original *CoNLL2003* dataset [4]. In the *CoNLLpp* version, 5.38% of the sentences in the test set have been manually corrected compared to the original version. *CoNLL2003* is a widely used NER benchmark dataset. The whole dataset is already partitioned into a training set containing 14041 sentences, a validation set consisting of 3250 sentences, and a test set consisting of 3453 sentences. There are 4 types of named entities. The first entity is persons (PER), denoting the names of individuals or groups. The next type of named entities are locations (LOC), where the names of political or geographically defined places such as cities, provinces, states, international regions, bodies of water, mountains, etc. are included. The third group is organizations (ORG), which includes names of companies, agencies, institutions, etc. The last type of named entities is miscellaneous (MISC), which includes names of entities that do not fit into any of the previous three categories. They may include names of events, nationalities, products, artworks, etc.

The version of *CoNLLpp* from the HuggingFace portal that we use contains only data in English. In addition, each word is also given its corresponding part-of-speech tag. However, in our experiment we will not use this knowledge and will only focus on prediction based on word sequences. A single row consists of an array of words and an array of their associated named entity tags. During the experiment we used all available sets. The training set was used to train the models, the validation set was used to tune the hyperparameters, and the test set was used to evaluate the performance of the models on previously unseen data. We kept the individual sets in their original form, i.e., we did not change the order of the sentences during training.

3.2   Replacing Words with Pseudo Words

In our research, we focused on the effect of using pseudo words on the performance of models in a NER task. The sentences and the words occurring in them were sequentially walked through in each dataset. Each word was subject to a series of tests. If a word met any of the conditions, it was replaced in the sentence by the pseudo word corresponding to that condition. A word that already met one of the conditions was excluded from further consideration.

The whole process began with the creation of a dictionary of known words. This dictionary was created based on the training set only. The sentences occurring in the training set are flattened and a single array containing all the words occurring in the corpus is created. In the next stage of dictionary definition, there are two possible scenarios. In the first scenario, words that contain numbers or only consist of punctuations remain in the array. In the second scenario, such words are removed. Independently of the applied scenario, the frequency of occurrences of each word is computed based on the given array. Finally, words with occurrence frequency below a certain threshold are removed. From the remaining unique words, a dictionary of known words is created.

The next stage of the process involved the actual replacement of words in the sets by pseudo words. This phase is applied to all the sets (train, validation, test). The sets are sequentially walked through sentence by sentence and sentence by word. Each word is subjected to a series of conditions. The first condition is the occurrence of the word in a dictionary of known words. If the word is found in this dictionary, its form is kept and it is excluded from further processing. If the word is not found in the dictionary, it is subjected to further testing. In the second step, the conditions are applied to this unknown word in a well-defined order. The individual conditions and their order of application are listed in Table 1. If a word satisfies the corresponding condition, it is replaced by the corresponding pseudo word and is excluded from further processing. If the word does not satisfy the actual condition, the following condition is applied to it in order. If a word does not satisfy any of the conditions, it is placed on the last condition satisfied by each word.

Datasets processed in this way are used to train the model and evaluate its performance. In our experiment, we have tested the following scenarios:

- **No modification** - In this case, we have not applied any changes to the individual datasets and have used them in the format in which we got them from the source.
- **Removal of words containing numbers or consisting only of punctuation marks from the dictionary of known words** - All words from the test set are included in the list of known words, except for words containing digits or consisting only of non-alphanumeric characters. Thus, only these words are replaced by pseudo words in the training set. In the validation and test sets, words that did not appear in the training set are also replaced. However, the model had no opportunity to learn to recognize these pseudo words and thus they will only appear as unknown words for the model.
- **Removal of words from the dictionary of known words where the frequency of occurrences in the training set is less than a threshold** - This scenario contains four sub-scenarios for each frequency of occurrences (1, 2, 3, 4).
- **Remove those words from the dictionary of known words that contain numbers, consist only of punctuation marks, or have a frequency of occurrence less than the threshold** - This scenario is a combination of the two previous scenarios.

3.3 Model Training

The training understandably varied depending on the model. Each model required its specific training data format. Some of the models, namely CRF, GRU and BiLSTM, also required hyperparameter tuning.

The training process of Naïve model looks as follows. The sentences in the training set are flattened into an array of words and an array of associated named entity tags. For each word the most frequently used named tag for that word is defined. Also the most frequent tag in whole dataset is identified. It will be later assigned to unknown words.

As HMM model we have used the *HiddenMarkovModelTagger* implementation, available in the *NLTK* library. This implementation requires a collection of sentences for its training, where each sentence is represented by an array of pairs where the first position contains the word and the second position contains the associated named entity tag. We used the data prepared in this way as an argument to the model's training function.

From the *sklearn-crfsuite* library, we have used the *CRF* implementation. From a training dataframe, individual sentences represented by an array of words and a separate array of corresponding named entity tags are extracted. CRF requires defining a set of features and converting words to these features. We have taken the function that converts a word into a dictionary of features from the documentation page of the *sklearn-crfsuite* library. The original version of the function also produced features based on part-of-speech tags. Such features have been removed from the function to ensure equal conditions and available information across models. The model on its input for training and prediction requires every word in the sentence to be converted into dictionary of features. The CRF contained hyperparameters that needed to be determined. Using restricted grid search, we have tested different combinations of hyperparameters on the validation set. From the measured values, we have found that the best results are given by the combination of hyperparameters shown in Table 2. Remaining hyperparameters were left at their default values.

The training of the GRU and BiLSTM models looks identical in both cases. Since neural networks require only numerical data on their input, it is necessary to convert words and named entity tags to numbers. A special tag [**PAD**] is added to the list of named entity tags, which is used to represent padding. The sentences from the (preprocessed) training dataset are flattened and for each word the frequency of its occurrence within the whole dataset is determined. Since the size of the neural network inputs affects the number of parameters and hence the time required to train them, only a subset of the most frequently occurring words is selected from the list of unique words. The number of these words is determined by the *vocab_size* parameter. From the set of unique words, *vocab_size* − 2 most frequently occurring words are selected. This set of unique words constitutes our vocabulary. The value 2 is subtracted from the original *vocab_size* parameter, since two values are reserved for special tokens that represent unknown words and padding. Based on the vocabulary, a *Keras StringLookup* layer is created. This layer will provide the conversion of words to numbers. Within all datasets, word arrays representing sentences are converted to number arrays using this layer. Training dataset prepared in this fashion needs to be divided into equally sized mini-batches. The size of a single batch is determined by the *batch_size* parameter. Tensors are created from the training sentences and their associated named entity tags. These tensors are concatenated into equally sized mini batches (the exception is the last batch, which may be smaller). The tensors in each of these mini batches have the same size, which is equal to the number of words in the longest sentence within the mini batch. Shorter sentences are aligned to the required size using a special character, padding. The adjusted data is used as an argument to the fit method, which is used to train the model. Since padding is used, a custom loss function based on the *SparseCategoricalCrossentropy* loss function was created. This loss function only takes into account the error in positions corresponding to the original sentence, so any part with padding is ignored when computing the error. The described loss function as well as the preprocessing of the input data is a modification of an example taken from the official documentation page of the *Keras* library, where the NER task using the transformer model has been solved. During training we used early stopping in order to reduce overtraining. This monitored the loss on the validation set and if the loss increased for two consecutive epochs, training was terminated. The structure of the GRU network can be seen in Table 3 and the structure of the BiLSTM is shown in Table 5. GRU and BiLSTM require hyperparameter tuning for their proper functioning. Using restricted grid search, we tested different combinations of hyperparameters on the validation set. From the measured values, we have found that the best results for GRU model are given by the combination of hyperparameters

shown in Table 4 and best hyperparameters combination for BiLSTM is shown
in Table 6. The other hyperparameters were left at their default value.

Table 2. CRF hyperparameters

Hyperparameter	Value
algorithm	lbfgs
c1	0.1
c2	0.1
max_iterations	200
all_possible_transitions	True

Table 3. GRU model structure

Layer type	Parameters
Embedding	input_dim=lookup_layer.vocabulary_size()+1, output_dim=100
GRU	units=50, return_sequences=True
Dense	units=10, activation='sigmoid'

3.4   Performing a Test Scenario

The flow of each test scenario consists of several steps. In the first step, a dictionary of known words is generated based on the training data. Based on scenario, words that contain digits or that consist entirely of punctuation marks are
retained or removed. Next, the frequency of occurrences for each word is calculated and words that have a frequency lower than a given threshold are removed
from the dictionary. In the next step, word replacement with pseudo words is

Table 4. GRU hyperparameters

Hyperparameter	Value	Note
vocab_size	20000	Upper bound for number of words in string lookup layer
output_embedding	100	Each word is converted into 100 dimensional numeric vector
units	50	Number of GRU units
batch_size	32	Each training mini-batch consist of 32 padded sequences (except the last)
epochs	100	Early stopping was used, so this number is the upper limit

Table 5. BiLSTM model structure

Layer type	Parameters
Embedding	input_dim=lookup_layer.vocabulary_size()+1, output_dim=100
Bidirectional(LSTM)	units=100, return_sequences=True
Dense	units=10, activation='sigmoid'

Table 6. BiLSTM hyperparameters

Hyperparameter	Value	Note
vocab_size	20000	Upper bound for number of words in string lookup layer
output_embedding	100	Each word is converted into 100 dimensional numeric vector
units	100	Number of LSTM units in one direction
batch_size	32	Each training mini-batch consist of 32 padded sequences (except the last)
epochs	100	Early stopping was used, so this number is the upper limit

handled. Depending on the scenario, this phase can be skipped. If the replacement should be performed, the series of conditions is applied to each dataset (training, validation, test). For each word, it is first checked for its occurrence in the dictionary of known words. If the word occurs in the dictionary, it is left unchanged and excluded from further processing, otherwise, given the condition it satisfied, it is replaced by the corresponding pseudo word. Next comes the initialization of the model. The processed training data is sent to the model initialization method. This initialization step is used by the GRU and BiLSTM models to create the *StringLookup* layer. This is followed by transformation of training data (and, in the case of GRU and BiLSTM, also validation data) into the format needed for model training. Next, all three datasets are transformed into the format required for prediction and performance evaluation. In the last step, training and performance evaluation of the model takes place. This step is performed N times, storing the result in the result list. These N results are then used to calculate the average value of the observed metrics.

3.5 Performance Evaluation

The most commonly used metrics for evaluating NER models are precision, recall, and F1 score. These metrics provide a broad view of model performance. Their use is widespread as they provide a balance between the model's ability to correctly identify entities (precision) and its ability to not miss any real entities (recall). The F1 score provides a single metric that balances both considerations. Because of this, we provide the F1 score as the main metric.

To evaluate the performance of the models, we used the *seqeval* framework, which is a Python framework available through the *evaluate* library, designed to evaluate labeled sequences. In the context of NER, *seqeval* provides values for metrics such as accuracy, precision, recall, F1 score for the entire dataset and also provides the same metrics for individual named entity categories. The *seqeval* provides two evaluation modes, **default** and **strict**. The **default** mode aims to mimic *conlleval*, while the **strict** mode evaluates inputs based on the specified schema. Since our data uses the IOB2 scheme, we used **strict** mod in our evaluation.

For each scenario we performed 5 runs, i.e. in each run we re-created and re-trained the model. Using *seqeval*, we have evaluated the individual metrics and stored them in a list of results. The final result for a given metric is calculated as the average of all runs.

3.6 Experiment Limitations

There are several limitations in our experiment that can be potential sources of error. The first limitation is related to the dataset used. In the experiment, we only used the *CoNLLpp* dataset, containing data from English-language newspaper articles. In order to be able to make general conclusions applicable to different domains and languages, it would be necessary to perform experiments with datasets containing data from different domains and also in different languages.

Another limitation relates to individual sets. For training, validation, and testing, we used prepared sets that were directly available within *CoNLLpp*. For more accurate results, it would be appropriate to combine the individual parts into a whole, which would then be randomly used to create training, validation, and testing sets. We also did not perform random shuffling of the training data as part of the experiment. This is not a problem in case of the Naïve, HMM and CRF models, but in case of neural network based models, training on different mini batches could lead to slightly different results.

Rules that replace words with pseudo words can also be a source of distortions. We create the above regular expressions, and since we are not linguistic experts, we may have created expressions that inadequately capture some of the categories.

Another source of error may be the models themselves. We designed the GRU and BiLSTM models ourselves based on our knowledge and experience. These models are probably not achieving their maximum potential.

Technical limitations were the reason why model tuning was performed only on a subset of all available hyperparameters. Also, due to technical limitations and lack of computational power, we did not conduct model tuning during the experiments themselves, which means that the same hyperparameters are also used for models trained on data in which some words have been replaced by pseudo words.

4 Conclusion

To summarize, the study aimed to systematically investigate the impact of preprocessing based on pseudo word substitution on selected NER models, namely Naïve, HMM, CRF, GRU and BiLSTM. We introduced in detail the concept of pseudo word substitution and provided the Table 1 listing the pseudo words used along with the conditions that must be satisfied for a word to be replaced by a corresponding pseudo word. The remaining part of this work was devoted to a detailed description of the methodology used in the experiment.

The insights gained from this research not only advance our understanding of the interrelationship between preprocessing techniques and NER outcomes, but also have practical relevance for researchers and practitioners who would like to further investigate substitution as a preprocessing technique. We suggest that proper preprocessing techniques may be key to obtaining models capable of better generalization.

As an extension for the future, we propose to extend the set of existing pseudo words with new elements that will allow finer word discrimination and hence more fine-grained feature encoding.

Acknowledgment. It was supported by the Erasmus+ project: Project number: 2022-1-SK01-KA220-HED-000089149, Project title: Including EVERyone in GREEN Data Analysis (EVERGREEN) funded by the European Union. Views and opinions expressed are however those of the author(s) only and do not necessarily reflect those of the European Union or the Slovak Academic Association for International Cooperation (SAAIC). Neither the European Union nor SAAIC can be held responsible for them.

References

1. Bikel, D.M., Schwartz, R., Weischedel, R.M.: An algorithm that learns what's in a name. Mach. Learn. **34**, 211–231 (1999)
2. Hickman, L., Thapa, S., Tay, L., Cao, M., Srinivasan, P.: Text preprocessing for text mining in organizational research: review and recommendations. Organ. Res. Methods **25**(1), 114–146 (2022)
3. Naseem, U., Razzak, I., Eklund, P.W.: A survey of pre-processing techniques to improve short-text quality: a case study on hate speech detection on twitter. Multimedia Tools Appl. **80**, 35239–35266 (2021)
4. Sang, E.F., De Meulder, F.: Introduction to the CoNLL-2003 shared task: Language-independent named entity recognition (2003). arXiv preprint cs/0306050
5. Situmeang, S.: Impact of text preprocessing on named entity recognition based on conditional random field in Indonesian text. Jurnal Mantik **6**(1), 423–430 (2022)
6. Wang, Z., Shang, J., Liu, L., Lu, L., Liu, J., Han, J.: CrossWeigh: training named entity tagger from imperfect annotations (2019). arXiv preprint arXiv:1909.01441

Correlation n-ptychs of Multidimensional Datasets

Adam Dudáš[✉]

Faculty of Natural Sciences, Matej Bel University, Banská Bystrica, Slovakia
adam.dudas@umb.sk

Abstract. Correlation analysis studies the predictive potential stored in attribute pairs of multidimensional datasets, which is subsequently summarized through a correlation matrix or visualized using a correlation heatmap. Both of these approaches are suitable for examining pairs of attributes and the correlation values between them but are difficult to use when working with a larger number of attributes. Datasets containing several dozen or hundreds of attributes are, in the era of big data, gaining importance rapidly and are often the basis for building analytical models using machine or deep learning models. This paper focuses on the visual representation of parts of multidimensional datasets that carry a significant part of the prediction potential - correlation n-ptychs, while we consider the values $2 \leq n \geq 6$. The proposed concept is subsequently presented in a case study conducted on a dataset focused on renewable energy and weather conditions.

Keywords: Correlation structures · Visual data analysis · Prediction potential · Pattern recognition

1 Introduction

The current era is characterized by the large amounts of data which is being generated on a daily basis. Data is commonly referred to as Big Data in the case it takes on properties referred to as $3V$ - the number of these "V"s increases over time, in some literature $5V$ is the most basic model of view of big data. Out of all of these properties the volume, variety and velocity of the data constitute core of modern large and commonly dynamic datasets [1,2]. These properties bring forth several problems related to their processing and analysis - processes, which need to be supported by the use of methods of statistical analysis, visualization of the data and other methods of exploratory or predictive data analysis [3,4].

In this paper, we are working within the area of pattern recognition in large datasets - the area which is lively in regards to building and optimizing machine and deep learning models. Yet, our work focuses on the use of approaches and methods originating from correlation data analysis and visual analysis of data - the methods more interpretable by human observers. The combination of these two areas brings interesting results which are useful in knowledge, pattern or trend discovery and are applicable to number of practical problems.

Á. Rocha et al. (Eds.): WorldCIST 2024, LNNS 990, pp. 151–160, 2024.
https://doi.org/10.1007/978-3-031-60328-0_15

For example, the authors of [5] explore three techniques of correlation visualization for multidimensional geo-temporal datasets. The study presents visualization's effectiveness depending on the task to be carried out. Based on the findings authors present a set of design guidelines for geo-temporal visualization techniques and identification of prediction potential in such data.

On the other hand the work [6] focuses on visualization of prediction potential stored in datasets through a correlation colour map of the transformed data used to identify true and false alarms. In the correlation colour map correlation and redundancy information can be easily found and used to improve the alarm settings, and statistical methods such as singular value decomposition techniques can be applied within each cluster to help design multivariate alarm strategies.

The main objective of this work is the visual representation of parts of multidimensional datasets that carry a significant part of the prediction potential. These parts of a dataset are called correlation n-ptychs, where n denotes the number of considered attributes for visualizing - in the paper, we consider the values $2 \leq n \geq 6$. The main contribution of this study is a proposal of a novel visualization method of correlation n-ptychs which is presented on a case study conducted on a dataset focused on renewable energy and weather conditions.

The body of the paper is structured as follows - the second section of the work focuses on the description of correlation n-ptychs nested in multidimensional datasets, metrics to ensure minimization of n-ptych which are constructed and evaluation metric. Section 3 consists of the case study of the renewable energy dataset, construction of correlation n-ptych on this dataset and subsequent evaluation of this visualization model. In the last section of the paper, we present the conclusion of the work, offer a scenario in which the proposed model can be used and define some possible future work ideas.

2   Correlation n-ptychs Nested in Multidimensional Datasets

A pair of attributes A_i and A_j is correlated with each other when attribute A_i has predictive potential for attribute A_j or vice versa. We measure the correlation of two attributes using the correlation coefficient, which points to how much attribute A_j is a function of attribute A_i. There are several types of correlation coefficients (eg. Pearson, Spearman, Kendall) that are aimed at different types of relationships between attributes, but regardless of the type of correlation, this coefficient can take on values from the $[-1, 1]$ interval while [7]:

- $corr(A_i, A_j) = 1$ denotes complete correlation and significant prediction potential between values of attributes $A_i and A_j$.
- $corr(A_i, A_j) = 0$ denotes no correlation between values of attributes A_i $and A_j$.
- $corr(A_i, A_j) = -1$ denotes complete anticorrelation and - similar to complete correlation - significant prediction potential between values of attributes $A_i and A_j$.

With the use of correlation coefficients, we examine the predictive potential stored in attribute pairs of a dataset, which is, then, summarized through a correlation matrix and visualized using a correlation heatmap [8].

Correlation n-ptych is a complete graph composed of n vertices interconnected via a set of edges. In the correlation n-ptych, vertices represent attributes of a dataset, while edges are weighted by a value of correlation coefficient measured between a pair of attributes. Since n-ptych is a complete graph, for each n-ptych N, the number of correlation coefficient values to be measured is computed as follows:

$$\#(corr)_N = \frac{n(n-1)}{2} \tag{1}$$

The number of correlation coefficients can be problematic mainly in methods similar to the Kendall rank correlation coefficient, which takes into account a high number of attribute value combinations. Number of possibly constructible n-ptychs nested in dataset A can be computed as:

$$\#(N) = \binom{\#(A)}{n} \tag{2}$$

where $\#(A)$ denotes number of attributes of dataset A. Since the construction of all n-ptych can be a problem for datasets with a high number of attributes, we need to truncate the state space of the problem. This can be done by application of border of admissibility σ, which denotes minimal user-defined admissible absolute value of correlation coefficient measured between pair of attributes in a n-ptych. The value of σ border is set on the basis of the use of correlation n-ptych - eg. in the case, we are building a linear prediction model on a dataset, the literature recommends not using values of correlation coefficients $|corr(A_i, A_j)| < 0.8$, for non-linear models this border can be relaxed significantly [3,8].

Each correlation n-ptych N is constructed and then scored by its' correlational contribution η:

$$\forall N, \ \eta_N = (\sum_{i,j=1}^{n} |corr(A_i, A_j)|) - n \tag{3}$$

This measure represents the correlational contribution of n-ptych N to the dataset the N is nested in and can be used to evaluate the importance of n-ptych to the studied dataset.

In the scope of this paper, we consider only n-ptychs of size $2 \leq n \geq 6$ - the diptychs, triptychs and tetraptychs are only planar n-ptychs (can be visualized in a plane without crossing of edges of the graph), the pentaptychs and hexaptychs are not planar, but still legible (see Table 1). The n-ptychs of higher dimension are hard to read.

Table 1. Diagram, example of possible attribute combinations and number of correlations for correlational n-ptychs of size $n = \{2, 3, 4, 5, 6\}$

n	n-ptych	Diagram	Attribute combinations	#(corr)
2	diptych		$(atr_1, \ atr_2, \ atr_3, \ atr_4, \ atr_5, \ atr_6)$	1
3	triptych		$(atr_1, \ atr_2, \ atr_3, \ atr_4, \ atr_5, \ atr_6)$	3
4	tetraptych		$(atr_1, \ atr_2, \ atr_3, \ atr_4, \ atr_5, \ atr_6)$	6
5	pentaptych		$(atr_1, \ atr_2, \ atr_3, \ atr_4, \ atr_5, \ atr_6)$	10
6	hexaptych		$(atr_1, \ atr_2, \ atr_3, \ atr_4, \ atr_5, \ atr_6)$	15

3 Pattern Recognition in Renewable Energy Dataset with the Use of Correlation n-ptychs

The proposed visualization method of correlation n-ptychs can be applied to any multidimensional dataset. For the purposes of this work, we present the concept on a case study conducted on a dataset focused on Renewable energy and weather conditions (Renewable energy dataset for short) with the use of Python programming language.

The Renewable energy dataset [9] is focused on the examination of the effects of environmental variables on energy consumption in Poland from 2017 to 2022. This dataset consists of 17 attributes measured over 196 776 records, specifically: Time of measurement, Energy delta (in W/h) - change of energy consumption between two consecutive times, Global Horizontal Irradiance (GHI), Temperature, Pressure, Humidity, Wind speed, Rain and Snow - the amount of rain/snow in milimeters measured between two consecutive times, Clouds - situation of cloudiness, binary attribute Is Sun, Sunlight Time, Day Length, Sunlight Time/Day Length for the proportion of sunlight time compared to the length of the day, Wheather type, Hour of measurement and Month of measurement.

Since correlation coefficient values are measured on numerical values only, the correlation analysis takes into account only 16 of these attributes (the attribute

Time of measurement is in the timestamp format). In Fig. 1 we present a correlation heatmap for the Spearman rank correlation for the Renewable energy dataset.

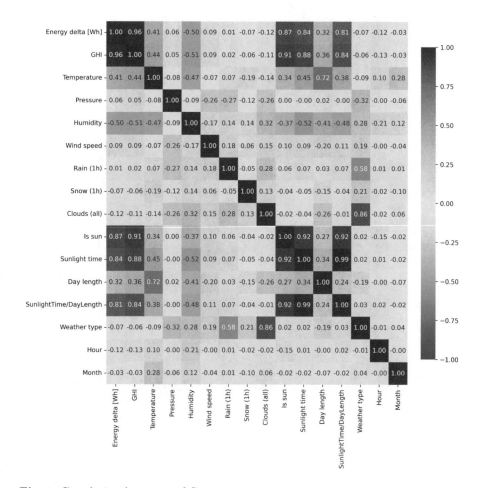

Fig. 1. Correlation heatmap of Spearman type measured on the Renewable energy dataset

In the correlation map in Fig. 1, we see several interesting correlation coefficient values but also number of weaker correlations, so first we truncate the correlation matrix (or map) using σ border. In this case study, we choose the value $\sigma = 0.6$ and thus the minimum absolute value of the correlation coefficient between two attributes will be 0.6. With this operation, we significantly reduce the required number of assembled n-ptychs. The internal representation of the truncated correlation heatmap is presented in Fig. 2. In this heatmap, we set all values of correlation coefficients lower than the specified σ border to 0. In the case that all values in a row or column of the matrix $corr(A_i, A_j) = 0$ (except for $corr(A_i, A_i)$, naturally), the entire row of the truncated correlation heatmap is dropped.

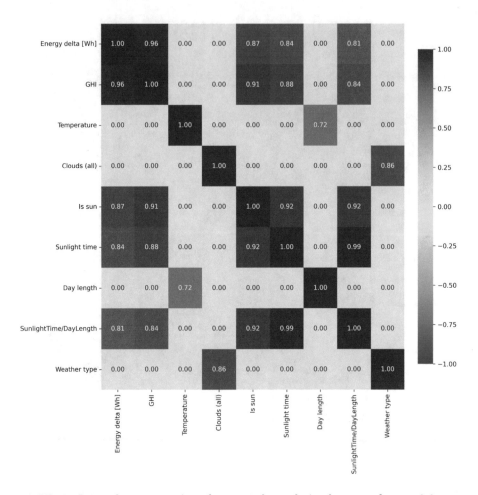

Fig. 2. Internal representation of truncated correlation heatmap for $\sigma = 0.6$

This truncated correlation heatmap (matrix) serves as the data structure for the construction of correlation n-ptych nested in the truncated dataset. We present the visualization of all correlational diptychs (Fig. 3), triptychs (Fig. 4), tetraptychs (Fig. 5) and the pentaptych (Fig. 6) which are constructable on the truncated dataset. In these figures, the attributes are labelled with the use of the first capital letters of each attribute title.

The use of the σ border set for correlation coefficient values truncated the considered state space of the problem considerably. Specifically, there is

T	DL	C	WT	ED	GHI	ED	IS
$\eta = 0.72$		$\eta = 0.86$		$\boldsymbol{\eta = 0.96}$		$\eta = 0.87$	

ED	ST	ED	ST/DL	GHI	IS	GHI	ST
$\eta = 0.84$		$\eta = 0.81$		$\eta = 0.91$		$\eta = 0.88$	

GHI	ST/DL	IS	ST	IS	ST/DL	ST	ST/DL
$\eta = 0.84$		$\eta = 0.92$		$\eta = 0.92$		$\boldsymbol{\eta = 0.99}$	

Fig. 3. Correlation diptychs of renewable energy dataset for $\sigma = 0.6$

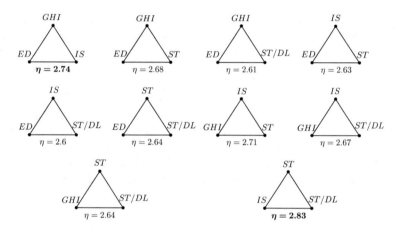

Fig. 4. Correlation triptychs of renewable energy dataset for $\sigma = 0.6$

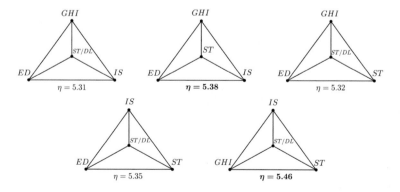

Fig. 5. Correlation tetraptychs of renewable energy dataset for $\sigma = 0.6$

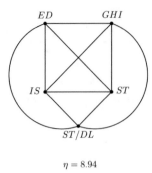

$$\eta = 8.94$$

Fig. 6. Correlation pentaptych of renewable energy dataset for $\sigma = 0.6$

$\sum_{n=2}^{5} \binom{16}{n} = 6868$ possible n-ptychs nested in the untruncated dataset (counting only n-ptychs of size $2 \leq n \geq 5$). The use of the σ border lowered the number of possible combinations of all n-ptychs to 28. Therefore, we truncated at least 99.59% of state space which would produce n-ptychs with inadmissible values of correlation coefficients.

From the experiment, we see that we are looking for a small n-ptych (small number of attributes) with a high η score - such subsets of attributes of the studied dataset should be important in decision-making processes. In the figures, we mark the strongest η scores in bold lettering. We can see that we identified a pentaptych containing attributes Energy Delta, Global Horizontal Irradiance, Is Sun, Sunlight Time and Sunlight Time/Day Length. The tetraptych with

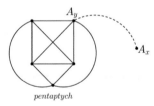

Fig. 7. Example of use of pentaptych values to estimate the values of attributes outside of the structure

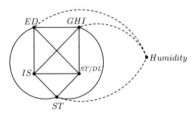

Fig. 8. Example of predictive potential transitivity between attributes of correlation pentaptych and Humidity attribute for $\sigma = 0.5$

the highest η score contained the same attributes except for the Energy Delta attribute and in the strongest triptych we also excluded Global Horizontal Irradiance from the pentaptych.

4 Conclusion

The research presented in the scope of this work presents a method of identifying correlation n-ptychs using methods of visual data analysis based on K_n graphs. Since it is possible to compile a large number of such graphs of different sizes on a multidimensional dataset, we also created a visualization method, which significantly reduces the number of combinations of attributes visualized in correlation n-ptychs.

Since correlation n-ptychs represent a set of attributes of the studied dataset carrying a large portion of its predictive potential, they significantly influence the quality of decisions made using machine learning models. In the case we understand this predictive potential as a transitive measure, we can perceive the correlation n-ptychs as sets of attributes that can be computed precisely based on each other. In such a case, it is possible to impute the values of attributes that do not have to have a relationship with all of these attributes (one of the n-ptych attributes is sufficient) on the basis of n-ptych correlation values. Figure 7 shows an example of such imputation - even though the attribute A_x outside of the correlation pentaptych has a strong enough correlation coefficient value only with attribute A_y of correlation pentaptych, in the case the value of A_y is missing, we can use any of the remaining attributes of pentaptych to compute the value of A_y and then A_x. Naturally, the greater the η score of the n-ptych, the better the prediction.

If we were to apply this concept to the case study from Sect. 3, when setting $\sigma = 0.5$, we have this type of relationship between the identified pentaptych and the Humidity attribute (see Fig. 8).

The presented correlation structures can also be used as a means of meta-analysis or interpreting of decision models and can be considered as an alternative to approaches such as the Garson and Olson method [10–12]. Correlation n-ptychs differ from these methods in that they are not parametric - the method does not need a trained model or a specific task to build correlation n-ptychs on. This feature and the use of the presented model need to be verified more deeply in future work. In the following research, we are also aiming at the preparation of a tool package for the Python language, which will be aimed at identifying correlation structures (such as n-ptychs) in multidimensional datasets.

Acknowledgements. The support of Advtech_AirPollution project (Applying some advanced technologies in teaching and research, in relation to air pollution, 2021–1-RO01-KA220-HED-000030286) funded by European Union within the framework of Erasmus+ Program is gratefully acknowledged.

References

1. Kvet, M., et al.: Master index access as a data tuple and block locator. In: Conference of Open Innovation Association FRUCT, pp. 176–183 (2019). https://doi.org/10.23919/FRUCT48121.2019.8981531
2. de Espona Pernas, L., Vichalkovski, A., Steingartner, W., Pustulka, E.: Automatic indexing for MongoDB. In: Abelló, A., et al. (eds.) New Trends in Database and Information Systems: ADBIS 2023 Short Papers, Doctoral Consortium and Workshops: AIDMA, DOING, K-Gals, MADEISD, PeRS, Barcelona, Spain, September 4–7, 2023, Proceedings, pp. 535–543. Springer Nature Switzerland, Cham (2023). https://doi.org/10.1007/978-3-031-42941-5_46
3. Xue, L., Jiang, D., Wang, R., Yang, J., Hu, M.: Learning semantic dependencies with channel correlation for multi-label classification. Vis. Comput. **36**(7), 1325–1335 (2019). https://doi.org/10.1007/s00371-019-01731-5
4. Li, X., Fan, Y., Lv, G., et al.: Area-based correlation and non-local attention network for stereo matching. Vis. Comput. (2022). https://doi.org/10.1007/s00371-021-02228-w
5. Pena-Araya, V., Pietriga, E., Bezerianos, A.: A comparison of visualizations for identifying correlation over space and time. IEEE Trans. Visual Comput. Graphics **26**(1), 375–385 (2019). https://doi.org/10.48550/arXiv.1907.06399
6. Yang, F., Shah, S.L., Xiao, D., Chen, T.: Improved correlation analysis and visualization of industrial alarm data. ISA Trans. **51**(4), 499–506 (2021). https://doi.org/10.1016/j.isatra.2012.03.005
7. Nettleton, D.: Commercial data mining. Elsevier (2014). ISBN 978-0-12-416602-8
8. Bon-Gang, H.: Performance and improvements of green construction projects. Elsevier (2018). ISBN 978-0-12-815483-0
9. Renewable Energy and Weather Dataset. https://www.kaggle.com/datasets/samanemami/renewable-energy-and-weather-conditions
10. Pineiro, L.J., Portillo, W.L.: Web architecture for URL-based phishing detection based on random forest, classification trees, and support vector machine. Inteligencia Artif. **25**(69), 107–121 (2022). https://doi.org/10.4114/intartif.vol25iss69
11. Michalíková A., Pažický B.: Classification of tire tread images by using neural networks. In: 15th International Scientific Conference on Informatics (2019). https://doi.org/10.1109/Informatics47936.2019.9119306
12. Al-Azawi, M.: Symmetry-based brain abnormality detection using machine learning. Inteligencia Artif. **24**(68), 138–150 (2022). https://doi.org/10.4114/intartif.vol24iss68

Performance Analysis of the Data Aggregation in the Oracle Database

Michal Kvet[✉]

Faculty of Management Science and Informatics, University of Žilina, Univerzitná 8215/1,
010 26 Žilina, Slovak Republic
Michal.Kvet@fri.uniza.sk

Abstract. The amount of data for processing and evaluation is growing enormously. Temporal database monitoring object state evolution is an inseparable part of the analytics. Oracle Database provides a robust environment for treating complex datasets by offering indexing, partitioning, and other enhancements allowing to create scalable systems. Aggregate and analytic functions in SQL are associated with the data groups or partitions, characterizing the data set over which individual functions are evaluated. Commonly, groups are not defined only by the attributes, instead, function results and transformations form the groups. This paper deals with the performance of the group definition associated with the function calls, referenced by the pure definition or aliases, introduced in Oracle 23c.

Keywords: data analytics · aggregate functions · group definition · expression reference · column alias

1 Introduction

Data to be handled has enormously risen over the decades. The structure of the data became more complex, requiring storing the whole data evolution in the temporal database [9]. Database systems form the data layer holding the data, which are then evaluated by the aggregate and analytic functions. Relational databases are still most often used and widespread, because of the strict data structures, references, and integrity [5]. Database transactions ensure the data to be loaded pass the requirements defined by the data model, as well as all the constraints defined for the data [6]. Thus, the data are correct, ensured by consistency and integrity. Online transaction systems are mostly used for operational data, which are then grouped, evaluated, and shifted to warehouses for complex data analytics [5, 7, 12]. Autonomous data warehouses benefit from the ability to be deployed in the Oracle Cloud Infrastructure [1, 7, 12], so the whole maintenance and administration is left to the cloud vendor. Furthermore, it offers many enhancements allowing to optimize the data access by using auto-indexing features, dynamic partitioning, and scalability options.

Advanced analytics is required almost in every sphere. Intelligent information systems require temporal data to be evaluated by monitoring the changes [2, 3]. Intelligent

© The Author(s), under exclusive license to Springer Nature Switzerland AG 2024
Á. Rocha et al. (Eds.): WorldCIST 2024, LNNS 990, pp. 161–170, 2024.
https://doi.org/10.1007/978-3-031-60328-0_16

decision-making systems [4] are based on complex data offering to predict the evolution using machine learning methods and AI [10, 11]. There are many tools for data analytics getting reports, charts, correlations, tables, or any other forms of the result representation. PowerBI, R, and Oracle Analytics are the most popular tools allowing to make reports and outputs easily using wizards, so the process does not need to be done by professionals. Whatever system is used, there are always data behind them to be acquired, processed, evaluated, and subsequently represented in the form defined by the user. Data and their representations are therefore critical.

This paper deals with the analytical and aggregate functions by defining the data set ranges – groups of data, which are processed by those functions. Well, typically, a group is not defined of only attributes, but also expressions and functions are used. In the past, it was impossible to use column aliases in the Where clause, as well as group definitions (Group by clause), forcing the system to process the function multiple times, even by calculating the output of the function several times, even with the same parameters, degrading the performance of the whole system.

Oracle Database 23c [14] offers the function result alias to be placed in almost all clauses of the SQL Select statement. Thanks to that, it can be directly referred to the function usage. This paper evaluates the impacts of using aliases in the Group by clause forming the groups for the data aggregations and analytics. Please note, that this paper focuses exclusively on the database system Oracle. The goal is not to compare individual implementations and features of multiple database systems and to provide a universal solution applicable in any environment. Instead, only the Oracle database is taken into consideration for multiple reasons. Oracle Database is the most complex and most powerful database type. It offers autonomous database types, so the administration, patching, and optimization are strongly reduced. Finally, this paper forms the output of the Erasmus + project EverGreen emphasizing green data analytics [13], in which Oracle Analytics Cloud forms the infrastructure.

2   SQL Statement Execution

To process an SQL statement and produce the results, the database system must perform the following steps:

- Parsing SQL statement
- Validating SQL statement
- Generating various execution plans and selecting the most suitable, based on the data statistics and heuristics.
- Executing the statement by using the selected access plan.

Besides, the database system collects many additional statistics to perform optimization and provides recommendations for using additional features, optimization, tuning, and extended numbers of statistics. The execution plan is identified by the hash value of the original statement and temporarily stored in the instance memory, so if it is executed multiple times, it is not necessary to process all the above steps, just the pre-calculated plan is taken.

Select statements can consist of multiple clauses defining the data set and processing, which significantly influence the processing steps. The following clauses can be used, Select and From clauses are mandatory [5]:

- *Select* clause defining the list of values produced in the result set.
- *From* clause defining the data sources and methods for interconnecting them.
- *Where* clause specifying conditions limiting the data set.
- *Group by* clause defining data groups and splits used for the aggregate and analytic function calculations.
- *Having* clause for specifying conditions based on aggregate functions and analytics (window functions).
- *Order by* clause offering to sort the result data set. Generally, the data are not sorted at all, defined by the order in which individual results were obtained
- *Limit* (*Fetch*) clause allowing to get only a portion of data, instead of the full result set, based on the evaluation and conditions.

Figure 1 shows the process of the execution and statement evaluation. First, source tables are extracted, joined, and merged, followed by applying conditions specified in the Where clause attempting to reduce the amount of data for further steps. Data are filtered, so it´s time to group by the data, optionally reduced by the Having clause specifying conditions based on the aggregate functions and analytics. Then, the Select clause is treated, and enhanced by the column, expression or function call aliases. Finally, the pre-prepared result set is ordered, or limit & offset can be applied.

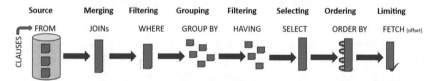

Fig. 1. SQL query execution order

The order of individual steps is significant. It is always preferred to reduce the cardinality as much as possible. Thus, if there are multiple conditions, the order of their execution is defined by the reduction factor - the ratio between the original input and the expected output from the point of view of the data amount. Aliases are defined in the Select clause. As evident, Where conditions and group definitions are considered before attempting to process Select clause. Consequently, SQL norm does not allow to use column, expression, and function aliases in the Group by and Where clauses of the same Select statements. The limiting factor of the performance is defined in the next section.

3 Problem

Aliases defined in the Select clause are primarily used for specifying column references in the provided output. As stated, they cannot be used in other clauses of the same statement, because they are not known at those stages, except for the Order by clause, which is evaluated at the end of the processing, in which the data set itself and the contained data in it are clear. In principle, the Group by clause must state all data attributes from the Select statement clause, except of the aggregate function. Additional data attributes can be stated, mostly determined by the primary keys. Furthermore, function calls can be present in Select, Where, and Group by clause (Order by clause can be also enhanced by the function calls, but it is not significant from the data access processing point of view). By digging deeper into SQL tuning and performance accelerations, the impossibility of using function call aliases brings additional costs and processing time. One function is evaluated multiple times. Even caching the function results does not provide sufficient power, while the function must be explicitly stated as deterministic, and obtaining the result from the memory cache also requires some time and resources. Figure 2 shows the problem by determining one function in three positions. The same function is called in multiple positions and thus evaluated multiple times. Furthermore, calling the PL/SQL function requires a context switch. Even optimization of the function for the SQL calls (by using User Defined Function (UDF) pragma) does not provide relevant improvement.

```
select FUNC(params) as function_result, agg_function
from table_list+joining
    where FUNC(params) > limit_val
        group by FUNC(params)
        [order by FUNC(params)];
```

Fig. 2. Function calls in a Select statement

In conclusion, the inability to use function call aliases impact and limit the performance of the query. In the current standard, aliases cannot be used in the clauses part of the same statement, as specified. Proposed solutions, techniques, and workarounds, compared to the Oracle 23c enhancement are discussed in the next section.

4   Proposed Solutions and Enhancements

In this paper, three new solutions and enhancements are discussed, denoted by the SOL1-SOL3 display marks. The reference solution (marked as REF) uses the original approach defined by the SQL norm, which does not use aliases, and a particular function is listed in each clause separately.

- **REF** – Original solution calling a function multiple times. The improvement of this reference method lies in the optimization of the called method for SQL language usage through the UDF pragma usage (REF_UDF).
- **SOL1** - Nested statement – This solution is based on calculating the function results in the first phase by forming a nested query. Then, the function result is called in the outer query, referred to by the aliases specified in the inner query. Thanks to that, a particular function is called only once for each row/group and stored temporarily in the inline view.

- **SOL2** - Pre-processing using dynamic view – With clause extensions have been introduced in Oracle Database 12c Release allowing to use of subquery factoring clause enhancements, like PRAGMA UDF, deterministic hint, and subquery caching.
- **SOL3** - Direct usage of the alias specified in the Select clause is offered in the Oracle Database 23c. Groups can be enhanced by the function calls defined in the same statement and referenced by the aliases. The extended Select statement evaluation process is depicted in Fig. 3.

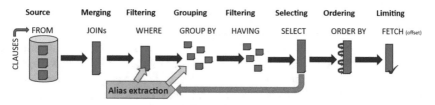

Fig. 3. Function calls (alias) extraction module

The next section deals with the computational study using environmental data impact analysis.

5   Computational Study

The computational study was run on the server with the following parameters:

- Processing unit: AMD Ryzen 5 PRO 5650U, 2.30 GHz, Radeon Graphics
- Memory: Kingston, DDR4 type, 2x 32 GB, 3200 MHz, CL20
- Storage: 2 TB, NVMe disc type, PCIe Gen3 x 4, 3500 MB/s for read/write operations
- Operating system: Windows Server 2022, x64
- Database system: Oracle Database 23c, release bundle Oracle 23c Free, Developer Release Version 23.2.0.0.0.

European region flight monitoring data set was used to determine environmental data by treating planned and real routes of the airplanes, compared to the optimal plans reducing the environmental impacts and emissions [8]. Flight companies attempt to reduce costs by using regions, which are cheaper and avoid expensive territories. This results in longer journeys, greater fuel consumption, and thus greater consequences for the environment and climate. For each flight, location data were monitored, delimited by the 50 parameters monitoring the flight and temporal assignment to the flight information regions (FIR). These provided values were compared to the planned routes and calculated journey optimized for reducing environmental burdens. The whole data set consisted of 5 million rows for the positional data and 1000 of rows for the FIR assignment. The structural data example of the FIR assignment is in Fig. 4. Each flight is identified by the ECTRL_ID. Individual FIR assignments (identified by the AUA_ID) are sequentially referenced (Sequence_number) over the timeline (Entry_time, Exit_time).

```
"ECTRL ID","Sequence Number","AUA ID","Entry Time","Exit Time"
"186858226","1","EGGXOCA","01-06-2015 04:55:00","01-06-2015 05:57:51"
"186858226","2","EISNCTA","01-06-2015 05:57:51","01-06-2015 06:28:00"
"186858226","3","EGTTCTA","01-06-2015 06:28:00","01-06-2015 07:00:44"
"186858226","4","EGTTTCTA","01-06-2015 07:00:44","01-06-2015 07:11:45"
"186858226","5","EGTTICTA","01-06-2015 07:11:45","01-06-2015 07:15:55"
```

Fig. 4. FIR assignment data [8]

The computational study deals with the costs of the processing and processing time.

The evaluation study is divided into two parts. In the first experiment, the groups are defined by the pure attributes, however, they are referred to by the aliases to highlight the additional workload of extracting and translating aliases into the original definition. The aim is to calculate the flight efficiency, delimited by the FIR references comparing optimal and real routes, while external circumstances like weather are taken into consideration. The efficiency is calculated for each flight separately, so the group is formed by the flight identifiers – ECTRL_ID attribute. Table 1 shows the results.

Table 1. Results considering attribute alias

	REF	SOL1	SOL2	SOL3
Costs	3660 (24% CPU)			
Processing time (ss. ff)	23.25	27.33	21.78	18.24

The worst solution was obtained by the nested query, it is caused by the execution plan calculation necessity. By transforming the nested query to the dynamic view, a 20.31% improvement was detected, caused by the optimization, pre-fetching, and data optimization, offered by the parallelism options. Using aliases references lowers the processing time demands to 18.24 s, which refers to a 21.55% improvement compared to the REF. Furthermore, it does not require any query nesting. Compared to SOL1, it provides a 33.26% improvement and 16.25% for SOL2. Figure 5 shows the processing time demand results graphically.

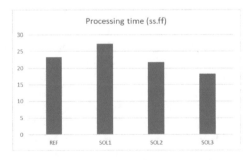

Fig. 5. Results – group definition by attributes

The second experiment deals with the function calls, which are to be calculated only once and consecutively referenced by the internal views, nested query, or pointed by the function result aliases. In this case, efficiency is calculated across all flights together. The month and year extractions are considered as the forming values for the groups. Each flight refers to the takeoff, as long as the flight is at the border of the months. Standard function to_char, bundled in the Standard package is used, optimized for the SQL calls. The main advantage of that function is that it can treat multiple date and time elements in one call, compared to the extract function, which is also available, however, such a function can proceed only one temporal element, consequencing in the necessity to call it twice – once for getting the month and once for year element. Table 2 shows the results. The first part deals with the to_char function usage, and the second part highlights extract function usage.

Table 2. Results considering function references

	REF	SOL1	SOL2	SOL3	Used function
Costs	3660 (24% CPU)				to_char
Processing time (ss. ff)	23.05	26.12	20.46	17.01	
Costs	3660 (24% CPU)				extract
Processing time (ss. ff)	23.67	26.41	21.03	16.46	

The number of functions defining the groups does not play a significant role. On the one hand, we have the to_char function, which processes both elements in one call, but it results in a more complex output in string format. Opposite, there is an extract function, which needs to be called twice, once for the month and the second time for the year components. Thus, the results are indeed two values, but in numerical format, which reduces the cost of further processing. So, summarizing, in both cases, the achieved results are almost identical from the time requirements point of view. The number of groups also impacts the performance. Namely, too many groups can cause tiny data frames and a small amount of data. Although the groups can be processed in parallel, it is impossible to establish separate threads for each group, if the number is too high. Reflecting on the results, the function alias is the most beneficial and reduces the processing time demands by 26.20% for the to_char function and 30.46% for the extract function (referencing REF). Similarly to the previous results obtained in experiment 1, the most demanding solution is based on query nesting (SOL1) (Fig. 6).

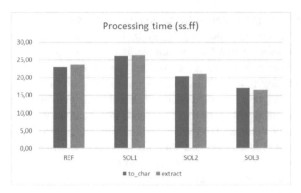

Fig. 6. Results – group definition by functions

6 Results and Conclusions

Data aggregation is a core element and prerequisite of the analysis by shifting the data from the transaction-oriented databases into data warehouses, marts, lakes, etc. To calculate aggregated data, it is necessary to form the groups, for which the aggregate functions are calculated, followed by the analytic function calls defined by the partitions. The groups can be formed by an unlimited number of attributes, expressions, and functions. Generally, function calls can be present in any clause of the Select statement. By emphasizing the execution plan and order of individual clause processing, it is evident that the Where clause, Data set merging, and Group by definitions are processed before evaluating the Select clause, which can consist of the expressions and function calls, enhanced by the aliases. In the past, column aliases were mostly intended for the definition and naming of the attributes of the output set, for hiding the definition and easier expression of the results. Users had to be aware of the aliases and their visibility across the definition, forcing the system to nest queries to make them more complicated, or to refer to the same expressions multiple times. If something had to be changed in the definition, it had to be applied in multiple places, which could potentially introduce errors. Later, the queries became more and more complex, so they were necessary for consecutive references in the outer queries. Data analytics is a typical example of defining complex queries. Oracle Database 23c brought significant change in terms of alias management. Namely, individual aliases are identified and extracted in the first phase making them referencable in any clause of the statement. It brings not only simplification of the code, better readability, but as shown in this paper, it also brings improvement of performance characteristics and reduction of costs and processing time.

In this paper, four solutions are referenced and compared. REF solution is a reference, which does not use any alias to focus on the impacts. Usage of the nested query brings additional demands, caused by the result set fetching and loading necessity for the inner query. This fact can have a significant impact on the instance memory, whereas the data need to be placed there, for the consecutive processing by the outer queries. If the result set is huge and does not fit the available instance memory, the problem is even huge, because of the disc-swapping necessity. For the discussed environment, nested querying processing time demands are higher by 4.08 s, which expresses an additional

17.55%. Vice versa, using a function alias lowers the demands compared to the referenced (REF) solution by 5.01 s, just because of the function reference extraction and reducing content switches. Namely, even if the function call parameters differ, they can always be processed in bulk, instead of separating them for each statement clause separately. Pre-calculation of the function results is also appropriate because the dynamic view already contains the results of the function calls, which are treated as stored results. Thus, in the data analysis itself, the functions are no longer called, they only use the calculated results in the previous phase.

The limitation of the discussed alias extraction module is its strict interconnection with the Select statement and reference to the steps of the execution plan. Therefore, individual clauses can be executed only if there is no not yet processed alias. Otherwise, the processing has to wait.

In further research, we will focus on our own extended extraction methods of columns, expressions, and functions aliases and their placement in a collection defined in the form of an index, instead of the flat structure, that is currently used. Thanks to that, complex queries will benefit, because the validity of the alias will be for the entire processing, not only a specific block and its nested subqueries.

Acknowledgment. It was supported by the Erasmus+ project: Project number: 2022-1-SK01-KA220-HED-000089149, Project title: Including EVERyone in GREEN Data Analysis (EVER-GREEN) funded by the European Union. Views and opinions expressed are however those of the author(s) only and do not necessarily reflect those of the European Union or the Slovak Academic Association for International Cooperation (SAAIC). Neither the European Union nor SAAIC can be held responsible for them.

Co-funded by
the European Union

References

1. Anders, L.: Cloud Computing Basics. Apress (2021)
2. Dudáš, A., Škrinárová, J.: Edge coloring of set of graphs with the use of data decomposition and clustering. IPSI Trans. Internet Res. **16**(2), 67–74 (2020). ISSN 1820-4503
3. Dudáš, A., Škrinárová, J., Vesel, E.: Optimization design for parallel coloring of a set of graphs in the high-performance computing. In: Proceedings of 2019 IEEE 15th International Scientific Conference on Informatics, pp 93–99. ISBN 978-1-7281-3178-8
4. Jánošíková, Ľ., Jankovič, P., Kvet, M., Zajacová, F.: Coverage versus response time objectives in ambulance location. Int. J. Health Geogr. **20**, 1–16 (2021). ISSN 1476-072X
5. Kuhn, D., Kyte, T.: Expert Oracle Database Architecture: Techniques and Solutions for High Performance and Productivity. Apress (2021)
6. Kuhn, D., Kyte, T.: Oracle Database Transactions and Locking Revealed: Building High Performance Through Concurrency. Apress (2020)
7. Kumar, Y., Basha, N., Kumar, K., Sharma, B., Kerekovski, K.: Oracle High Availability, Disaster Recovery, and Cloud Services: Explore RAC, Data Guard, and Cloud Technology. Apress (2019)

8. Kvet, M.: Dangling predicates and function call optimization in the oracle database. In: 2023 Communication and Information Technologies (KIT) (2023)
9. Kvet, M.: Developing Robust Date and Time Oriented Applications in Oracle Cloud: A comprehensive guide to efficient Date and time management in Oracle Cloud. Packt Publishing (2023). ISBN: 978-1804611869
10. Png, A., Helskyaho, H.: Extending Oracle Application Express with Oracle Cloud Features: A Guide to Enhancing APEX Web Applications with Cloud-Native and Machine Learning Technologies. Apress (2023)
11. Riaz, A.: Cloud Computing Using Oracle Application Express. Apress (2019)
12. Sarkar, P., Ruiy, G.: Oracle Cloud Infrastructure for Solutions Architects: A practical guide to effectively designing enterprise-grade solutions with OCI services. Packt Publishing (2021)
13. EverGreen project. https://evergreen.uniza.sk/
14. Oracle 23c enhancements. https://oracle-base.com/articles/23c/articles-23c

BipartiteJoin: Optimal Similarity Join for Fuzzy Bipartite Matching

Ondrej Rozinek[1]([✉]), Monika Borkovcova[2], and Jan Mares[1,3]

[1] Department of Process Control, University of Pardubice, Studentska 95, 532 10 Pardubice, Czech Republic
ondrej.rozinek@gmail.com
[2] Department of Information Technology, University of Pardubice, Studentska 95, 532 10 Pardubice, Czech Republic
[3] Department of Mathematics, Informatics and Cybernetics, University of Chemistry and Technology Prague, Technicka 5, 166 28 Prague, Czech Republic

Abstract. Set similarity join, crucial for data cleaning, integration, and recommendation systems, identifies set pairs exceeding a similarity threshold. Our approach combines a count Q-gram filter with maximum weighted bipartite matching, balancing accuracy and efficiency. The Q-gram filter, based on the relationship between Q-gram similarity and edit distance, reduces the number of comparisons, operating in constant time on a pre-built index. This enables real-time processing, as only a minimal number of pairs are verified through Fuzzy Bipartite Matching, significantly enhancing the efficiency of similarity joins.

Keywords: similarity join · Q-gram filter · record linkage · entity resolution · similarity space · bipartite matching

1 Introduction

Set similarity join identifies all pairs of sets within a single record collection or across two different collections when the similarity score is above a certain threshold α [3,5,7]. This process is an essential operation in many applications, such as data cleaning and integration, personalized recommendation and record deduplication.

As shown in the Fig. 2, our focus is on a two-step method for similarity join: firstly, using a count Q-gram filter, and secondly, employing maximum weighted bipartite matching as the best approach for solving the combinatorial assignment problem [10]. This method involves fuzzy token similarity in bipartite matchings, recognized state-of-the-arts for its high accuracy in classifying records as matches or non-matches in an error-tolerant manner. However, this approach results in polynomial time complexity, $\mathcal{O}(n^3)$, which is managed by the Kuhn-Munkres algorithm (Fig. 1).

To improve the efficiency of similarity joins, we avoid exhaustive pair comparisons by applying a count Q-gram filter. In real-world applications, Ukkonen's lemma [9,12] is used to establish a direct relation between Q-gram similarity

Á. Rocha et al. (Eds.): WorldCIST 2024, LNNS 990, pp. 171–180, 2024.
https://doi.org/10.1007/978-3-031-60328-0_17

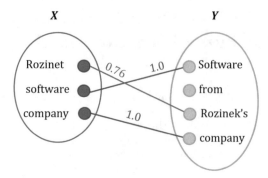

Fig. 1. Maximum weighted bipartite matching of two records $\mathcal{X}$ and $\mathcal{Y}$.

and edit distance (Levenshtein), which acts as a rigorous mathematical filter to reduce the number of comparisons without missing any true matches. As indicated in the figure, the Q-gram filter operates in constant time, $\mathcal{O}(1)$ on a pre-built inverted Q-gram index. This significant reduction in comparison pairs enables real-time processing, as only a small fraction of candidate pairs is further verified for actual matches in the second stage of Fuzzy Bipartite Matching (Fig. 2).

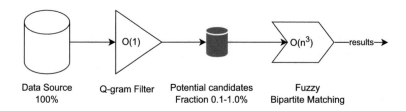

Fig. 2. Block diagram of the processing of records from the source in real-time by a two-step system of Q-gram filter and Fuzzy Bipartite Matching

According to Ukkonen's lemma [9,12], let X and Y be tokens with the edit distance $d(X, Y)$. Then, the Q-gram similarity $|Q_X \cap Q_Y|$ of the tokens X and Y is at least $t = \inf_d\{|Q_X \cap Q_Y|\} = \max\{|X|, |Y|\} - q + 1 - qd(X, Y)$, where t is a Q-gram similarity threshold with respect to $d(X, Y)$ and q is the Q-gram length.

The problem often arises with the assumption that this constraint is applied to the entire record [12]. Consider two records $\mathcal{X}$ and $\mathcal{Y}$, split into sets of tokens $X_i \in \mathcal{X}$ and $Y_j \in \mathcal{Y}$. This simplification is demonstrated in Example 1, leading to differing results of the Q-gram filter as a constraint of fuzzy bipartite matching. We address this gap in our article's goal.

Example 1. Consider two records $\mathcal{X}$ and $\mathcal{Y}$, and their token sets, $\{X_1, X_2\}$ and $\{Y_1, Y_2\}$. Assume a matching set of token pairs $\mathcal{M}$, such that $M = \{\{X_1, Y_1\}, \{X_2, Y_2\}\}$. Then, for the entire record, we have $t = \max\{|X_1 \cup X_2|, |Y_1 \cup Y_2|\} - q + 1 - qd(X_1 \cup X_2, Y_1 \cup Y_2)$, which is different from the calculation on the matching pairs separately $t_\mathcal{M} = \max\{|X_1|, |Y_1|\} - q + 1 - qd(X_1, Y_1) + \max\{|X_2|, |Y_2|\} - q + 1 - qd(X_2, Y_2)$, given by edges in maximum weighted bipartite matching. Hence, we prove that $t \neq t_\mathcal{M}$ and a more precise filter for $t_\mathcal{M}$ must exist.

2 Related Work

Recent research surveys, such as the one by Papadakis et al. (2020) [7], have identified state-of-the-art methods for set similarity joins based on edit constraints, including FastJoin, SilkMoth, and MF-Join.

FastJoin [10], the initial method in this domain, is based on the principle that two sets with a bipartite matching score of at least α must have at least $\lceil \alpha \rceil$ shared tokens. For a set $\mathcal{X}$ with n tokens, any $n - \lceil \alpha \rceil + 1$ tokens form its signature. If another set $\mathcal{Y}$ lacks these signature tokens, a match with $\mathcal{X}$ is improbable. FastJoin essentially adapts prefix filtering to fuzzy criteria.

SilkMoth [2] enhances FastJoin's signature scheme, reducing token count to limit candidate sets. It applies refinement filters like Check Filter (CF) and Nearest Neighbor Filter (NNF) for sets $\mathcal{Y}$ containing a signature token. CF calculates a similarity threshold for each element in $\mathcal{X}$, pruning $\mathcal{Y}$ if no element pair meets this threshold. NNF pairs each $\mathcal{X}$ element with its closest in $\mathcal{Y}$, setting an upper limit for matching scores.

MF-Join [11] diverges by using two thresholds: δ_1 for sets and δ_2 for elements. It considers element pairs above δ_2 for matching, but manually setting these thresholds is challenging.

TokenJoin [13] proposes a lightweight and effective token-based filtering approach that accelerates the speed of similarity joins.

Unlike other methods, our approach aims to use a count Q-gram filter [7] to make the similarity join process quicker in maximum weighted bipartite matching. Our main goal is to have a precise filter that is reliable in the pruning process, rather than a faster filter that is less accurate.

3 Optimal Count Q-Gram Filter

Consider the edit distance between two strings X and Y, denoted as $d(X, Y)$, and the worst case of the expected distance function represented as $d(\alpha, |X|, |Y|)$. If we use the floor function, denoted by $\lfloor . \rfloor$, we find that the normalized similarity metric $s_n(X, Y)$, with $\alpha \in [0, 1] \subset \mathbb{R}$ as a threshold, satisfies the following relationship:

$$s_n(X, Y) \geq \alpha \iff d(X, Y) \leq d(\alpha, |X|, |Y|) = \left\lfloor \frac{1 - \alpha}{1 + \alpha}(|X| + |Y|) \right\rfloor. \quad (1)$$

This relationship is derived from the general understanding of distance and similarity metrics [8]. When substituting self-similarities, which equal the corresponding cardinality of the sets, as $s(X,X) = |X|$ and $s(Y,Y) = |Y|$, the expected distance $d(\alpha, |X|, |Y|)$ is maximized when the similarity is at its minimum, bound by the threshold α. This condition is met if and only if $s_n(X,Y) = \alpha$:

$$d(X,Y) = \lfloor d_R \rfloor = \left| \frac{1 - s_n(X,Y)}{1 + s_n(X,Y)}(s(X,X) + s(Y,Y)) \right|$$

$$\leq \sup_\alpha d(X,Y) = \left| \frac{1-\alpha}{1+\alpha}(|X| + |Y|) \right| = d(\alpha, |X|, |Y|). \tag{2}$$

Given that the edit distance is an integer, the floor function is used to discard the fractional part, ensuring the measure remains defined and accurate.

In our refinement of the count Q-Gram filter model, we introduce more precise assumptions about the token sets $\mathcal{X}$ and $\mathcal{Y}$, as detailed in the Example 1 provided. This leads to the derivation of an optimized filter, specifically designed to retain every comparison pair where Fuzzy Bipartite Matching exceeds the threshold α.

Central to our approach is the assumption that the matching token pairs from set $\mathcal{M}$ are known to us, but their exact edit distances remain undetermined. Despite this, we can approximate these edit distances by relying on predictions based on expected edit distances. This predictive approach allows us to effectively gauge similarity without the need for precise edit distance values, aligning with our refined model's objectives.

The optimality of the filter is understood from the perspective that it represents the infimum of Q-gram similarity, and consequently, it is the optimal edit constraint for maximum weighted bipartite matching.

Theorem 1 (Optimal Count Q-gram Filter for Bipartite Matching).
Let $\mathcal{X}$ and $\mathcal{Y}$ be records representing a set of tokens. Then the Q-gram similarity in bipartite matching of $\mathcal{X}$, $\mathcal{Y}$ and cardinality $|\mathcal{M}|$ for a given threshold $s_n(\mathcal{X},\mathcal{Y}) \geq \alpha$ is at least

$$t_{\mathcal{M}} = \inf_\alpha \{|Q_{\mathcal{X}} \cap Q_{\mathcal{Y}}|\}$$

$$= \underbrace{\sum_{(i,j)\in\mathcal{M}} \max\{|X_i|, |Y_j|\} - |\mathcal{M}|q + |\mathcal{M}|}_{\text{maximum shared Q-grams}} - q \max_\alpha \underbrace{\sum_{(i,j)\in\mathcal{M}} d(\alpha_{i,j}, |X_i|, |Y_j|)}_{\text{loss function}}, \tag{3}$$

containing a linear combination of

$$d(\alpha_{i,j}, |X_i|, |Y_j|) = \frac{1 - \alpha_{i,j}}{1 + \alpha_{i,j}}(|X_i| + |Y_j|) \tag{4}$$

under the constraint α for which the linear combination is maximized.

Proof. Consider the sum over connected pairs of tokens with cardinality $|\mathcal{M}|$

$$
\inf_{\alpha}\{|Q_{\mathcal{X}} \cap Q_{\mathcal{Y}}|\} = \inf_{\alpha}\left\{ \sum_{(i,j)\in\mathcal{M}} |Q_{X_i} \cap Q_{Y_i}| \right\} = \sum_{(i,j)\in\mathcal{M}} \inf_{\alpha_{i,j}}\{|Q_{X_i} \cap Q_{Y_j}|\}
$$

$$
= \sum_{(i,j)\in\mathcal{M}} \inf_{\alpha_{i,j}}\{\max\{|X_i|,|Y_j|\} - q + 1 - qd(X,Y)\}
$$

$$
= \sum_{(i,j)\in\mathcal{M}} \{\max\{|X_i|,|Y_j|\} - q + 1 - q\sup_{\alpha_{i,j}} d(X,Y)\}
$$

$$
= \sum_{(i,j)\in\mathcal{M}} \max\{|X_i|,|Y_j|\} - |\mathcal{M}|q + |\mathcal{M}| - q\max_{\alpha} \sum_{(i,j)\in\mathcal{M}} d(\alpha_{i,j},|X_i|,|Y_j|).
$$

(5)

Each $\alpha_{i,j}$ is the distributed minimum similarity for each token, giving a threshold vector that should maximize the sum of the expected distances $d(\alpha_{i,j},|X_i|,|Y_j|)$ so that $s_n(\mathcal{X},\mathcal{Y}) \geq \alpha$ holds for t_M. Formalizing this, we get the task

$$
\text{maximize} \quad \sum_{(i,j)\in\mathcal{M}} d(\alpha_{i,j},|X_i|,|Y_j|),
$$

$$
\text{subject to} \quad \sum_{(i,j)\in M} \alpha_{i,j} \geq (\text{Table 1}) \quad \alpha \in [0,1],\, i = 1,\dots,|\mathcal{M}|,
$$

$$
\alpha_{i,j} \in [0,1],\, j = 1,\dots,|\mathcal{M}|.
$$

This leads to an integer linear programming task equivalent to the Knapsack problem, solvable in $\mathcal{O}(nb)$ time. The optimization algorithm determines the maximum expected edit distance distribution across tokens, maintaining the similarity threshold $s_n(\mathcal{X},\mathcal{Y}) \geq \alpha$.

Table 1. Overview of fuzzy token similarity functions and their corresponding constraints [6] on the integer linear programming problem.

Similarity Measure	Subject to $\sum_{(i,j)\in M} \alpha_{i,j} \geq$
Fuzzy Dice	$\frac{\alpha}{2}(|\mathcal{X}| + |\mathcal{Y}|)$
Fuzzy Cosine	$\alpha\sqrt{|\mathcal{X}||\mathcal{Y}|}$
Fuzzy Jaccard	$\frac{\alpha}{1+\alpha}(|\mathcal{X}| + |\mathcal{Y}|)$
Fuzzy Overlap	$\alpha\min\{|\mathcal{X}|,|\mathcal{Y}|\}$

4 Approximate Count Q-Gram Filter

Let F_x and F_Y be discrete distribution functions of ascending sorted lengths $|X_i|$ and $|Y_i|$. Then the Q-gram similarity in bipartite matching for unknown connected edges of records $\mathcal{X}$, $\mathcal{Y}$ and cardinality $|\mathcal{M}|$ is at least

$$
t_M \approx \hat{t}_M = \frac{2q\alpha + \alpha - 2q + 1}{2 + \alpha}(F_X[|\mathcal{M}|] + F_Y[|\mathcal{M}|])
$$

$$
+ \frac{1}{2}\Big|F_X[|\mathcal{M}|] - F_Y[|\mathcal{M}|]\Big| - |\mathcal{M}|q + |\mathcal{M}|
$$

(6)

for a classification Fuzzy Bipartite Matching threshold $s_n(\mathcal{X}, \mathcal{Y}) \geq \alpha$.

The derivation results from a new approximation method that involves using several techniques to establish a less tight lower bound for certain terms. To aid the reader's understanding, we first present the following equations that are integral to the final deduced formula.

$$\max\left\{\sum_{(i,j)\in\mathcal{M}} |X_i|, \sum_{(i,j)\in\mathcal{M}} |Y_j|\right\} \leq \sum_{(i,j)\in\mathcal{M}} \max\{|X_i|, |Y_j|\}. \tag{7}$$

We can also express the maximum of any two variables $a, b \in \mathbb{R}$ in another analytical form:

$$\max\{a, b\} = \frac{1}{2}(a + b + |a - b|), \tag{8}$$

and now define the cumulative sum (discrete distribution function) of ascending sorted length F_X and F_Y. Finally, we obtain the inequality

$$\max\{F_{\mathcal{X}}[|\mathcal{M}|], F_{\mathcal{Y}}[|\mathcal{M}|]\} \leq \max\left\{\sum_{(i,j)\in\mathcal{M}} |X_i|, \sum_{(i,j)\in\mathcal{M}} |Y_j|\right\}. \tag{9}$$

With Eqs. (7), (8), and (9), we proceed to the full derivation, assuming the constancy of $\alpha_{i,j} = \alpha$ for simplicity. We also apply the floor function $\lfloor . \rfloor$ to maintain integer values for shared Q-grams. The entire derivation is as follows:

$$
\begin{aligned}
t_{\mathcal{M}} &= \sum_{(i,j)\in\mathcal{M}} \max\{|X_i|, |Y_j|\} - |\mathcal{M}|q + |\mathcal{M}| - q\max_{\alpha} \sum_{(i,j)\in\mathcal{M}} d(\alpha, |X_i|, |Y_j|) \\
&= \sum_{(i,j)\in\mathcal{M}} \max\{|X_i|, |Y_j|\} - |\mathcal{M}|q + |\mathcal{M}| - q\max_{\alpha} \sum_{(i,j)\in\mathcal{M}} \frac{1-\alpha_{i,j}}{1+\alpha_{i,j}}(|X_i| + |Y_j|) \\
&\approx \sum_{(i,j)\in\mathcal{M}} \max\{|X_i|, |Y_j|\} - |\mathcal{M}|q + |\mathcal{M}| - q\frac{1-\alpha}{1+\alpha} \sum_{(i,j)\in\mathcal{M}} (|X_i| + |Y_j|) \\
&\geq \max\left\{\sum_{(i,j)\in\mathcal{M}} |X_i|, \sum_{(i,j)\in\mathcal{M}} |Y_j|\right\} - |\mathcal{M}|q + |\mathcal{M}| - \frac{q-q\alpha}{1+\alpha} \sum_{(i,j)\in\mathcal{M}} (|X_i| + |Y_j|) \\
&= \frac{1}{2}\sum_{(i,j)\in\mathcal{M}} (|X_i| + |Y_j|) + \frac{1}{2}\left|\sum_{(i,j)\in\mathcal{M}} |X_i| - \sum_{(i,j)\in\mathcal{M}} |Y_k|\right| - |\mathcal{M}|q + |\mathcal{M}| - \frac{q-q\alpha}{1+\alpha} \sum_{(i,j)\in\mathcal{M}} (|X_i| + |Y_j|) \\
&= \frac{2q\alpha + \alpha - 2q + 1}{2 + \alpha} \sum_{(i,j)\in\mathcal{M}} (|X_i| + |Y_j|) + \frac{1}{2}\left|\sum_{(i,j)\in\mathcal{M}} |X_i| - \sum_{(i,j)\in\mathcal{M}} |Y_k|\right| - |\mathcal{M}|q + |\mathcal{M}| \\
&\geq \frac{2q\alpha + \alpha - 2q + 1}{2 + \alpha}(F_X[|\mathcal{M}|] + F_Y[|\mathcal{M}|]) + \frac{1}{2}\left|F_X[|\mathcal{M}|] - F_Y[|\mathcal{M}|]\right| - |\mathcal{M}|q + |\mathcal{M}| \\
&\geq \left\lfloor \frac{2q\alpha + \alpha - 2q + 1}{2 + \alpha}(F_X[|\mathcal{M}|] + F_Y[|\mathcal{M}|]) + \frac{1}{2}\left|F_X[|\mathcal{M}|] - F_Y[|\mathcal{M}|]\right| \right\rfloor - |\mathcal{M}|q + |\mathcal{M}| \\
&= \hat{t}_{\mathcal{M}} \\
&\implies t_{\mathcal{M}} \approx \hat{t}_{\mathcal{M}}.
\end{aligned}
\tag{10}
$$

Consequently, when utilizing a pre-built inverted Q-gram index that includes the distribution of ascending sorted lengths, we can achieve a time complexity of $\mathcal{O}(1)$.

5   Experiments

We computed the *non-interpolated average precision* of this ranking. According to the papers [1,4], we calculate the precision and recall as:

$$\text{Precision} = \frac{c(i)}{i}, \tag{11}$$

$$\text{Recall} = \frac{c(i)}{m}, \tag{12}$$

where $c(i)$ is the number of correct matching pairs ranked before position i, and m is total number of correct matches. Consequently *interpolated precision* at recall r is the $\max_i \frac{c(i)}{i}$, where the max is taken over all ranks i such that $\frac{c(i)}{m} \geq r$. The overall relative performance of the compared similarity functions is calculated using the maximum F1-score as:

$$\text{F1-score} = 2 \times \frac{(\text{Precision} \times \text{Recall})}{\text{Precision}+\text{Recall}}, \tag{13}$$

and shown in the Table 3. The table shows that the best results, with an accuracy of 85.09%, were achieved using the Fuzzy Overlap method based on maximum weighted bipartite matching on its own, and slightly lower at 85.01% when combined with the Q-gram filter. These results support the idea that the approximated optimal Q-gram filter is highly accurate and confirm that our way of approximating it mathematically is correct. However, it's worth mentioning that even though this is an approximation, a few records were found in a detailed analysis that didn't pass through the filter. Still, the difference in the F-score is very small, only 0.08% (Table 2).

The Q-gram filter and Fuzzy Overlap, when used together, enable real-time fuzzy matching. This combination operates on a pre-built inverted Q-gram index and completes the task in just 220 ms, which is a speed increase by an order of magnitude compared to running Fuzzy Overlap alone, which takes 13 s:426 ms. Note that the speed improvement is influenced by the α threshold parameter, affecting the size of the potential candidate pool in the second stage of Fuzzy

Table 2. Datasets used in experiments from original sources [1]

Name	Number of strings	Name	Number of strings
Animal	5,709	Game	911
Bird Kunkel	336	Park	654
Bird Nybird	982	Restaurant	863
Bird Scott1	38	Ucd-people	90
Bird Scott2	719	Census	841
Business	2,139		

Table 3. Comparison of selected similarity functions ranked in descending order of F1-score

Similarity	F1-score	Similarity	F1-score
Fuzzy Overlap	85.09 %	Smith-Waterman	75.71 %
approx. 3-gram filter+Fuzzy Overlap	85.01 %	Smith-Waterman-Gotoh	75.54 %
approx. 2-gram filter+Fuzzy Overlap	84.88 %	Jaro	75.29 %
Fuzzy Jaccard (Levenshtein $\delta = 0.8$))	84.17 %	Overlap 3-gram	73.21 %
Jaro-Winkler	81.45 %	Jaccard 2-gram	71.05 %
L2 Monge-Elkan (Levenshtein)	80.80 %	Dice 2-gram	71.05 %
Damerau-Levenshtein	76.86 %	Jaccard 3-gram	70.86 %
Levenshtein	76.83 %	Dice 3-gram	70.86 %
Needleman-Wunsch	76.25 %	Overlap 2-gram	66.92 %

Overlap (refer to Fig. 2). Testing to evaluate the relative time complexity was performed on a single-core Intel i7 11370H processor with a maximum turbo frequency of 4.80 GHz and 16 GB of RAM (Table 4).

Table 4. Relative Time Complexity

Similarity	Elapsed Time	Similarity	Elapsed Time
Levenshtein	13 s:426 ms	L2 Monge-Elkan (Levenshtein)	14 s:209 ms
Damerau-Levenshtein	22 s:824 ms	Jaccard 2-gram	10 s:542 ms
Jaro	3 s:902 ms	Jaccard 3-gram	9 s:829 ms
Jaro-Winkler	3 s:772 ms	Dice 2-gram	11 s:95 ms
Needleman-Wunsch	28 s:170 ms	Dice 3-gram	10 s:717 ms
Smith-Waterman	28 s:600 ms	Overlap 3-gram	10 s:251 ms
Fuzzy Overlap	13 s:474 ms	Overlap 2-gram	11 s:549 ms
Q-Gram Filter+Fuzzy Overlap	0 s:220ms	Fuzzy Jaccard ($\delta = 0.8$)	12 s:824 ms

6   Conclusion

In this research, we conducted a detailed analysis of different similarity functions. We compared these functions using metrics such as precision, recall, and F1-score. Specifically, our focus was on evaluating our similarity join method, which is based on a two-step approach. This approach combines an approximated Q-gram count filter with the Fuzzy Overlap technique. Through this analysis, we aimed to understand how our method performs in comparison to other existing similarity functions, particularly in terms of accuracy and efficiency.

Our tests, conducted using a range of datasets, showed that the standalone Fuzzy Overlap method achieved the best accuracy at 85.09%, with a slight decrease to 85.01% when combined with the Q-gram filter. These results highlight the accuracy of our approximated Q-gram filter and affirm the correctness of our mathematical approach.

An important finding of our study is the significant speed improvement observed when combining the Q-gram filter with Fuzzy Overlap. This combination was able to complete tasks in only 220 ms, much faster than the 13 s:426 ms needed by Fuzzy Overlap alone. This speed is especially notable because it brings the process into the realm of real-time capability, a critical factor for applications requiring immediate data processing. The α threshold parameter, which influences the number of potential matches considered in the second stage of Fuzzy Overlap, plays a key role in this efficiency.

Overall, our research provides useful insights into the use of similarity functions for large datasets, showing a promising balance between accuracy and speed. This balance is essential for practical applications like data analysis and integration, where real-time data processing can be crucial. Future research could focus on further optimizing these functions for even faster and more accurate real-time applications.

Acknowledgment. It was supported by SGS FEI UPCE 2024 and the Erasmus+ project: Project number: 2022-1-SK01-KA220-HED-000089149, Project title: Including EVERyone in GREEN Data Analysis (EVERGREEN) funded by the European Union. Views and opinions expressed are however those of the author(s) only and do not necessarily reflect those of the European Union or the Slovak Academic Association for International Cooperation (SAAIC). Neither the European Union nor SAAIC can be held responsible for them.

References

1. Cohen, W.W., Ravikumar, P., Fienberg, S.E., et al.: A comparison of string distance metrics for name-matching tasks. In: IIWeb. vol. 3, pp. 73–78 (2003)
2. Deng, D., Kim, A., Madden, S., Stonebraker, M.: Silkmoth: An efficient method for finding related sets with maximum matching constraints. arXiv preprint arXiv:1704.04738 (2017)
3. Elmagarmid, A.K., Ipeirotis, P.G., Verykios, V.S.: Duplicate record detection: a survey. IEEE Trans. Knowl. Data Eng. **19**(1), 1–16 (2007)
4. Gali, N., Mariescu-Istodor, R., Hostettler, D., Fränti, P.: Framework for syntactic string similarity measures. Expert Syst. Appl. **129**, 169–185 (2019)
5. Li, B.H., Liu, Y., Zhang, A.M., Wang, W.H., Wan, S.: A survey on blocking technology of entity resolution. J. Comput. Sci. Technol. **35**, 769–793 (2020)
6. Okazaki, N., Tsujii, J.: Simple and efficient algorithm for approximate dictionary matching. In: Proceedings of the 23rd International Conference on Computational Linguistics (Coling 2010), pp. 851–859 (2010)
7. Papadakis, G., Skoutas, D., Thanos, E., Palpanas, T.: Blocking and filtering techniques for entity resolution: a survey. ACM Comput. Surv. (CSUR) **53**(2), 1–42 (2020)

8. Rozinek, O., Mareš, J.: The duality of similarity and metric spaces. Appl. Sci. **11**(4) (2021). https://www.mdpi.com/2076-3417/11/4/1910
9. Ukkonen, E.: Approximate string-matching with q-grams and maximal matches. Theoret. Comput. Sci. **92**(1), 191–211 (1992)
10. Wang, J., Li, G., Fe, J.: Fast-join: an efficient method for fuzzy token matching based string similarity join. In: 2011 IEEE 27th International Conference on Data Engineering, pp. 458–469. IEEE (2011)
11. Wang, J., Lin, C., Zaniolo, C.: Mf-join: efficient fuzzy string similarity join with multi-level filtering. In: 2019 IEEE 35th International Conference on Data Engineering (ICDE), pp. 386–397. IEEE (2019)
12. Yang, Z., Yu, J., Kitsuregawa, M.: Fast algorithms for top-k approximate string matching. In: Twenty-Fourth AAAI Conference on Artificial Intelligence (2010)
13. Zeakis, A., Skoutas, D., Sacharidis, D., Papapetrou, O., Koubarakis, M.: Tokenjoin: efficient filtering for set similarity join with maximumweighted bipartite matching. Proc. VLDB Endowment **16**(4), 790–802 (2022)

Scalable Similarity Joins for Fast and Accurate Record Deduplication in Big Data

Ondrej Rozinek[1(✉)], Monika Borkovcova[2], and Jan Mares[1,3]

[1] Department of Process Control, University of Pardubice, Studentska 95,
532 10 Pardubice, Czech Republic
ondrej.rozinek@gmail.com
[2] Department of Information Technology, University of Pardubice, Studentska 95,
532 10 Pardubice, Czech Republic
[3] Department of Mathematics, Informatics and Cybernetics, University of Chemistry
and Technology Prague, Technicka 5, 166 28 Prague, Czech Republic

Abstract. Record linkage is the process of matching records from multiple data sources that refer to the same entities. When applied to a single data source, this process is known as deduplication. With the increasing size of data source, recently referred to as big data, the complexity of the matching process becomes one of the major challenges for record linkage and deduplication. In recent decades, several blocking, indexing and filtering techniques have been developed. Their purpose is to reduce the number of record pairs to be compared by removing obvious non-matching pairs in the deduplication process, while maintaining high quality of matching. Currently developed algorithms and traditional techniques are not efficient, using methods that still lose significant proportion of true matches when removing comparison pairs. This paper proposes more efficient algorithms for removing non-matching pairs, with an explicitly proven mathematical lower bound on recently used state-of-the-art approximate string matching method - Fuzzy Jaccard Similarity. The algorithm is also much more efficient in classification using Density-based spatial clustering of applications with noise (DBSCAN) in log-linear time complexity $\mathcal{O}(|\mathcal{E}| \log(|\mathcal{E}|))$.

Keywords: record deduplication · Q-gram filter · record linkage · entity resolution · similarity space · bipartite matching · similarity join

1 Introduction

Record deduplication, a critical process in data management, refers to the identification and removal of duplicate records in databases. This process is essential for maintaining data quality and integrity, especially in large databases where duplicate records can lead to inconsistent, misleading, or erroneous data analysis [1,4].

© The Author(s), under exclusive license to Springer Nature Switzerland AG 2024
Á. Rocha et al. (Eds.): WorldCIST 2024, LNNS 990, pp. 181–191, 2024.
https://doi.org/10.1007/978-3-031-60328-0_18

The growing volume of digital data poses significant challenges to data dedu-plication. Traditional methods, which rely primarily on rule-based and exact matching techniques, are often unable to cope with the complexity of today's datasets, which include natural language variations.

Recent advances in machine learning and natural language processing have opened up new opportunities for more sophisticated deduplication strategies. [1,8,9]. These approaches use complex algorithms to identify duplicates with higher accuracy, even in data sets with high variability and noise.

This article aims to extend the model of similarity join methods for bipartite record matching, incorporating a newly developed count Q-gram filter, with an application to record deduplication.

2 Related Work

Recent developments in Q-gram count filters have had a significant impact on methods for approximate string matching and data deduplication. Ukkonen's seminal work [12] introduced Q-grams to string processing and laid the founda-tion for subsequent algorithmic advances. This approach has been further refined, as shown by Yang et al. [17], to be suitable for large-scale data environments.

The introduction of the Ed-join algorithm by Xiao et al. (2008) [16] marked a notable advance in algorithmic development for similarity joins with edit dis-tance constraints, using Q-grams to optimise performance. At the same time, the survey by Yu et al. (2016) [18] emphasized the effectiveness of Q-gram-based techniques in string similarity joins, highlighting their role in balancing accuracy and computational efficiency.

Hybrid approaches that integrate Q-gram count filters with other computa-tional methods have demonstrated improved effectiveness in data deduplication. Jiang et al. (2014) [6] illustrated the benefits of combining Q-gram filters with token-based methods, showing improved results on datasets characterised by typographical variations.

Challenges of scalability and data noise continue to drive innovation in the field. Vernica et al. (2010) [13] addressed scalability with a parallelized approach to set-similarity joins using Q-gram methods. Koudas et al. (2006) focused on the use of Q-gram filters in flexible string matching against large, diverse databases.

Recent research, such as that of Papadakis et al. (2020) [8], indicates a trend towards integrating machine learning with Q-gram count filters to improve entity resolution in deduplication processes. This integration represents a potential shift towards more sophisticated data processing capabilities.

3 Problem Formulation

At its core, deduplication involves the concept of entities (or entity profiles) [8], which provide a uniquely identified description of a real-world object in the form of name-value pairs. Two entities e_i and e_j match, $e_i \equiv e_j$, if they refer to the same real-world entity. Matching entities are also called duplicates. The task of entity resolution is to find all matching entities within an entity collection or across two or more entity collections. The term entity is also interchangeable with the term record, which is mainly used in the fields of databases and data storage.

Definition 1 (Deduplication). *Deduplication is a process represented by a function $\mathcal{D}\colon \mathcal{E} \to \mathcal{C}$, where $\mathcal{E}$ is a collection of entities, and $\mathcal{C}$ is a collection of clusters of duplicate entities within $\mathcal{E}$. Each cluster in $\mathcal{C}$ consists exclusively of entities that are considered equivalent (duplicates) under a specified equivalence relation $\equiv$. Formally, the function is defined as:*

$$\mathcal{D}(\mathcal{E}) = \mathcal{C} = \{\{e_i, \ldots, e_j\}\colon e_i, \ldots, e_j \in \mathcal{E}, \forall e_i \equiv e_j, i \neq j\}. \tag{1}$$

The definition based on the family of sets $\mathcal{C}$ imposes the necessity to have the output as a cluster of entities with the same entity resolution. This is the main difference with the introduced definition [8].

The use of clusters requires the use of Euclidean or metric spaces, which provide the fundamental properties of most clustering methods. These spaces define distance metrics, which are crucial optimization criteria for many clustering algorithms. Working in non-metric spaces presents challenges such as difficulties in point localization, distance measurement, algorithm convergence, and identifying cluster shapes. However, a wide range of similarity functions, such as the Jaccard index, Tanimoto coefficient and edit similarity, have been shown to be dual to metric spaces [11]. We refer to this dual construct as a 'similarity space', which shapes a different axiomatic system. We therefore focus on similarity spaces with proven duality to metric spaces [11].

Definition 2 (Similarity Space [10,11]). *Given a non-empty set $\mathcal{X}$, a function $s\colon \mathcal{X} \times \mathcal{X} \to \mathbb{R}$ is a similarity metric if for all elements $x, y, z \in \mathcal{X}$, it satisfies the following axioms:*

(S1) $s(x,y) = s(y,x)$ (symmetry),
(S2) $s(x,z) + s(y,y) \geq s(x,y) + s(y,z)$ (triangle inequality),
(S3) $s(x,x) = s(x,y) = s(y,y) \iff x = y$ (identity of indiscernibles),
(S4) $s(x,y) \geq 0$ (non-negativity),
(S5) $s(x,y) \leq \min\{s(x,x), s(y,y)\}$ (bounded by self-similarity).

A similarity space is an ordered pair $(\mathcal{X}, s)$.

Theorem 1 (Generalized Rozinek Distance [11]). *Suppose given a normalized similarity metric s_n and an arbitrary similarity metric s. The Generalized Rozinek distance $d_R\colon \mathcal{X} \times \mathcal{X} \to \mathbb{R}$ is the distance metric derived from an arbitrary normalized similarity metric $s_n(x,y)$ and self-similarities $s(x,x)$ and $s(y,y)$ by*

$$d_R(x,y) = \frac{1 - s_n(x,y)}{1 + s_n(x,y)}(s(x,x) + s(y,y)). \tag{2}$$

Proof. [11]

Definition 3 (Self-Join in Similarity Space). *Given an entity collection* $\mathcal{E}$, *a similarity metric* $s\colon \mathcal{E}^2 \to \mathbb{R}$, *and a similarity threshold* α, *a similarity join identifies all pairs of entity in* $\mathcal{E}$ *that have similarity at least* α

$$\mathcal{E} \bowtie_\alpha \mathcal{E} = \{(e_i, e_j) \in \mathcal{E}^2 \colon s(e_i, e_j) \geq \alpha, i \neq j\}. \tag{3}$$

The effectiveness of deduplication lies in its ability to detect true positive duplicates, while the efficiency is related to the computational cost of these detections, usually measured by the number of comparisons or the computational time complexity $\mathcal{O}(D(\mathcal{E}))$. The brute-force approach, involving all pairwise comparisons within an entity collection $\mathcal{E}$, leads to a quadratic complexity $\mathcal{O}(D(\mathcal{E})) = \mathcal{O}(|\mathcal{E}|^2)$. For deduplicating a single structured data source with $|\mathcal{E}|$ entities, the maximum number of comparisons is equal to half the entries in a $|\mathcal{E}| \times |\mathcal{E}|$ symmetric matrix, excluding the diagonal. This results in a final time complexity of $\mathcal{O}(D(\mathcal{E})) = (|\mathcal{E}|^2 - |\mathcal{E}|)/2$, since each entity may have to be compared with all others.

To avoid an exhaustive pairwise comparison, the similarity join typically consists of two steps

- *Filtering* is a function $\mathcal{F}_\alpha\colon \mathcal{E}^2 \to \mathcal{E}^2$ that returns a set of candidates for each entity e_i, excluding all those that do not match e_i.
- *Matching* is a function $\mathcal{M}_\alpha\colon \mathcal{E}^2 \to \mathcal{E}^2$.

Self-Join in similarity space could be decomposed into functional composition as follows

$$\mathcal{E} \bowtie_\alpha \mathcal{E} = \mathcal{M}_\alpha \circ \mathcal{F}_\alpha. \tag{4}$$

4   Self-Join in Similarity Space

In this generalization, we consider two entities e_i, e_j each comprising token sets $\mathcal{X}$ and $\mathcal{Y}$, which are matched in a bipartite graph using Fuzzy Jaccard Similarity. The variable $|\mathcal{M}|$ represents the maximum number of connected token pairs in the optimal combinatorial assignment problem.

Definition 4 (Fuzzy Jaccard Similarity [14,15]). *At our disposal are two sets of tokens,* $\mathcal{X}$ *and* $\mathcal{Y}$. *Write* $\mathcal{X} \widetilde{\cap} \mathcal{Y}$ *for the fuzzy overlap of* $\mathcal{X}$ *and* $\mathcal{Y}$:

Fuzzy Jaccard Similarity, $s_n(\mathcal{X}, \mathcal{Y})$,

$$s_n(\mathcal{X}, \mathcal{Y}) = \frac{|\mathcal{X} \widetilde{\cap} \mathcal{Y}|}{|\mathcal{Y}| + |\mathcal{X}| - |\mathcal{X} \widetilde{\cap} \mathcal{Y}|}. \tag{5}$$

In the articles [14,15], the incident edge for the token pair is considered only if $s_n(X_i, Y_j) \geq \delta$. In our definition, we have dropped the second threshold δ for a more fuzzy approach, because applying δ would essentially create a binary classifier determining whether a token pair (X_i, Y_j) is classified as a match. Furthermore, the original paper violates the triangle inequality $S2$ due to the normalization of edit distance $d_n(X_i, Y_j) = \frac{d(X_i, Y_j)}{\max|x|, |y|}$.

Example 1. Consider the strings $X = $ "ab", $Y = $ "abc", and $Z = $ "bc". Then we obtain

$$d(X, Z) \leq d(X, Y) + d(Y, Z),$$

$$\frac{d(X, Z)}{max\{|X|, |Z|\}} \leq \frac{d(X, Y)}{max\{|X|, |Y|\}} + \frac{d(Y, Z)}{max\{|Y|, |Z|\}}, \qquad (6)$$

$$\frac{2}{2} \nleq \frac{1}{3} + \frac{1}{3}.$$

Theorem 2 (Threshold of Normalized Edit Similarity Metric). *Let the edit distance be $d(X, Y)$, the worst case of the expected distance function be $d(\alpha, |X|, |Y|)$, and write the floor function by $\lfloor . \rfloor$. Then*

$$s_n(X, Y) \geq \alpha \iff d(X, Y) \leq d(\alpha, |X|, |Y|) = \left\lfloor \frac{1 - \alpha}{1 + \alpha}(|X| + |Y|) \right\rfloor, \qquad (7)$$

where $\alpha \in [0, 1] \subset \mathbb{R}$ is a threshold of the normalized similarity metric given by $s_n(X, Y) \geq \alpha$.

Proof. According to Theorem 1 and substituting for self-similarities which equal the corresponding cardinality of the sets, $s(X, X) = |X|$, $s(Y, Y) = |Y|$, the expected distance $d(\alpha, |X|, |Y|)$ reaches a maximum just when the similarity is minimal under the lowest similarity given by the threshold α. This happens if and only if we substitute $s_n(X, Y) = \alpha$:

$$d(X, Y) = \lfloor d_R \rfloor = \left\lfloor \frac{1 - s_n(X, Y)}{1 + s_n(X, Y)}(s(X, X) + s(Y, Y)) \right\rfloor$$

$$\leq \sup_\alpha d(X, Y) = \left\lfloor \frac{1 - \alpha}{1 + \alpha}(|X| + |Y|) \right\rfloor = d(\alpha, |X|, |Y|). \qquad (8)$$

The edit distance is an integer, hence we use the floor function to remove the undefined fractional part.

4.1   Optimal Count Q-Gram Filter

In [7], a lower bound relationship is established between edit distance and the Q-gram method for a pattern string X of length $|X|$ and a text string Y of length $|Y|$. This lower bound is crucial for string similarity search and similarity join algorithms, which are widely applied in data cleaning, search engines, and data integration [18]. These algorithms, going beyond traditional exact search methods, handle data errors and inconsistencies. Their importance lies in speeding up similarity joins and minimizing exhaustive pairwise comparisons.

Theorem 3 (Q-gram Count Filtering [7,17]). *Let X and Y be strings with the edit distance $d(X, Y)$. Then, the Q-gram similarity $|Q_X \cap Q_Y|$ of the token X and Y is at least*

$$t = \inf_d \{|Q_X \cap Q_Y|\} = max\{|X|, |Y|\} - q + 1 - qd(X, Y), \qquad (9)$$

where t is a Q-gram similarity threshold with respect to $d(X, Y)$.

We refine the Q-Gram Count filter model by introducing more precise assumptions for the token sets $\mathcal{X}, \mathcal{Y}$ and derive an optimal filter that ensures no comparison pair with a Fuzzy Jaccard Similarity higher than α is lost.

We assume that the matching token pairs provided by $\mathcal{M}$ are known, while their edit distances are unknown. However, these distances can be predicted based on expected edit distances.

Theorem 4 (Optimal Count Q-gram Filter for Bipartite Matching).
Let $\mathcal{X}$ and $\mathcal{Y}$ be records representing a set of tokens. Then the Q-gram similarity in bipartite matching of $\mathcal{X}$, $\mathcal{Y}$ and cardinality $|\mathcal{M}|$ for a given threshold $s_n(\mathcal{X}, \mathcal{Y}) \geq \alpha$ is at least

$$t_{\mathcal{M}} = \inf_{\alpha}\{|Q_{\mathcal{X}} \cap Q_{\mathcal{Y}}|\}$$

$$= \underbrace{\sum_{(i,j)\in\mathcal{M}} \max\{|X_i|, |Y_j|\} - |\mathcal{M}|q + |\mathcal{M}|}_{\text{maximum shared Q-grams}} \underbrace{-q \max_{\alpha} \sum_{(i,j)\in\mathcal{M}} d(\alpha_{i,j}, |X_i|, |Y_j|),}_{\text{loss function}} \quad (10)$$

containing a linear combination of

$$d(\alpha_{i,j}, |X_i|, |Y_j|) = \frac{1 - \alpha_{i,j}}{1 + \alpha_{i,j}}(|X_i| + |Y_j|) \quad (11)$$

under the constraint $\alpha = \frac{\sum_{(i,j)\in\mathcal{M}} \alpha_{i,j}}{|\mathcal{X}|+|\mathcal{Y}|-\sum_{(i,j)\in\mathcal{M}} \alpha_{i,j}}$ for which the linear combination is maximized.

Proof. Consider the sum over connected pairs of tokens with cardinality $|\mathcal{M}|$

$$\inf_{\alpha}\{|Q_{\mathcal{X}} \cap Q_{\mathcal{Y}}|\} = \inf_{\alpha}\left\{\sum_{(i,j)\in\mathcal{M}} |Q_{X_i} \cap Q_{Y_i}|\right\} = \sum_{(i,j)\in\mathcal{M}} \inf_{\alpha_{i,j}}\{|Q_{X_i} \cap Q_{Y_j}|\}$$

$$= \sum_{(i,j)\in\mathcal{M}} \inf_{\alpha_{i,j}}\{\max\{|X_i|, |Y_j|\} - q + 1 - qd(X, Y)\}$$

$$= \sum_{(i,j)\in\mathcal{M}} \{\max\{|X_i|, |Y_j|\} - q + 1 - q\sup_{\alpha_{i,j}} d(X, Y)\}$$

$$= \sum_{(i,j)\in\mathcal{M}} \max\{|X_i|, |Y_j|\} - |\mathcal{M}|q + |\mathcal{M}| - q\max_{\alpha} \sum_{(i,j)\in\mathcal{M}} d(\alpha_{i,j}, |X_i|, |Y_j|).$$

$$(12)$$

Each $\alpha_{i,j}$ is the distributed minimum similarity for each token, giving a threshold vector that should maximize the sum of the expected distances $d(\alpha_{i,j}, |X_i|, |Y_j|)$ so that $s_n(\mathcal{X}, \mathcal{Y}) \geq \alpha$ holds for t_M. Formalizing this, we get the task

$$\text{maximize} \sum_{(i,j)\in\mathcal{M}} d(\alpha_{i,j}, |X_i|, |Y_j|),$$

$$\text{subject to} \sum_{(i,j)\in\mathcal{M}} \alpha_{i,j} \geq \frac{\alpha}{1+\alpha}(|\mathcal{X}| + |\mathcal{Y}|) \quad \alpha \in [0,1], i = 1, \ldots, |\mathcal{M}|,$$

$$\alpha_{i,j} \in [0,1], j = 1, \ldots, |\mathcal{M}|.$$

This leads to an integer linear programming task equivalent to the Knapsack problem, solvable in $\mathcal{O}(nb)$ time. The optimization algorithm determines the maximum expected edit distance distribution across tokens, maintaining the similarity threshold $s_n(\mathcal{X}, \mathcal{Y}) \geq \alpha$.

We aim to merge the objective function and its constraint into a single expression using the Lagrange multiplier method

$$\mathcal{L}(\alpha_{i,j}, ..., \alpha_{|\mathcal{M}|}, \lambda)$$

$$= \sum_{(i,j)\in\mathcal{M}} \frac{1-\alpha_{i,j}}{1+\alpha_{i,j}}(|X_i|+|Y_j|) - \lambda\left(\frac{\alpha}{1+\alpha}(|\mathcal{X}|+|\mathcal{Y}|) - \sum_{(i,j)\in\mathcal{M}} \alpha_{i,j}\right).$$

and solve $\nabla_{\alpha_{i,j},...,\alpha_{|\mathcal{M}|},\lambda}\mathcal{L}(\alpha_{i,j}, ..., \alpha_{|\mathcal{M}|}, \lambda) = 0$. Differentiating with respect to a specific $\alpha_{i,j}$ and setting the derivative to zero:

$$\frac{\partial\mathcal{L}}{\partial\alpha_{i,j}} = -\frac{2(|X_i|+|Y_j|)}{\alpha_{i,j}^2 + 2\alpha_{i,j} + 1} - \lambda = 0. \tag{13}$$

Differentiating with respect to λ and setting this derivative to zero:

$$\frac{\partial\mathcal{L}}{\partial\lambda} = \frac{\alpha}{1+\alpha}(|\mathcal{X}|+|\mathcal{Y}|) - \sum_{(i,j)\in\mathcal{M}} \alpha_{i,j} = 0. \tag{14}$$

Solving these equations will yield the values of $\alpha_{i,j}$ and the optimal λ for the given optimization problem. From this, we can solve for λ as follows

$$\lambda = -\frac{2(|X_i|+|Y_j|)}{\alpha_{i,j}^2 + 2\alpha_{i,j} + 1}. \tag{15}$$

4.2   Approximate Count Q-gram Filter

Considering the analytical intractability of the optimal count Q-gram filter, our goal is to establish a suitable approximation that maintains accuracy while ensuring computational efficiency, achieving a constant time complexity of $O(1)$.

Our approach is predicated on the following assumptions:

- The cardinalities of the sets $\mathcal{X}$, $\mathcal{Y}$, and the matching set $\mathcal{M}$ are equivalent, i.e., $|\mathcal{X}| = |\mathcal{Y}| = |\mathcal{M}|$.
- For every token pair $(i, j) \in \mathcal{M}$, the length of Y_j is assumed to be the expected value of the lengths of tokens in $\mathcal{Y}$, represented as $|Y_j| = \mathbb{E}[Y_j]$.
- The similarity threshold α is uniformly applied across all token pairs, such that $\mathbb{E}[\alpha_{i,j}] = \alpha$, and hence $\mathbb{E}[\mathcal{L}(\alpha_{i,j}, ..., \alpha_{|\mathcal{M}|}, \lambda)] = \mathcal{L}(\alpha, \lambda)]$.
- The expected number of destroyed Q-grams, $\mathbb{E}[q_{i,j}]$, is estimated based on the assumption that the edit distance $d = 1$ is a uniformly distributed random variable over the token. This estimation is represented by the formula:

$$\mathbb{E}[q_{i,j}] = q \cdot \frac{\sup|Q_{X_i} \cap Q_{Y_j}|}{\max\{|X_i|, \mathbb{E}[|Y_j|]\}} = q \cdot \frac{\max\{|X_i|, \mathbb{E}[|Y_j|]\} - q + 1}{\max\{|X_i|, \mathbb{E}[|Y_j|]\}} < q. \tag{16}$$

Specifically, by establishing that $\mathbb{E}[q_{i,j}] < q$, the filter criteria become more stringent and so enhancing the selectivity of the filter.

- The mean value of the expected number of destroyed Q-grams across all tokens $\mathcal{X}$ is denoted $\mathbb{E}[\mathbb{E}[q_{i,j}]]$.
- Due to certain simplifications and estimations made across the collection of records, we introduce the filter sensitivity factor γ within the range $[0, 1]$. This factor is empirically set to be slightly weaker, allowing a balance between high efficiency and precision of the filter.

Under these considerations, the approximation for the Q-gram similarity is derived as follows:

$$
\begin{aligned}
\mathbb{E}[t_{\mathcal{M}}] &= \sum_{i \in \mathcal{X}} \max\{|X_i|, \mathbb{E}[Y_j]\} - |\mathcal{X}|q + |\mathcal{X}| - \mathcal{L}(\alpha, \lambda) \\
&= \sum_{i \in \mathcal{X}} \max\{|X_i|, \mathbb{E}[Y_j]\} - |\mathcal{X}|q + |\mathcal{X}| - \sum_{i \in \mathcal{X}} \mathbb{E}[q_{i,j}] d(\alpha, |X_i|, \mathbb{E}[Y_j]) \\
&\quad - \mathbb{E}[\mathbb{E}[q_{i,j}]] \cdot \gamma \cdot \lambda \cdot \left(\frac{\alpha(1 - \alpha)}{1 + \alpha} |\mathcal{X}| \right).
\end{aligned}
\tag{17}
$$

The loss function $d(\alpha, |X_i|, |Y_j|)$, considering the expected value of Y_j, is defined as $d(\alpha, |X_i|, \mathbb{E}[Y_j]) = \frac{1-\alpha}{1+\alpha}(|X_i| + \mathbb{E}[Y_j])$ and Lagrange multiplier $\lambda = -\frac{2(\mathbb{E}[|X_i|]+\mathbb{E}[Y_j])}{\alpha^2+2\alpha+1}$.

This approximation greatly simplifies the original problem. By standardizing the length of tokens in $\mathcal{Y}$ to their expected value and using a consistent similarity threshold, we optimize the Q-gram similarity calculation to $\mathcal{O}(1)$. This method is especially useful when the exact lengths of $|Y_j|$ are unknown or when computational efficiency is a priority.

5   Experiments

In our experiments, we estimate the expected token length as the average length across all $\mathcal{E}$ records, denoting $\mathcal{Y} \in \mathcal{E}$ as $\mathbb{E}[Y_j] = \overline{Y}$ and $\mathcal{X}$ as $\mathbb{E}[X_i] = \overline{X}$. We set the factor γ to 0.75 empirically.

The precision and recall metrics are based on the concepts of true positives, false positives, and false negatives:

- *True Positives (TP):* Pairs of records that are correctly placed in the same clusters and belong to the same entity.
- *False Positives (FP):* Pairs of records that are incorrectly placed in the same cluster but belong to different entities.
- *False Negatives (FN):* Pairs of records that belong to the same entity but are incorrectly placed in different clusters.

Based on the defined terms, Precision and Recall are calculated using the formulas: Precision $= \frac{TP}{TP+FP}$ and Recall $= \frac{TP}{TP+FN}$. The F-Score, which is the harmonic mean of Precision and Recall, is given by: $F - Score = 2 \times \frac{\text{Precision} \times \text{Recall}}{\text{Precision} + \text{Recall}}$.

These equations represent the standard approach to calculating the accuracy of a process such as clustering, balancing the trade-off between precision and recall. The maximum F-Score is calculated over all thresholds α in range $[0, 1]$. For our clustering analysis, we employ DBSCAN [5] and Nearest Neighbors (NN) as demonstrated in Table 1. The Q-Gram filter achieved a precision of 100%, indicating that no true positive (TP) comparison pairs were erroneously removed, while maintaining a recall of 52%. This high efficiency and filtering capability of the Q-Gram filter are evident, as filter passes only twice as many candidate pairs compared to the fraction of TP comparison pairs.

Table 1. Sorted Comparison of Max F-scores for DBSCAN and Nearest Neighbor Clustering Algorithms on Labelled Vauniv Dataset (116 Records and 15 Clusters) [2]

Similarity	DBSCAN Max F-score	NN Max F-score
2-Gram Filter + Fuzzy Jaccard	**0.8520**	0.6979
3-Gram Filter + Fuzzy Jaccard	**0.8520**	0.6979
Fuzzy Jaccard	**0.8520**	0.6979
2-Gram Jaccard	0.8248	0.6061
Jaro	0.7917	0.7163
2-Gram Overlap	0.6586	0.7917
3-Gram Jaccard	0.7886	0.6619
3-Gram Overlap	0.6720	0.7405
Jaro-Winkler	0.7333	0.6169
Levenshtein	0.6859	0.5516

6   Conclusion

In this work, we extend Ukkonen's lemma by incorporating an edit constraint for bipartite matching, a notable field contribution. Our main contribution outperforms existing models with an optimal Q-Gram Count filter for bipartite matching. This development, from a mathematically derived lower bound, ensures no loss of true positive (TP) comparison pairs. Given its analytical intractability, we propose a precise estimation method for this filter that operates in constant time complexity $\mathcal{O}(1)$. In our tests, this approach achieved 100% precision in filtering with a high filtering capability. Altogether, the proposed extended count Q-gram filter significantly speeds up the process of similarity join while maintaining high filter efficiency and precision. The record deduplication was efficiently conducted using DBSCAN clustering, which has the significant advantage of being able to form clusters of arbitrary shape while maintaining fast performance, characterized by a time complexity of $\mathcal{O}(|\mathcal{E}| \log(|\mathcal{E}|))$ [5]. It is worth noting that scalability can be achieved by using one of the parallel versions of DBSCAN, as discussed in [3].

Acknowledgment. It was supported by SGS FEI UPCE 2024 and the Erasmus+ project: Project number: 2022-1-SK01-KA220-HED-000089149, Project title: Including EVERyone in GREEN Data Analysis (EVERGREEN) funded by the European Union. Views and opinions expressed are however those of the author(s) only and do not necessarily reflect those of the European Union or the Slovak Academic Association for International Cooperation (SAAIC). Neither the European Union nor SAAIC can be held responsible for them.

References

1. Christen, P.: A survey of indexing techniques for scalable record linkage and deduplication. IEEE Trans. Knowl. Data Eng. **24**(9), 1537–1555 (2012)
2. Cohen, W.W., Ravikumar, P., Fienberg, S.E., et al.: A comparison of string distance metrics for name-matching tasks. In: IIWeb, vol. 3, pp. 73–78 (2003)
3. Dafir, Z., Lamari, Y., Slaoui, S.C.: A survey on parallel clustering algorithms for big data. Artif. Intell. Rev. **54**, 2411–2443 (2021)
4. Elmagarmid, A.K., Ipeirotis, P.G., Verykios, V.S.: Duplicate record detection: a survey. IEEE Trans. Knowl. Data Eng. **19**(1), 1–16 (2007)
5. Ester, M., Kriegel, H.P., Sander, J., Xu, X., et al.: A density-based algorithm for discovering clusters in large spatial databases with noise. In: KDD, vol. 96, pp. 226–231 (1996)
6. Jiang, Y., Li, G., Feng, J., Li, W.S.: String similarity joins: an experimental evaluation. Proc. VLDB Endow. **7**(8), 625–636 (2014)
7. Jokinen, P., Ukkonen, E.: Two algorithms for approxmate string matching in static texts. In: Tarlecki, A. (ed.) MFCS 1991. LNCS, vol. 520, pp. 240–248. Springer, Heidelberg (1991). https://doi.org/10.1007/3-540-54345-7_67
8. Papadakis, G., Skoutas, D., Thanos, E., Palpanas, T.: Blocking and filtering techniques for entity resolution: a survey. ACM Comput. Surv. (CSUR) **53**(2), 1–42 (2020)
9. Papadakis, G., Svirsky, J., Gal, A., Palpanas, T.: Comparative analysis of approximate blocking techniques for entity resolution. Proc. VLDB Endow. **9**(9), 684–695 (2016)
10. Rozinek, O., Borkovcova, M.: Theorems for boyd-wong contraction mappings on similarity spaces. Mathematics **11**(20), 4359 (2023)
11. Rozinek, O., Mareš, J.: The duality of similarity and metric spaces. Appl. Sci. **11**(4) (2021). https://www.mdpi.com/2076-3417/11/4/1910
12. Ukkonen, E.: Approximate string-matching with q-grams and maximal matches. Theoret. Comput. Sci. **92**(1), 191–211 (1992)
13. Vernica, R., Carey, M.J., Li, C.: Efficient parallel set-similarity joins using mapreduce. In: Proceedings of the 2010 ACM SIGMOD International Conference on Management of Data, pp. 495–506 (2010)
14. Wang, J., Li, G., Fe, J.: Fast-join: an efficient method for fuzzy token matching based string similarity join. In: 2011 IEEE 27th International Conference on Data Engineering, pp. 458–469. IEEE (2011)
15. Wang, J., Li, G., Feng, J.: Extending string similarity join to tolerant fuzzy token matching. ACM Trans. Database Syst. (TODS) **39**(1), 1–45 (2014)
16. Xiao, C., Wang, W., Lin, X.: Ed-join: an efficient algorithm for similarity joins with edit distance constraints. Proc. VLDB Endow. **1**(1), 933–944 (2008)

17. Yang, Z., Yu, J., Kitsuregawa, M.: Fast algorithms for top-k approximate string matching. In: Twenty-Fourth AAAI Conference on Artificial Intelligence (2010)
18. Yu, M., Li, G., Deng, D., Feng, J.: String similarity search and join: a survey. Front. Comput. Sci. **10**(3), 399–417 (2016)

Impact of Preprocessing Using Substitution on the Performance of Selected NER Models - Results

Miroslav Potočár[✉]

University of Žilina, Žilina, Slovakia
`Miroslav.Potocar@fri.uniza.sk`

Abstract. This paper investigates the impact of preprocessing using substitution in word sequences on selected models of named entity recognition. The study is focused on evaluating the results of the performed experiments. It briefly describes the concept of substitution using pseudo words and the methodology used in performing the experiments. Based on the outputs of the experiments, it discusses in detail the implications of substitution on the selected models and provides possible explanations for the results. In the end, conclusions and recommendations for the use of substitution as a preprocessing technique are made based on the observed results.

Keywords: named entity recognition · preprocessing · substitution · pseudo words

1 Introduction

Nowadays, the number of data increases considerably over time [2]. Much of this data is in the form of unstructured text, so a lot of attention is currently being paid to natural language processing tasks. Named Entity Recognition (NER) task is focused on identifying, locating and classifying important objects in text data [5,8]. NER is a fundamental task that is addressed in other tasks such as information extraction, question answering and knowledge ontology construction. In these tasks, NER is indispensable [3].

Data preprocessing is an important aspect in solving many natural language processing tasks. Despite its importance, it has received little attention in the literature. In this paper, we investigate a specific preprocessing strategy - substitution - and its impact on the performance of selected NER models. Substitution is the strategic replacement of words in sequences by pseudo words that encode a certain feature of the replaced word. This type of substitution may reduce overfitting and improve the model's generalization abilities for some models under certain conditions.

This paper is dedicated to revealing the results of a large set of experiments designed to evaluate the impact of preprocessing using substitution on selected NER models. Specifically, we investigate the effects on the hidden Markov model

Á. Rocha et al. (Eds.): WorldCIST 2024, LNNS 990, pp. 192–202, 2024.
https://doi.org/10.1007/978-3-031-60328-0_19

(HMM), conditional random fields (CRF), gated recurrent unit (GRU), bidirectional long short-term memory network (BiLSTM) and our Naïve model. With respect to the experiment results, we aim to provide a comprehensive understanding of how substitution affects the prediction capabilities of the models.

2 Concept of Pseudo Word Substitution

The idea to investigate the impact of word substitution in the sequence by pseudo words arose while studying the work of Bikel et al. [1] where pseudo words appeared as one of the features entering the process of solving the NER task. We took this concept of pseudo words and focused on researching how the use of pseudo words affects NER models whose input is only a sequence of words. The idea is to replace unknown or rarely occurring words with a pseudo word that encodes one of the features. We have taken the word features from the original work, with their order, examples, and the intuition behind each feature, and extended them to include a representative pseudo word and a condition that must be fulfilled in order to replace the word with the pseudo word. We substitute words that:

- consist exclusively of two numbers,
- consist exclusively of four numbers,
- containing numbers and letters,
- contain numbers and dashes,
- contain numbers and slashes,
- contain numbers and commas,
- contain numbers and a period,
- represent other numbers,
- consist entirely of capital letters,
- have a capital initial letter and a period,
- are the first word in a sentence,
- have a capital initial letter,
- consist of lower case letters only,
- contains any alphanumeric and non-alphanumeric characters.

A detailed list of pseudo words along with rules, examples and intuition can be found in [6].

An example of the functioning and impact of substitution is best illustrated with words containing numbers. We often encounter this type of words in texts. If we wanted to include all numbers in the vocabulary, this would not be possible due to their infinite nature. Words such as '22/11/2023' and '23/11/2023', would appear as unique words within the vocabulary, but we can infer from their structure that they are words representing a date. For NER models to be able to identify that a given word is a date, they would need to encounter the particular word in multiple possible contexts. However, in this way they would only learn to recognize one particular date. Clearly, this method of learning would lead to overfitting and an inability to generalize in NER models. If we replace each word that has the shape of a date with a pseudo word [CDS], the model will have more opportunities to learn the context in which dates occur, and as a result the model will be capable to better handle new, unique dates as well. This improves the model's ability to generalize and prevents overfitting of the model. Similar logic can be applied to words containing a capital letter. The set of possible company names is theoretically unlimited. However, most names share a common feature, which is the first initial letter. Replacing this name with the pseudo word [IC] allows the model to better learn the context in which the company names may occur.

3   Methodology

In the following section we briefly describe the methodology used. A detailed explanation of our experimental procedure can be found in the paper [6].

3.1   Data

As test data, we used the dataset *CoNLLpp* [9], which is a modification of the original dataset *CoNLL2003* [7]. The dataset uses the IOB2 labeling scheme. It distinguishes four types of entities, persons (PER), locations (LOC), organizations (ORG) and miscellaneous (MISC). There are three sets already pre-prepared in the dataset namely training, validation and test sets. A summary of the data in each part of the dataset can be seen in Table 1. For each part, we have listed the number of sentences, the number of words, the number of unique words. For each IOB category, we have listed the number of words associated with this category and also the number of unique words for this category.

3.2   Models Implementations

We have used *Python* in our research, so the model implementations used are influenced by this.

In the case of the Naïve model, we self-implemented a simple class that stored information about words and the named entity tag that occurred most frequently with a given word in the training set. It assigns the most frequent tag in train dataset to unknown words.

For the HMM, we have used the *HiddenMarkovModelTagger* implementation available in *NLTK* library. This implementation can be used in sequence prediction, and is also able to handle unseen tokens.

As a CRF model, we used the *CRF* implementation available in the *sklearn-crfsuite* library, which is a library focused specifically on the CRF model. Unfortunately, this library is no longer active. Because of this, we have encountered several issues when we were using it, related to incompatibility with newer versions of the *numpy* library. However, due to the library being open source, it was possible to fix these issues and make use of the implementation.

To implement the GRU and BiLSTM models, we used the tools provided by the *Keras* and *TensorFlow* libraries.

Table 1. Datasets summary

	Train	Validation	Test
Sentences	14041	3250	3453
Words	203621	51362	46435
Unique words	23623	9966	9488
B-LOC words	7140	1837	1646
B-LOC unique words	1223	511	476
B-MISC words	3438	922	723
B-MISC unique words	707	304	279
B-ORG words	6321	1341	1715
B-ORG unique words	1767	547	684
B-PER words	6600	1842	1618
B-PER unique words	2275	919	858
I-LOC words	1157	257	259
I-LOC unique words	294	90	105
I-MISC words	1155	346	254
I-MISC unique words	334	148	115
I-ORG words	3704	751	882
I-ORG unique words	1124	334	399
I-PER words	4528	1307	1161
I-PER unique words	2398	952	797
O words	169578	42759	38177
O unique words	15704	6806	6401

3.3 Replacing Words with Pseudo Words

The process of word substitution with pseudo words consists of several steps. At the beginning, a dictionary of known words is created based on the training data. From this dictionary, depending on the scenario, words containing numbers, composed entirely of non-alphanumeric characters, or words whose number of occurrences within the training dataset satisfy a certain threshold are removed or kept. In the next step, the individual words in each of the datasets are sequentially reviewed and it is determined whether they should be replaced by a pseudo word. First, the occurrence of the word in the dictionary of known words is checked. If it is found in this dictionary, the original form is retained and it is discarded from further processing. If it is not found in the dictionary, it means that this word will be replaced by a pseudo word. For words not found in the dictionary of known words, the conditions are successively applied in the exact order. Depending on which condition the word satisfies, it is replaced by the corresponding pseudo word. The datasets modified in this way are used to train, validate and also test the models. The detailed word substitution process is described in the paper [6].

3.4   Test Scenarios

As described in detail in the paper [6], we performed several experimental scenarios with each model. For each model, we determined its raw performance, i.e., the overall F1 score that the model achieved when no preprocessing was being applied. We then moved on to the actual scenarios where preprocessing was used.

We tested two types of preprocessing scenarios. In the first type of scenario, we did not remove words consisting entirely of punctuation or words containing numerical values from the set of all words that were later used to build the dictionary of known words. We only removed those words from the set whose number of occurrences was below a given threshold. Specifically, we focused on scenarios where we removed words that had a frequency of occurrences less than 1, 2, 3, 4, and 5. Each frequency threshold represented one test scenario. In the second type of scenario, we removed words containing numbers or composed entirely of non-alphanumeric characters from the set of all words used to build the dictionary of known words. As in the first type of scenario, we also removed words whose number of occurrences was less than 1, 2, 3, 4, and 5 from the rest of the words. From the remaining words, a dictionary of known words was created and used during the substitution process.

3.5   Evaluation

To evaluate the performance of the models, we used the F1 score metric, which is widely used in NER task evaluation. F1 score is a combination of precision and recall metrics. Thus, it expresses the balance of both metrics and indicates how well the model is able to correctly identify entities and at the same time what is the ability of the model to detect all real entities.

We used the *seqeval* [4] framework available through the *evaluate* library, which is designed to evaluate labeled sequences. This framework provides values for precision, accuracy, recall, and F1 score for the entire dataset and also provides the same metrics with respect to specific categories of named entities. The *seqeval* provides two evaluation modes, **default** and **strict**. The default mode simulates *conlleval* and **strict** evaluates the inputs based on a specific schema. Our dataset used the IOB2 schema, so we used the **strict** mode where we defined the use of the IOB2 schema.

For each test scenario, we performed 5 runs, which means that we recreated and retrained the model 5 times and evaluated its performance on each dataset. The final value of each metric is calculated as the average of the measured values in each run.

4   Results

The results of the experiments are shown in Table 2. One row represents the averaged results for one tested scenario. Each row consists of the following values:

- **Model** - This column specifies the tested model. In our case, we performed the experiments with the Naïve, HMM, CRF, GRU and BiLSTM models.
- **S** - This column provides information on if substitution was (T) or was not (F) performed over the individual datasets.
- **N&P** - This column contains information about whether words that contained a number and words that consisted entirely of punctuation were removed (T) or kept (F) in a given scenario.
- **FT** - This column informs about the threshold value of the frequency of occurrence. Words that had a frequency of occurrence equal or less than this value were excluded from the dictionary of known words.
- **P, L, O, M** - Provides the F1 score for the specified named entity type (P = PER, L = LOC, O = ORG, M = MISC).
- **ΔP, ΔL, ΔO, ΔM** - Indicates how the F1 score for a given named entity type has changed from the value without substitution (P = PER, L = LOC, O = ORG, M = MISC).
- **OA** - Indicates the overall value of the F1 score.
- **Δ** - Indicates how the overall F1 score changed from the value without substitution.

The grey rows indicate the scenarios where the best result was achieved for a particular model with respect to the overall F1 score.

Considering the increasing trend of the overall F1 score with increasing threshold frequency of occurrence for the BiLSTM model, we performed an additional set of experiments. In these experiments, we observed how the model performed when substitution was used, words containing numbers and words consisting entirely of punctuation were removed, and the threshold varied from 3 to 6. We combined the measured values with the original values for the BiLSTM model and calculated the average values in this case as well. All values were calculated based on 5 runs. The exceptions were the scenarios with a threshold equal to 3 and 4, where the value was calculated based on 10 runs (5 original runs and 5 additional runs). We called the model *BiLSTM-6* and the resulting values are shown in Table 3.

Table 4 contains a subset of the rows from Table 2, giving for each model the row where the best overall F1 scores were obtained. It also includes a row from Table 3, for the BiLSTM model that was tested on an additional set of experiments. We have created an additional Table 5 to this Table 4, which for each model with the best F1 score on the test set, shows how that model performed on the training data.

The Table 4 shows that the use of substitution led to an improvement in F1 scores on the test data for each of the tested models. The exception may be the CRF model, where the difference between the overall F1 score for the with and without substitution scenario is very small. Considering the Table 2, we can see

Table 2. Experiment results - test data predictions

Model		Processing		F1 score per class(%)								F1 score(%)	
	S	N&P	FT	P	ΔP	L	ΔL	O	ΔO	M	ΔM	OA	Δ
Naïve	F	F	0	21.68	0.0	75.22	0.0	54.05	0.0	66.12	0.0	54.62	0.0
Naïve	T	F	1	45.96	24.27	75.25	0.03	51.54	−2.51	64.16	−1.96	59.8	5.18
Naïve	T	F	2	44.49	22.81	74.66	−0.56	46.14	−7.92	62.75	−3.36	57.62	3.01
Naïve	T	F	3	40.48	18.79	73.28	−1.94	44.52	−9.53	60.21	−5.91	55.4	0.79
Naïve	T	F	4	37.12	15.43	71.7	−3.52	39.3	−14.75	56.7	−9.42	52.24	−2.37
Naïve	T	T	0	21.68	0.0	75.22	0.0	53.9	−0.16	66.12	0.0	54.57	−0.04
Naïve	T	T	1	45.96	24.27	75.25	0.03	51.38	−2.68	64.16	−1.96	59.75	5.14
Naïve	T	T	2	44.49	22.81	74.66	−0.56	46.04	−8.02	62.75	−3.36	57.6	2.98
Naïve	T	T	3	40.48	18.79	73.28	−1.94	44.42	−9.64	60.21	−5.91	55.38	0.76
Naïve	T	T	4	37.12	15.43	71.7	−3.52	39.19	−14.87	56.7	−9.42	52.22	−2.4
HMM	F	F	0	67.72	0.0	66.81	0.0	60.98	0.0	55.17	0.0	63.71	0.0
HMM	T	F	1	74.8	7.08	76.75	9.94	65.31	4.33	69.12	13.95	72.06	8.34
HMM	T	F	2	72.76	5.04	80.96	14.14	60.74	−0.24	66.15	10.98	70.45	6.73
HMM	T	F	3	71.23	3.51	79.39	12.58	59.1	−1.88	63.84	8.67	68.75	5.03
HMM	T	F	4	71.54	3.82	78.05	11.23	55.75	−5.23	61.59	6.42	67.04	3.33
HMM	T	T	0	71.38	3.66	80.4	13.59	64.16	3.18	58.91	3.74	70.32	6.61
HMM	T	T	1	75.87	8.15	76.92	10.1	64.56	3.58	67.48	12.32	72.02	8.3
HMM	T	T	2	64.74	−2.98	81.01	14.2	60.3	−0.68	66.87	11.71	68.19	4.47
HMM	T	T	3	63.42	−4.3	79.46	12.65	58.27	−2.71	64.4	9.24	66.41	2.69
HMM	T	T	4	65.35	−2.37	77.9	11.09	51.05	−9.93	62.11	6.94	63.27	−0.44
CRF	F	F	0	83.32	0.0	84.05	0.0	72.36	0.0	75.87	0.0	79.48	0.0
CRF	T	F	1	82.2	−1.12	82.61	−1.44	69.79	−2.57	72.26	−3.61	77.49	−1.99
CRF	T	F	2	83.2	−0.12	82.36	−1.7	67.96	−4.4	73.89	−1.98	77.37	−2.11
CRF	T	F	3	82.38	−0.95	82.51	−1.54	67.0	−5.36	72.93	−2.94	76.79	−2.69
CRF	T	F	4	82.04	−1.28	83.24	−0.81	67.66	−4.7	71.56	−4.31	76.91	−2.57
CRF	T	T	0	35.93	−47.39	84.56	0.51	69.61	−2.76	74.27	−1.6	66.41	−13.07
CRF	T	T	1	79.84	−3.48	85.0	0.95	72.65	0.28	71.75	−4.12	78.41	−1.08
CRF	T	T	2	79.82	−3.5	86.15	2.1	73.43	1.07	72.71	−3.15	79.07	−0.42
CRF	T	T	3	80.21	−3.11	86.29	2.23	74.74	2.38	72.77	−3.1	79.58	0.09
CRF	T	T	4	79.7	−3.62	85.96	1.91	75.15	2.79	71.05	−4.81	79.23	−0.25
GRU	F	F	0	66.51	0.0	75.52	0.0	59.07	0.0	62.88	0.0	67.03	0.0
GRU	T	F	1	73.87	7.36	79.34	3.82	62.07	2.99	64.73	1.85	70.94	3.91
GRU	T	F	2	74.94	8.43	78.86	3.34	59.14	0.06	62.59	−0.29	69.92	2.89
GRU	T	F	3	74.03	7.52	77.92	2.4	57.75	−1.32	61.32	−1.56	68.97	1.94
GRU	T	F	4	72.53	6.03	76.52	1.0	55.32	−3.75	58.89	−3.98	67.26	0.22
GRU	T	T	0	57.28	−9.23	77.83	2.3	62.44	3.37	62.7	−0.17	64.82	−2.21
GRU	T	T	1	74.61	8.1	78.55	3.02	63.35	4.28	65.25	2.37	71.39	4.36
GRU	T	T	2	74.34	7.83	78.85	3.33	60.82	1.75	63.48	0.6	70.42	3.39
GRU	T	T	3	73.4	6.89	77.18	1.66	60.58	1.51	62.98	0.1	69.65	2.62
GRU	T	T	4	71.45	4.95	76.18	0.66	56.3	−2.77	60.01	−2.87	67.14	0.11
BiLSTM	F	F	0	66.44	0.0	75.22	0.0	65.12	0.0	67.29	0.0	69.07	0.0
BiLSTM	T	F	1	78.59	12.15	82.46	7.23	70.85	5.73	65.56	−1.74	75.84	6.78
BiLSTM	T	F	2	79.29	12.85	82.18	6.96	67.35	2.23	65.08	−2.21	74.79	5.72
BiLSTM	T	F	3	79.62	13.17	82.12	6.9	68.6	3.49	65.78	−1.51	75.3	6.23
BiLSTM	T	F	4	79.12	12.68	81.61	6.38	69.12	4.01	63.5	−3.8	74.8	5.74
BiLSTM	T	T	0	72.52	6.08	81.23	6.01	66.13	1.02	68.42	1.13	72.95	3.88
BiLSTM	T	T	1	76.46	10.02	83.78	8.56	68.41	3.29	66.58	−0.71	75.16	6.09
BiLSTM	T	T	2	77.14	10.7	83.91	8.69	70.05	4.94	67.64	0.35	75.98	6.91
BiLSTM	T	T	3	78.02	11.57	83.6	8.38	71.16	6.04	65.81	−1.48	76.2	7.13
BiLSTM	T	T	4	78.02	11.58	84.02	8.79	71.83	6.72	65.52	−1.77	76.42	7.35

in the CRF case that the application of substitution led to a deterioration in performance on the test set in all scenarios except one.

Referring to the Table 4, we can notice the major change compared to the scenario without substitution in the case of the HMM model. There is an 8.34% increase in the F1 score. Significant improvements also occurred for GRU, BiL-STM and a non-negligible improvement also occurred for our Naïve model.

Looking at the Table 4, we see that for BiLSTM, the overall F1 score increases as the threshold increases. An additional set of experiments presented in the Table 3 indicate that this trend is random and ends at a threshold of 3.

Table 3. Additional BiLSTM experiments - test data predictions

Model	S	N&P	FT	P	ΔP	L	ΔL	O	ΔO	M	ΔM	OA	Δ
		Processing		F1 score per class(%)								F1 score(%)	
BiLSTM-6	F	F	0	66.44	0.0	75.22	0.0	65.12	0.0	67.29	0.0	69.07	0.0
BiLSTM-6	T	T	0	72.52	6.08	81.23	6.01	66.13	1.02	68.42	1.13	72.95	3.88
BiLSTM-6	T	T	1	76.46	10.02	83.78	8.56	68.41	3.29	66.58	−0.71	75.16	6.09
BiLSTM-6	T	T	2	77.14	10.7	83.91	8.69	70.05	4.94	67.64	0.35	75.98	6.91
BiLSTM-6	T	T	3	77.99	11.55	83.95	8.73	71.29	6.18	66.22	−1.08	76.36	7.29
BiLSTM-6	T	T	4	77.91	11.46	84.22	9.0	71.6	6.49	65.38	−1.91	76.34	7.28
BiLSTM-6	T	T	5	77.42	10.98	83.81	8.59	71.19	6.08	65.18	−2.11	75.93	6.87
BiLSTM-6	T	T	6	75.5	9.06	83.64	8.42	70.21	5.09	62.3	−5.0	74.51	5.44

Table 4. Best overall F1 model - test data predictions

Model	S	N&P	FT	P	ΔP	L	ΔL	O	ΔO	M	ΔM	OA	Δ
		Processing		F1 score per class(%)								F1 score(%)	
Naïve	T	F	1	45.96	24.27	75.25	0.03	51.54	−2.51	64.16	−1.96	59.8	5.18
HMM	T	F	1	74.8	7.08	76.75	9.94	65.31	4.33	69.12	13.95	72.06	8.34
CRF	T	T	3	80.21	−3.11	86.29	2.23	74.74	2.38	72.77	−3.1	79.58	0.09
GRU	T	T	1	74.61	8.1	78.55	3.02	63.35	4.28	65.25	2.37	71.39	4.36
BiLSTM	T	T	4	78.02	11.58	84.02	8.79	71.83	6.72	65.52	−1.77	76.42	7.35
BiLSTM-6	T	T	3	77.99	11.55	83.95	8.73	71.29	6.18	66.22	−1.08	76.36	7.29

When we look at the Table 4 and Table 5 we see that the overall F1 score of the best models on the training set is significantly higher than that on the test set. This indicates that the models are overfitted. The decrease in the overall F1 score on the training data indicates that the use of substitution helps to reduce overfitting and improves the ability of the model to generalize.

Table 5. Models with the best overall F1 score on test data - train data predictions

Model	Processing			F1 score per class(%)								F1 score(%)	
	S	N&P	FT	P	ΔP	L	ΔL	O	ΔO	M	ΔM	OA	Δ
Naïve	T	F	1	69.56	−5.69	85.22	−3.79	74.33	−5.45	78.2	−4.4	76.93	−4.8
HMM	T	F	1	91.84	−1.53	90.46	−4.17	85.51	−3.3	87.26	−0.4	89.06	−2.63
CRF	T	T	3	93.57	−6.12	95.45	−3.97	92.17	−6.97	92.82	−5.86	93.66	−5.65
GRU	T	T	1	95.33	−2.56	92.59	−2.27	87.64	−4.39	87.94	−3.84	91.37	−3.14
BiLSTM	T	T	4	93.21	−5.26	95.48	−2.59	90.26	−6.54	88.98	−6.67	92.5	−4.99
BiLSTM-6	T	T	3	94.5	−3.97	96.21	−1.86	92.25	−4.55	90.17	−5.48	93.8	−3.69

Considering the results in the Table 4, we see that most models performed best when words containing number or consisting only of punctuation were replaced by pseudo words (N&P is T). The exceptions are the Naïve and HMM models, where the model performed best when such words were kept. However, when we look at the Table 2, we see that the difference between the best model with omitted and retained words is very small for the Naïve and HMM models.

From the Table 4, we can infer that in most cases, the substitution of words that occurred only once in the training data (FT = 1) has the greatest benefit. We can notice this for models that, during tag prediction for a processed word, do not have access to the words that follow after it (Naïve, HMM, GRU). In contrast, words that can peek at least one word into the future (CRF, BiLSTM) perform best when more frequent words are replaced by pseudo words (FT > 1).

5 Discussion and Conclusions

Considering the results of our experiments, we conclude that word substitution using pseudo words has a beneficial impact on the performance of most of the tested models. The models for which we applied the substitution before the training phase performed better on the test data in terms of F1 scores.

The most significant difference appeared in the HMM model, where an 8.34% increase in F1 score was observed compared to the no-substitution scenario. The least increase was observed in the CRF model, where the F1 score rose by 0.09%. We suggest that substitution has the greatest impact on models that do not use sophisticated features to predict tags in a sequence, relying only on the current word being processed, or using knowledge about the position of that word in the sentence. For these models, pseudo words supply a significant amount of information that they use in their predictions. We consider this to be the reason why the CRF model did not achieve significant improvement. CRF also made its predictions from the beginning based on features, which were, for example, information about whether a word consists entirely of capital letters, whether it has a capitalized first letter, whether it is a number, whether it is at the beginning or at the end of a sentence. Much of this information was encoded just through the pseudowords we used. For the CRF model, the pseudo words did not provide any additional information. Indeed, by replacing the word with a pseudo word, the model lost information such as the last 2 and 3 characters of the word. This assumption is to some extent confirmed by the values in Table 2, where for the CRF model we see for almost every scenario a deterioration compared to the case without substitution. The improvement in one of the cases may be due to coincidence rather than causality.

When we compare the F1 scores on the training and test data, we see that there is overfitting in the models. After applying the substitution, the difference in F1 scores is reduced. Thus, substitution appears to be a good way to reduce model overfitting for some models and also as a way to improve the ability of the model to generalize.

Considering the dataset used, the types of named entities recognized within it, and the pseudowords along with the intuition behind them, we suggest that substitution may improve the prediction ability for similar entities in certain models. Based on the intuition behind the pseudo words used, we also predict a possible improvement if dates, monetary amounts, product names, etc. will be recognized in the text. However, the currently used list of pseudo words emphasizes primarily on features such as letter case and the presence of numbers in a specific form. These may not provide useful information in domains where the recognized named entities do not have these features. We assume that in such domains, substitution would lead to a deterioration of the overall performance for some models.

In future research, we propose to investigate the impact of pseudo word substitution on datasets in different languages and domains. We also propose to investigate the impact of pseudo word substitution in the case of models that will not suffer from overfitting. Some attention should also be given to the design of new pseudo words that would be able to more finely discriminate between cases and also encode additional information.

Acknowledgment. It was supported by the Erasmus+ project: Project number: 2022-1-SK01-KA220-HED-000089149, Project title: Including EVERyone in GREEN Data Analysis (EVERGREEN) funded by the European Union. Views and opinions expressed are however those of the author(s) only and do not necessarily reflect those of the European Union or the Slovak Academic Association for International Cooperation (SAAIC). Neither the European Union nor SAAIC can be held responsible for them.

References

1. Bikel, D.M., Schwartz, R., Weischedel, R.M.: An algorithm that learns what's in a name. Mach. Learn. **34**, 211–231 (1999)
2. Kvet, M.: Relational data index consolidation. In: 2021 28th Conference of Open Innovations Association (FRUCT), pp. 215–221. IEEE (2021)

3. Li, X., Wang, T., Pang, Y., Han, J., Shi, J.: Review of research on named entity recognition. In: Sun, X., Zhang, X., Xia, Z., Bertino, E. (eds.) ICAIS 2022. CCIS, vol. 1587, pp. 256–267. Springer, Cham (2022). https://doi.org/10.1007/978-3-031-06761-7_21

4. Nakayama, H.: seqeval: a python framework for sequence labeling evaluation (2018). https://github.com/chakki-works/seqeval

5. Pakhale, K.: Comprehensive overview of named entity recognition: models, domain-specific applications and challenges. arXiv preprint arXiv:2309.14084 (2023)

6. Potočár, M., Kvet, M.: Impact of preprocessing using substitution on the performance of selected ner models - methodology (2023, manuscript submitted for publication)

7. Sang, E.F., De Meulder, F.: Introduction to the conll-2003 shared task: language-independent named entity recognition. arXiv preprint cs/0306050 (2003)

8. Wang, Y., Zhao, W., Wan, Y., Deng, Z., Yu, P.S.: Named entity recognition via machine reading comprehension: a multi-task learning approach. arXiv preprint arXiv:2309.11027 (2023)

9. Wang, Z., Shang, J., Liu, L., Lu, L., Liu, J., Han, J.: Crossweigh: training named entity tagger from imperfect annotations. arXiv preprint arXiv:1909.01441 (2019)

Oracle APEX as a Tool for Data Analytics

Ivan Pastierik$^{(\boxtimes)}$

University of Žilina, 010 26 Žilina, Slovakia
`pastierik2@stud.uniza.sk`

Abstract. In the contemporary landscape of data-driven decision-making, data analytics plays a crucial role in extracting meaningful insights from extensive datasets. Oracle Application Express (APEX) emerges as a robust platform for data analytics, offering a low-code environment that facilitates the development of powerful applications. Oracle APEX's declarative approach expedites the development cycle, allowing for swift creation tailored to specific data analytics needs. The seamless integration of Oracle APEX with the Oracle Database ensures optimal performance, security, and scalability for data-intensive applications, laying a robust foundation for analytics. This narrative explores the foundational aspects of Oracle APEX and its integration with the Oracle Cloud. Oracle APEX's data-centric design, customization capabilities, and support for RESTful APIs enhance its relevance in data analytics. The Oracle Cloud integration introduces unparalleled scalability, security, and performance optimization. The collaboration extends to Oracle Autonomous Database, an intelligent solution leveraging machine learning for autonomous operation, minimizing the workload on administrators. The practical application of Oracle APEX in climate anomaly detection exemplifies its analytical capabilities. By showcasing a real-world Climate Anomaly Detection Application, this narrative demonstrates how Oracle APEX analyzes historical weather data, identifies anomalies, and visualizes insights through interactive dashboards. The application's principle is grounded in standardized data and boxplot analysis, ensuring accessibility and effectiveness in anomaly detection. The inclusion of Oracle Cloud Infrastructure Services and advanced analytics tools further enhances Oracle APEX's capabilities, making it a versatile tool for data analysis in various domains.

Keywords: Oracle APEX · Data Analytics · Low-Code Development · Climate Anomaly Detection · Oracle Database Integration

1 Introduction

In the contemporary era of data-driven decision-making, the role of data analytics is paramount, serving as a fundamental tool to extract meaningful insights and knowledge from extensive datasets. This capability is crucial for organizations aiming to gain a competitive advantage and make well-informed decisions. Oracle Application Express (APEX) stands out as a robust platform for data analytics, harnessing the capabilities of the Oracle Database. Its low-code nature facilitates the development of powerful applications, enabling both IT professionals and business users to seamlessly analyze and visualize data without the need for extensive coding expertise [1].

© The Author(s), under exclusive license to Springer Nature Switzerland AG 2024
Á. Rocha et al. (Eds.): WorldCIST 2024, LNNS 990, pp. 203–214, 2024.
https://doi.org/10.1007/978-3-031-60328-0_20

The appeal of Oracle APEX for data analytics is underpinned by its ability to expedite the development cycle, allowing the swift creation of applications tailored to specific data analytics needs [1]. Its declarative approach simplifies application building, empowering users to focus on analysis rather than intricate coding [1]. Furthermore, the seamless integration of Oracle APEX with the Oracle Database ensures optimal performance, security, and scalability for data-intensive applications. This integration provides a robust foundation for data analytics, supporting the processing of large datasets. The platform's user-friendly interface, featuring drag-and-drop functionality and pre-built components, enhances accessibility, enabling a diverse range of users to engage in collaborative data exploration [2].

To illustrate the practical application of Oracle APEX in data analytics, this narrative showcases its use in detecting climate anomalies. By leveraging Oracle APEX, users can analyze historical weather data, identify abnormal temperature or humidity patterns, and visualize these anomalies through interactive dashboards. Subsequent sections will delve into the specific functionalities of Oracle APEX, exploring its features for data analytics and providing an overview of creating an application for climate anomaly detection.

2 Oracle APEX and Cloud Integration

In this section, we will explore the foundational aspects of Oracle APEX and its seamless integration with the Oracle Cloud. Providing a comprehensive overview, we will delve into the basic functionalities of Oracle APEX, emphasizing its relevance and capabilities within the Oracle Cloud environment. This knowledge serves as a crucial foundation for leveraging Oracle APEX in the realm of data analytics, offering insights into its core features and integration possibilities on the cloud platform.

2.1 Oracle APEX Overview

Oracle APEX is a groundbreaking force in low-code development, revolutionizing application building with a focus on data-centricity. Its declarative development approach allows users to articulate application requirements, translating them seamlessly into functional applications without extensive coding. A data-centric design, exemplified by features like the SQL Workshop, enables users to refine and model data directly within the platform [2]. Oracle APEX's robust support for RESTful APIs enhances its data-centricity, facilitating effortless interaction with external data sources for a dynamic and connected ecosystem [3]. The platform's customization capabilities, leveraging PL/SQL, HTML, CSS, and JavaScript, empower users to tailor applications to specific needs, incorporating intricate business logic [4]. Additionally, Oracle APEX's commitment to accessibility, demonstrated through integration with the Oracle Academy [5], fosters education and skill development, nurturing a new generation of proficient developers and data enthusiasts.

2.2 Oracle Cloud Integration

The integration of Oracle APEX with Oracle Cloud introduces unparalleled scalability, enabling seamless adaptation to evolving data analytics needs. Cloud resources are leveraged to efficiently manage increasing workloads and expanding datasets, laying a foundation for resource-efficient solutions. Security is prioritized, with Oracle Cloud reinforcing Oracle APEX with advanced security measures, ensuring the confidentiality, integrity, and availability of critical data. This integration adheres to industry standards, fostering a confident and secure environment for data exploration. Performance optimization is a key focus, utilizing the cloud's capabilities to enhance processing speed, reduce latency, and improve overall efficiency in data-intensive operations. Oracle Cloud's dynamic and scalable computing environment, seamlessly integrated with Oracle APEX, allows organizations to flexibly scale data analytics processes based on demand, optimizing costs and maximizing efficiency. The collaboration extends to robust storage solutions, addressing the challenges of handling vast amounts of data with scalable and secure storage options, providing a reliable foundation for in-depth analysis [6].

2.3 Oracle Autonomous Database

Oracle Cloud integrates seamlessly with Oracle APEX, offering a suite of integrated database services that align with Oracle APEX's data-centric approach. This integration provides Oracle APEX applications with access to high-performance databases within the Oracle Cloud Infrastructure, enabling efficient management and processing of large datasets. The collaboration enhances data retrieval and analysis capabilities, showcasing the synergy between Oracle APEX and Oracle Cloud databases [7].

The introduction of autonomous databases represents a significant advancement in cloud-based database management. Leveraging machine learning, autonomous databases automate various routine tasks traditionally handled by Database Administrators, such as tuning, security measures, backups, and updates. This autonomous operation eliminates the need for direct human intervention, addressing the challenges posed by the increasing complexity of databases. The autonomous database operates with key characteristics: self-driving, self-securing, and self-repairing. This intelligent automation minimizes the workload on Database Administrators, allowing them to focus on strategic tasks like data modeling and governance. The benefits include maximum uptime, performance, and security, reduction of manual tasks through automation, improved productivity, and the ability to reallocate database management resources for higher-value contributions. Autonomous databases emerge as a transformative solution, enabling organizations to shift from routine database management to innovation-focused IT efforts [7].

2.4 Oracle Cloud Infrastructure Services

The rich set of advanced analytics tools provided by Oracle Cloud complements Oracle APEX's capabilities, enhancing the analytical prowess of applications. From machine learning algorithms to predictive analytics, these tools empower users to derive deeper insights from their data, contributing to informed decision-making [6].

Oracle Machine Learning (OML). Oracle Machine Learning, an integral part of the Oracle Cloud ecosystem, augments Oracle APEX's analytics prowess by incorporating advanced machine learning algorithms. OML empowers users to unlock patterns, trends, and predictive insights from their data. The integration of machine learning capabilities enhances Oracle APEX applications, allowing users to delve deeper into data analytics with predictive modeling, anomaly detection, and classification. This collaborative approach ensures that Oracle APEX users can harness the full potential of machine learning to make data-driven decisions and uncover valuable insights [3].

Oracle Analytics. Oracle Analytics further enriches the data analytics landscape within the Oracle Cloud environment. With a user-friendly interface and robust visualization tools, Oracle Analytics seamlessly integrates with Oracle APEX, offering a comprehensive platform for data exploration and reporting. Users can create interactive dashboards, perform ad-hoc analyses, and share insights across the organization. The incorporation of Oracle Analytics enhances the overall user experience within Oracle APEX, providing a versatile toolkit for generating actionable insights and facilitating collaborative decision-making [8].

3   Climate Anomaly Detection Application

In this section, we delve into the practical application of Oracle APEX in the realm of climate anomaly detection. Harnessing the power of Oracle APEX, we demonstrate its ability to analyze historical weather data, identify abnormal temperature or humidity patterns, and visualize these anomalies through interactive visualizations. The Climate Anomaly Detection Application serves as an example of how Oracle APEX can be instrumental in addressing complex data analysis challenges, particularly in the context of climate science.

For this application we have decided to use Oracle APEX in Oracle Autonomous Transaction Processing Database inside Oracle Cloud Infrastructure, because it offers high flexibility and ease of use.

3.1   Climate Anomaly Detection Principle

The Climate Anomaly Detection Application is grounded in anomaly detection principles, serving as a crucial analytical tool for meteorological and environmental research. Notably, it strategically utilizes basic functions within Oracle APEX, deliberately excluding advanced tools like Oracle Machine Learning or Oracle Analytics. The application focuses on standardized data, transforming original weather data into a common scale for effective comparisons. Employing box plots, a statistical technique, the application visually represents data distribution, highlighting central tendencies, variability, and potential outliers.

Data can be standardized using formula (1), where z is resulting vector, x is input vector, function avg(x) computes average of input vector and function std(x) computes standard deviation of input vector [9].

$$z = \frac{1}{std(x)}(x - avg(x)1) \tag{1}$$

The core of the anomaly detection process involves analyzing standardized data through boxplots, where anomalies appear as points outside the typical range. This straightforward yet effective methodology ensures the accessibility of climate anomaly detection, avoiding complex algorithms and sophisticated analytics tools.

3.2 Receiving and Loading Climate Data

To conduct a meaningful climate anomaly analysis, the availability of authentic and comprehensive data is paramount. In the case of the Climate Anomaly Detection Application, the decision was made to utilize real-world data obtained from weather stations situated near key regional cities in Slovakia. These cities include Bratislava, Trenčín, Trnava, Nitra, Žilina, Banská Bystrica, Prešov, and Košice. The data retrieved from these weather stations encompass a diverse range of climatic conditions, capturing variations in temperature, relative humidity, snowfall, rain, and snow depth.

The source of this rich dataset is the open-meteo Free Weather API [10], which serves as a reliable repository for weather-related information. The hourly data received from the API spans over a decade from January 1, 2012, to December 31, 2022. This extensive timeframe enables a comprehensive analysis of climate patterns, allowing the Climate Anomaly Detection Application to discern long-term trends, seasonal fluctuations, and, most importantly, anomalies within the selected parameters.

By grounding the analysis in actual weather data from these strategically chosen locations, the application ensures that the insights derived are reflective of real-world climate conditions. The inclusion of multiple cities across Slovakia adds a layer of complexity, enabling a nuanced examination of regional climate variations. Ultimately, the utilization of this real and diverse dataset enhances the credibility and applicability of the anomaly detection results. Weather stations used from this dataset are visualized in Fig. 1.

Fig. 1. Locations of weather stations used in the application visualized on map.

Based on the data we have created a data model that can contain the data. This data model consists of two tables: WEATHER_DATA and WEATHER_LOCATIONS. WEATHER_LOCATIONS table contains data about position of weather stations, their altitudes and description, that represents the name of the regional city closest to the weather station. WEATHER_DATA table contains data of temperature, relative humidity, rain, snowfall and snow depth for these individual weather stations with timestamps, when were this data measured. Data model is shown in Fig. 2.

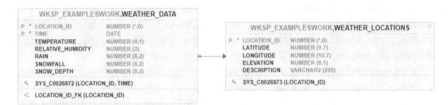

Fig. 2. Data model used in the application for storing weather stations data and locations.

We have loaded this data directly from Oracle APEX through data workshop into database from CSV files downloaded from open-meteo API. First 10 rows of downloaded data are shown in Fig. 3.

1	location_id	time	temperature_2m (°C)	relative_humidity_2m (%)	dew_point_2m (°C)	rain (mm)	snowfall (cm)	snow_depth (m)
2	0	2012-01-01T00:00	-2.1	99	-2.3	0.00	0.00	0.14
3	0	2012-01-01T01:00	-3.2	98	-3.5	0.00	0.00	0.14
4	0	2012-01-01T02:00	-3.8	98	-4.1	0.00	0.00	0.14
5	0	2012-01-01T03:00	-5.2	98	-5.5	0.00	0.00	0.14
6	0	2012-01-01T04:00	-6.1	97	-6.4	0.00	0.00	0.14
7	0	2012-01-01T05:00	-6.3	96	-6.8	0.00	0.00	0.14
8	0	2012-01-01T06:00	-6.1	95	-6.8	0.00	0.00	0.14
9	0	2012-01-01T07:00	-5.6	95	-6.3	0.00	0.00	0.14
10	0	2012-01-01T08:00	-4.2	95	-5.2	0.00	0.07	0.14
11	0	2012-01-01T09:00	-2.3	92	-3.4	0.00	0.14	0.14

Fig. 3. Raw data from dataset.

After successfully loading and preparing all data into database, we should check data for possible NULL values and add indexes where useful.

In Oracle Apex, the handling of NULL values and using indexes is pivotal for robust data processing [11]. Oracle Apex provides effective mechanisms to manage NULL values, though limitations arise, especially concerning indexes [11]. As NULL values cannot be part of the B + Tree index, the Table Access Full method is often used, impacting data access performance [11]. Proactive strategies, such as conditional expressions and functions like NVL, mitigate the impact of NULL values [12]. It is also possible to completely drop rows with missing data. In our case the data we used doesn't contain any NULL values and even if it contained, with the amount of data used, it won't be any problem do drop a couple of rows. Thanks to the fact that our dataset does not contain NULL values, we can effectively use indexes [12].

In this application we are most of the time only displaying data, so most common statement used in this application is SELECT statement. The main problem with using multiple indexes is, that when executing UPDATE, INSERT or DELETE statement, all indexes need to be modified according to updated, inserted or deleted data. When we are only displaying data, the index doesn't need to be modified, because data haven't changed. That's why in this application we are using multiple indexes, that allow us faster rendering of charts used in the application and their sorting.

3.3 Showcase of Climate Anomaly Detection Application

Now we would like to show you the finished application focused on climate anomalies detection, created in Oracle APEX. When you open the home page, you will see five charts displaying data of temperature, relative humidity, rain, snowfall and snow depth. You can show either only one of these charts or all of them. In Fig. 4 you can see how temperature graph looks like. In this graph are shown monthly temperatures from 1.1.2012 to 31.12.2022. As you can see in Fig. 4, it is possible to change the date span by changing the values in date fields From and To. It is also possible to change Sampling frequency to either yearly, monthly, daily or hourly. It is also possible to zoom the chart and scroll it in real time and show or hide individual weather stations.

Fig. 4. Application main page.

In Oracle APEX charts are generated with Oracle SQL Select statements. After writing the select statement you can map the selected columns to axes in chart. You can also change the type of chart. Oracle APEX supports various types of charts form bar charts, pie charts, to line charts, scatter plots or for example boxplots.

SQL Select statement for generating temperature chart:

```
SELECT location_id, description,
       TRUNC(time,
           DECODE(:P1_FREQUENCY, 0, 'YYYY', 1, 'MM', 2, 'DD', 'HH24')
       ) time,
       AVG(temperature) temperature
FROM weather_locations JOIN weather_data USING(location_id)
WHERE time >= to_date(:P1_FROM_DATE, 'DD.MM.YYYY HH24:MI:SS') and
      time <= to_date(:P1_TO_DATE, 'DD.MM.YYYY HH24:MI:SS')
GROUP BY location_id, description,
         TRUNC(time,
             DECODE(:P1_FREQUENCY, 0, 'YYYY', 1, 'MM', 2,
                     'DD', 'HH24')
         )
ORDER BY 3, 1;
```

In Fig. 5 you can see the same page with changed date span, sampling frequency and shown humidity chart that is zoomed in.

Fig. 5. Application main page after changing the date range and frequency.

If you click on specific weather station inside chart you will be able to view monthly analysis across all years in dataset of that weather station and that specific weather variable. So, in our case if we click on the line representing relative humidity for Žilina inside humidity chart, it will open a new Climate Analysis Across Years page non-modal dialog. Humidity chart will send information about id of weather station (location_id), name of chart (in our case Humidity) and name of city (value of description column).

This data will be stored in hidden items in Climate Analysis Across Years page. This linking process can be seen in Fig. 6.

Fig. 6. Link to Page 4: Area Climate Analysis Across Years

Figure 7 shows raw monthly data for each year in one chart, where each line represents one individual year. This chart serves as a tool for viewing the climate state for each month of specific weather station. This chart can also be zoomed in and scrolled as required.

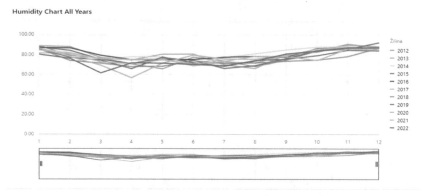

Fig. 7. Monthly humidity chart across all years in dataset.

Figure 8 shows boxplot created from standardized data. There we can see climate anomalies in form of outliers.

Figure 9 shows scatter plot that displays standardized data as points, where on x axis are individual months and on y axis are standardized values. In Fig. 9 are highlighted outliers based on boxplot from Fig. 8.

Standardized Humidity Boxplot

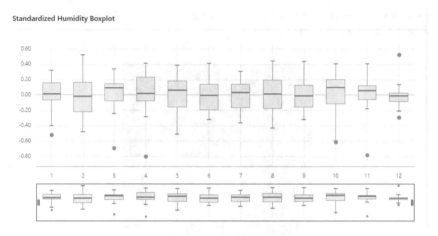

Fig. 8. Boxplot of standardized humidity across all years in dataset.

Standardized Humidity Chart All Years

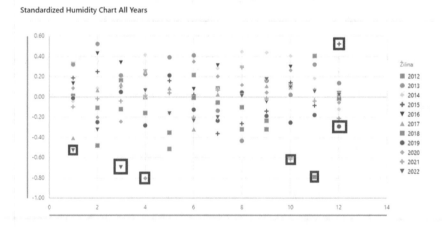

Fig. 9. Scatter plot of standardized humidity across all years in dataset with highlighted anomalies.

Thanks to all these charts it is possible to determine climate anomalies and if combined with other machine learning principles for example utilizing Oracle Machine Learning it is possible to create great tools for data analysis.

4 Conclusion

In conclusion, the integration of Oracle Application Express (APEX) with Oracle Cloud and its collaborative features with various Oracle Cloud services demonstrate a powerful framework for robust data analytics applications. Oracle APEX's low-code nature expedites the development cycle, making it accessible for both IT professionals and business users. The platform's data-centric design, declarative development approach, and support for RESTful APIs contribute to its efficiency in handling and analyzing diverse datasets.

The seamless integration of Oracle APEX with Oracle Cloud ensures optimal performance, security, and scalability for data-intensive applications. The collaboration between Oracle APEX and the Oracle Cloud Infrastructure Services provides advanced analytics tools, including Oracle Machine Learning (OML) and Oracle Analytics. These tools empower users to derive deeper insights from data, incorporating advanced machine learning algorithms, predictive analytics, and visualization capabilities.

The practical application of Oracle APEX in the Climate Anomaly Detection use case exemplifies its versatility in addressing complex data analysis challenges. The decision to utilize basic functions of Oracle APEX, without relying on Oracle Machine Learning or Oracle Analytics, highlights the platform's capabilities for anomaly detection using standardized data and boxplot techniques. This approach ensures accessibility and simplicity, making climate anomaly detection feasible for a diverse audience.

As organizations increasingly recognize the pivotal role of data analytics in decision-making, the collaboration between Oracle APEX, Oracle Cloud, and autonomous databases offers a strategic advantage. This synergy provides a foundation for innovative, scalable, and secure data analytics solutions, positioning organizations to stay competitive in the evolving landscape of data-driven insights.

Acknowledgment. It was supported by the Erasmus+ project: Project number: 2022-1-SK01-KA220-HED-000089149, Project title: Including EVERyone in GREEN Data Analysis (EVER-GREEN) funded by the European Union. Views and opinions expressed are however those of the author(s) only and do not necessarily reflect those of the European Union or the Slovak Academic Association for International Cooperation (SAAIC). Neither the European Union nor SAAIC can be held responsible for them.

References

1. Oracle APEX Documentation. https://apex.oracle.com/en/learn/documentation/. Accessed 25 Nov 2023
2. Sciore, E.: Understanding Oracle APEX 20 Application Development. Apress, New York (2020)
3. Png, A., Helskyaho, H.: Extending Oracle Application Express with Oracle Cloud Features. Apress, New York (2022)
4. Geller, A., Spendolini, B.: Oracle Application Express (APEX): Build Powerful Data-Centric Web Apps with APEX. McGraw Hill, New York (2017)
5. Oracle Academy. https://academy.oracle.com/en/oa-web-overview.html. Accessed 25 Nov 2023
6. Oracle Cloud Infrastructure Documentation. https://docs.oracle.com/en-us/iaas/Content/home.htm. Accessed 25 Nov 2023
7. Oracle Autonomous Database. https://www.oracle.com/autonomous-database/what-is-autonomous-database/. Accessed 25 Nov 2023

8. Oracle Analytics. https://www.oracle.com/business-analytics/analytics-platform/. Accessed 25 Nov 2023
9. Boyd, S., Vandenberghe, L.: Introduction to Applied Linear Algebra: Vectors, Matrices, and Least Squares. Cambridge University Press, Cambridge (2018)
10. Open-meteo weather API. https://open-meteo.com/. Accessed 25 Nov 2023
11. Kvet, M., Toth, Š., Kršák, E.: Concept of temporal data retrieval: undefined value management. Concurr. Comput. Pract. Exp. **32**(13) (2020). ISSN 1532-0634
12. Kvet, M.: Relational data index consolidation. In: 28th Conference of Open Innovations Association FRUCT, vol. 2021-January, pp. 215–221. FRUCT, Moscow (2021)

Phishing Webpage Longevity

Ivan Skula[iD] and Marek Kvet[(✉)][iD]

Faculty of Management Science and Informatics, University of Žilina,
Univerzitná 8215/1, 010 26 Žilina, Slovak Republic
skula@dobraadresa.sk, marek.kvet@uniza.sk

Abstract. Cybercriminals often spend considerable time preparing for phishing attacks, which surprisingly tend to last only a few hours. This short longevity is evident when older phishing links quickly become inaccessible, either taken down or flagged as malicious. While this benefits potential victims by shortening the effective risk period, it also poses a challenge for researchers aiming to track and analyze current phishing trends to develop effective countermeasures. Understanding the typical lifespan of phishing webpages is crucial for creating phishing data collection solutions. Our study involved monitoring phishing webpages from PhishTank and OpenPhish for three months while capturing their active/inactive status. Collected data had to undergo multiple steps - removing duplicate, incorrect, or non-relevant entries. The analysis focused on uncovering the phishing webpages' longevity, while the summary offered key findings. The study reveals that phishing webpages have a remarkably short active lifespan, with significant variations in the ratio of active to inactive pages across different periods. The initial rapid decrease by ≈12% in active phishing webpages is notable, dropping from 65% to 53% within the first five minutes and less than 40% remaining active after 24 h.

Keywords: phishing · webpage lifespan · statistics

1 Introduction

As organizations and governments advance on their journey of digitalization, the significance of online safety cannot be overstated. Over half (59%) of consumers are concerned about falling victim to a cyberattack and more than half of those who fall victim to a cyberattack report an impact on their personal finances [9]. Phishing, a prevalent form of online deception, harmfully impacts individuals and communities, making the study of its mechanisms - specifically, the lifespan of phishing pages - not just a technical pursuit. Phishing is, by number of victims, the most common type of cybercrime [5].

This study aims to explore the empirical evidence to understand how long phishing pages remain active post-deployment. Anti-Phishing Working Group (apwg.org) has been monitoring and reporting phishing webpages "uptime" from 2008 till 2014 in their half-yearly "Global Phishing Surveys". Though they

Á. Rocha et al. (Eds.): WorldCIST 2024, LNNS 990, pp. 215–226, 2024.
https://doi.org/10.1007/978-3-031-60328-0_21

reported the mean "uptime" value, they also reported the median "uptime" value as the distribution of the values is positively skewed. Skewness results from the fact that though a significant portion of phishing webpages have "uptime" in hours, some have it in weeks or even months, which shifts the mean "uptime" towards these extreme values. Median "uptime" values in the reported period (2008–2014) ranged from 5 h and 45 min to 19 h and 30 min, with both trends observed (decrease as well as increase). Mean "uptime" values were between ≈23 h and 73 h [1,2]. The last reported metric, from the second half of 2014, reported median "uptime" at ≈10 h (mean "uptime" was almost 30 h). Another research from 2007 reported 20 h as a median "lifetime", which was a figure very close to the number reported by APWG for H1/2008, at 19 h and 30 min. The mean "uptime" of 62 h was higher than the one reported by APWG (49 h 30 min) [8]. In regards to the webpage availability after a certain period, the figures vary, and researchers stated 66% [11] or 80% [3] of inactive webpages after the initial 24 h. Another study [7] reports only 25% of webpages being active after twelve days. The figures provided show the responsiveness and trends in phishing defense but don't fully capture the campaign's duration. Phishing campaigns often involve multiple domains or content migration between domains and IPs, especially after exposure and flagging [4].

The paper is organized as follows: Sect. 2 describes the approach taken to collect the phishing webpage data. In Sect. 3, we describe in detail the process of cleansing the collected data. 4th section summarizes the analysis performed and the gathered results before the last and concluding Sect. 5 summarizes the findings.

2   Data Collection Approach

The most important aspect of the feasible data source was the ability to gather suspicious phishing URLs as soon as they were observed and reported by internet users. The three most relevant sources identified were - PhishTank (phishtank.org), PhishStats (phishstats.info), and OpenPhish (openphish.com). All three sites provide a near real-time feed of phishing data. Due to the temporary unavailability of PhishStats, only PhishTank and OpenPhish data were collected and analyzed.

2.1   Solution for Collecting the Phishing Webpages Status

To collect and record the needed data, two separate tools were built *Identify* and *Verify*. The complete sequence of events from the initial phishing page deployment till its status capture by the tools is depicted in the diagram (Fig. 1).

1. Attacker deploys new phishing webpage (T_0)
2. A suspicious phishing webpage is observed by a user
3. User reports the suspicious webpage to OpenPhish or PhishStats (T_1)
4. *Identify* checks the phishing feed and records the new webpages (T_2)

5. *Verify* reads the new recorded suspicious URL in the database, navigates to the URL and captures its status - active or inactive (T_3)
6. *Verify* continues to read the suspicious webpage in a pre-determined time window until the page is observed as inactive.

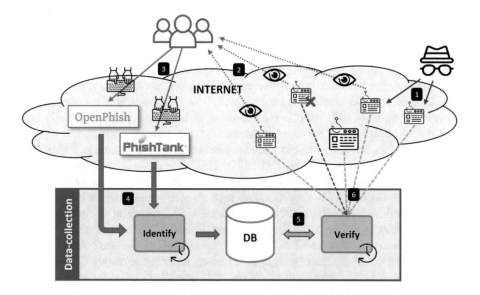

Fig. 1. Data collection process diagram.

Identify - was responsible for reading the data feeds every 2 min and recording new suspicious phishing URLs. This threshold was decided to balance the requirement to capture the status of phishing webpages as soon as possible with the impact of frequent reading on the underlying infrastructure. New records were saved into the database table, which utilized the uni-temporal structure of the destination table to capture the details while also separating inactive pages into the "HIST" table [6].

Verify - was periodically (every 10 s) monitoring the status of the records in the database and capturing the status if the record was new or the pre-defined time since the last status check has passed and during the last check, the webpage was still accessible (active). Verify stopped checking the status of those records that were inactive during the last status check (in practical terms, when the document size - HTML - read from the URL was zero, which meant it was removed or the hosting provider blocked the URL/site).

While in [8], authors were collecting the status every 30 min, APWG in [2] states that they were checking the status of the webpage several times per hour; we defined the following time windows at which the status was captured:

- immediately as the URL was recorded
- every hour within the first 24 h
- every day within the first 14 d
- 3rd and 4th week of first month
- 2nd - 6th, 9th and 12th month

Important Moments Within the Data Collection Process - from the (Fig. 1), four date and time points are observable, out of which three play an essential role in the analysis. The first is the moment(T_0) when the attacker deploys the phishing webpage and initiates the attack (via email, SMS, QR code, etc.). This exact moment is only known to the attacker and, for everyone else, remains unknown. It would be possible to capture this moment if we were able to capture the email or SMS used in the attack. This technique was successfully used in [10] and allows for a more precise measure of the attack duration. The next moment is when the user reports the observed suspicious webpage to PhishTank or OpenPhish (T_1), named SUBMISSION_DTTM. *Identify* can capture this moment only because PhishTank and OpenPhish share this information within their data feed. The following moment that *Identify* records is when it reads the data feed and identifies a new suspicious URL, which it records into the database table(T_2). This moment we named IDENTIFY_DTTM and serves as a primary reference moment for *Verify* when calculating the next time window for every individual phishing webpage. The last recorded moment (T_3) is when *Verify* reads and captures the status of the webpage after each pre-defined period passed. This moment was named VERIFY_DTTM and is recorded separately for each pre-defined time window.

3   Data Preparation

Data were collected during September, October, and November 2023 from two data sources - OpenPhish and PhishTank and only records with valid URLs were considered.

3.1   Confirmed Phishing URLs

As the users are reporting suspicious webpages, some are confirmed as phishing, and others are genuine webpages (also called False-Positives). PhishTank's final confirmation on the status of the reported phishing page requires manual review by more than one reviewer. So, many reported URLs are never closed as genuine or phishing. These undecided records, as well as those that were flagged as genuine webpages, had to be removed from the prepared dataset. OpenPhish also reports false positives, which were removed from the dataset but were rare. After this step the dataset shrank by nearly 23% from ≈219k records - ≈94K from OpenPhish and ≈125K from PhishTank - to ≈169k records out of which ≈93K originated from OpenPhish and ≈76K from PhishTank (Fig. 2).

3.2   Time-Delay Impact

As a result of the use of micro-batch execution in both components - Identify and Verify - varying delays in collecting the status of each recorder phishing webpage could have happened.

Delayed IDENTIFY_DTTM (T_2) occurred when OpenPhish or PhishtTank placed the suspicious URL into the feed with delay on their side or when Identify failed to ingest the feeds as per schedule. This causes a widening gap between when the URL was reported to OpenPhish or PhishTank and when Verify could start recording the webpage status. After a quick data analysis, a *maximum threshold time gap of 10 min was defined.* More than 85% of the records fell within this threshold. Any record saved into our database later than 10 min after being recorded by OpenPhish or PhishTank (SUBMISSION_DTTM) was removed from the analysis. This condition ensures the analysis results are coupled with the earliest captured date and time(SUBMISSION_DTTM) and IDENTITY_TM is captured within a maximum of 10 min since SUBMISSION_DTTM.

Delayed initial VERIFY_DTTM (T_3) occurred when Verify failed to capture the status shortly after the given time period for the webpage passed. Verify calculates the time passed since the IDENTIFY_DTTM only for those records seen as active in the previous check. If (for any reason) the status check doesn't happen, Verify will wait until it does. If, for example, Verify fails to record the webpage's status after two hours, it will not be able to update the status after three hours either, until the two-hour status is recorded - even if it happens many hours later. Verify can quickly recover and "catch up" with the current time window if the webpage is still active. However, if the webpage became inactive in between, such a record will not accurately capture when the page became inactive. To ensure consistency of the status capture, *a threshold of maximum delay was defined.* The maximum delay was breached if the Verify was supposed to capture the next time window, yet it is still missing the previous one. If, in this situation, inactive status is captured, it's impossible to decide whether the webpage became inactive in the current time window or the previous one. This ambiguity is relevant only if the most recent capture is inactive. Such records were, therefore, removed from analysis.

After applying the thresholds, the resulting dataset reduced by almost 27% to ≈124K records, out of which ≈77K were from OpenPhish and ≈47K from PhishTank (Fig. 2).

3.3   Duplicate Records Removal

Phishing attacks are often encountered by multiple users (Fig. 1), some of whom might report the attack to platforms like PhishTank or OpenPhish. As a result, the same domains are often reported repeatedly, leading to duplicate entries. To address this, we removed domains with identical elements up to the 5th-level subdomain (chosen based on a prior analysis of domain granularity levels in our phishing data [12]), reported within 24 h from their first occurrence. This step resulted in a significantly reduced dataset. The resulting dataset shrank by 33% to ≈83K records - ≈56K from OpenPhish and only ≈27K from PhishTank (Fig. 2).

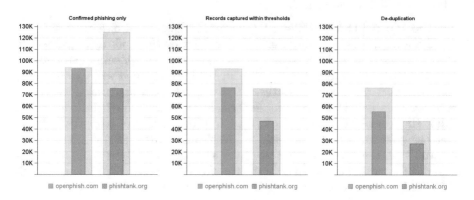

Fig. 2. Impact of data cleansing steps - confirmed phishing only (left), removal of records with delayed status capture (middle), and URL de-duplication (right). Before (bright) and after (dark).

3.4 Adjustment of Status for Small Pages

After a review of page sizes of captured records, we identified and decided to introduce a condition where all pages smaller than 200 bytes would be considered inactive. The size was based on an evaluation of a sample of pages of various sizes below 1000 bytes. The smallest phishing pages that used redirects were all slightly above 200 bytes. We found no webpage below 200 bytes, which was a valid phishing site. Most were pre-configured web server replies that the URL was unavailable or was taken down based on suspicion of being malicious. This rule improves the accuracy but doesn't fully correct the situation. There might still be webpages alerting users about the unavailability of the webpage on the domain, which might have more than 200 bytes.

The impact of this change was significant. Without adjustment, 31.2% of suspicious URLs collected by the Identify application were flagged as inactive during the initial status check. After the adjustment, this number grew to 40.2% (Fig. 3, last column 'ALL').

4 Phishing Webpages Longevity Analysis

Collected data were from a three-month period (Sep-Nov 2023), and so the volume of data decreases for weeks and months time-windows. The analysis was done in two parts to achieve the most accurate results based on the maximum number of applicable(mature) records. The *first* used all data with the age of at least *two weeks*. This way, the statistics used ≈82K records. The *second analysis* focused on the older data, *with the age of at least two months*. This analysis generated results while using approximately half of the records. Early results have shown that ≈40% (Fig. 4, row T) of suspicious webpages are inactive almost immediately after being captured from the data feed. To better understand this behavior, we conducted a separate analysis focusing on the first 10 min since the webpage is reported to OpenPhish or PhishTank (T_1=SUBMISSION_DTTM).

4.1 Longevity During the First Ten Minutes

The analysis compares the active vs. inactive webpages ratio within each of the first 10 min since SUBMISSION_DTTM. All records from the dataset had to be processed by the Identify application within 10 min of the SUBMISSION_DTTM as stated in data preparation section. This analysis grouped the records into buckets - "0.min" up to "9.min". Group "0.min" contains recorded suspicious URLs for which the initial status check happened within the 1st minute since the SUBMISSION_DTTM; group "1.min" contains URLs for which the status check happened between 1st and 2nd minute; and so on. Visualization of these groups (Fig. 3) and the ratio of active vs. inactive webpages in each group shows a visible trend where the ratio of inactive webpages grows slowly, minute by minute. In the first minute, 35.1% webpages are inactive; in the second, it's 38.3%; in the third, it's 39.4% and this trend continues up the fifth minute, topping at 44.2%. After that, the trends seem to stall, with some values higher and some lower. The important fact to consider here is the volume of records within each group. A green dotted line with correlating values on the right Y-axis represents this metric. The majority of the records are placed within the first five groups. Volume within the groups after "5.min" sharply decreases, which could explain why the trend doesn't copy the trend observed in the first five minutes.

4.2 Longevity Within First Two Weeks

A graphical representation of this analysis (Fig. 4) shows the gradual decrease of active webpages throughout the defined time windows. Only records older than two weeks were included in this analysis.

4.3 Longevity Within Months

This analysis could only consider records older than two months, as any "newer records" would still be yet to check their T+2m status. Collected data allowed us to analyze only three additional periods - T+3 weeks, T+4 weeks, and T+2 months. And since the set of data for this analysis would be only a subset of the data from previous analysis, we decided to build upon the figures from the above (first two weeks) analysis by applying a condition where only records with active status at the T+2 weeks(≈15K; 18.4% from 82K records) time-window

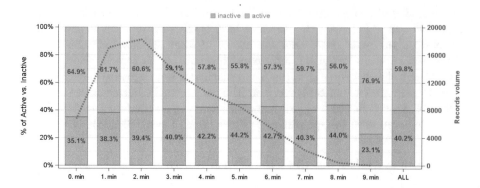

Fig. 3. Ratio of active vs. inactive phishing pages in each of initial 10 min and the overall ratio for all records

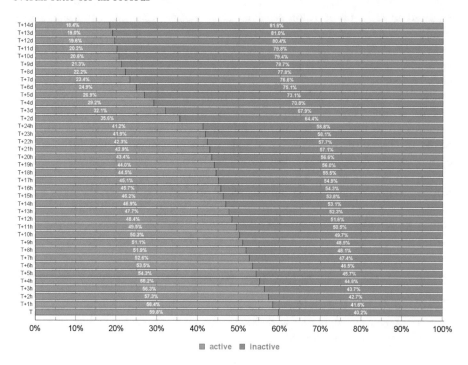

Fig. 4. Phishing webpage status - active vs. inactive ratio per time-window

check were considered and only those that were older than two months. After this filtering, the dataset had only ≈10K observations.

The final figures were as follows:

- T+3 weeks - 8.8% records became inactive from all active in T+2w
- T+4 weeks - 9.3% (additional 0.5%) inactive from all active in T+2w

– T+2 months - 15.5% (additional 6.2%) inactive from all active in T+2w

4.4   Volatility of Ratio Between Active and Inactive Webpages

As part of the review and validation process, we noticed that the ratio of active webpages varies a lot between different days. We, therefore, visualized the *daily ratio of active webpages through selected periods (T, T+24 h, and T+3 d)* to compare the change (Fig. 5). The bars on the graph are not stacked; each starts from 0% and reaches the height of the achieved ratio of active pages on a given day for a given time window. The blue dot within each day's column represents the volume of records on a given day to provide additional context. Volumes are linked to the values depicted on the right Y-axis of the graph. Daily mean number of records was ≈920, and all ≈83K records were considered.

The analysis confirmed the initial observation that the active webpages ratio at T (light-yellow color) varied from ≈44% (28th Of October) to ≈97% (3rd Of September - though the figure is based on less than 200 records) while mean value is ≈60% (Fig. 4). Figures at T+24 h (light-orange color) time window vary from ≈15% active webpages (26th Of October) to ≈71% (14th Of September) while mean value is ≈41%. Figures at T+3 d (dark-orange color) time window vary from ≈9% active webpages (7th Of November) to ≈68% (14th Of September) while mean value is ≈32%.

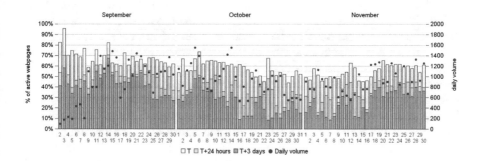

Fig. 5. Daily ratio of active webpages at T, T+24 h and T+3 d

4.5   Ratio of Active Webpages by the Source Data Feed

The analysis focused on comparing active webpages ratios between data collected from OpenPhish and PhishTank. Statistics were calculated for weekly as well as monthly periods. Both data feeds showed the same trends across the given period. The ratio for both PhishTank and OpenPhish gradually decreased from September to October and then gradually increased between October and November. Both data sources' figures copied this trend, and no opposing trends or notable differences were observed. Weekly active figures within the whole period ranged from ≈26% to ≈64%.

4.6   Median and Mean Lifespan of Active Webpage

Some articles that touched upon this topic calculated the mean and median lifespan of the phishing webpages, though they didn't provide a granular enough methodology to follow, which would allow for a direct comparison of figures. The data collection and cleansing process significantly impact the final (mean) figures, which is also why we described these steps in detail. All records - active and inactive were considered. The mean value was ≈224 h, which is approx. nine days and eight hours, but this number might increase as many records are older than two months and are still active. This figure is significantly higher than those reported by APWG in [1,2] and even higher than the high values mentioned in [8]. The median value, a more relevant metric, was only 10 h, the same as the last reported number by APWG [2] and lower than the figures reported in [8].

Comparing our figures with the stated webpages availability, we observed ≈60% of inactive pages after 24 h, similar to 66% [11]. After twelve days, we captured ≈20% webpages active, which is close to 25% [7].

5   Summary

Using the data from PhishTank and OpenPhish, the paper demonstrates that phishing webpages often have a very short active lifespan. To capture the details of phishing webpages, every minute matters, and availability drops rapidly (from 65% to 56% within the first five minutes). Approximately only 41% of reported phishing pages are active after 24 h, and only ≈36% after 48 h. However, the most significant finding is that 40% of the reported webpages are inactive when being collected from the data feed of PhishTank and OpenPhish. Focusing on the very first minute when the suspicious webpage is reported, already 35% of these URLs are inactive. Behind this relatively high number, there will be some that could be attributed to human error while reporting the suspicious webpage, some that could be a result of advanced techniques deployed on the phishing webpages to prevent scrapping or access from tools like curl. Some, which were removed as a result of tagging by anti-phishing solutions, but the question is how many were taken down by attacker as they achieved their objective?

Though the analysis is performed without the actual knowledge of the phishing webpage deployment date, it provides practical longevity figures for other researchers who plan to use data from OpenPhish or PhishTank. Based on this research, they can adapt their data collection strategy.

Not only having the actual real-world figures but also understanding the reasons behind them is crucial for optimizing the strategies for timely phishing prevention and detection. The transient nature of these sites poses challenges for real-time data collection and analysis. While the study provides a robust baseline for understanding phishing webpage longevity, it also emphasizes the need for continuous and rapid-response methodologies in cybersecurity research. Future studies could focus on extended periods beyond the two-month time window or assess the potential seasonal aspect of webpage longevity.

Acknowledgment. It was supported by the Erasmus+ project: Project number: 2022-1-SK01-KA220-HED-000089149, Project title: Including EVERyone in GREEN Data Analysis (EVERGREEN) funded by the European Union. Views and opinions expressed are, however, those of the author(s) only and do not necessarily reflect those of the European Union or the Slovak Academic Association for International Cooperation (SAAIC). Neither the European Union nor SAAIC can be held responsible for them.

References

1. Anti-Phishing Working Group: Global phishing survey: Trends and domain name use in 1h2009. Tech. rep., Anti-Phishing Working Group (2009). https://docs.apwg.org/reports/APWG_GlobalPhishingSurvey_1H2009.pdf
2. Anti-Phishing Working Group: Global phishing survey: Trends and domain name use in 2h2014. Tech. rep., Anti-Phishing Working Group (2014). https://docs.apwg.org/reports/APWG_GlobalPhishingSurvey_2H2014.pdf
3. Chu, W., Zhu, B., Xue, F., Guan, X., Cai, Z.: Protect sensitive sites from phishing attacks using features extractable from inaccessible phishing urls. In: IEEE International Conference on Communications (06 2013). https://doi.org/10.1109/ICC.2013.6654816
4. Cui, Q., Jourdan, G.V., Bochmann, G., Couturier, R., Onut, I.V.: Tracking phishing attacks over time. In: Proceedings of the 26th International Conference on World Wide Web Companion, pp. 667–676 (04 2017). https://doi.org/10.1145/3038912.3052654
5. FBI's Internet Crime Complaint Center: Internet Crime Report 2022. Tech. rep., FBI's Internet Crime Complaint Center (2022). https://www.ic3.gov/Media/PDF/AnnualReport/2022_IC3Report.pdf
6. Kvet, M., Matiasko, K., Vajsova, M.: Sensor based transaction temporal database architecture. 2015 IEEE World Conference on Factory Communication Systems (WFCS), pp. 1–8 (2015). https://api.semanticscholar.org/CorpusID:19918990
7. McGrath, D.K., Gupta, M.: Behind phishing: An examination of phisher modi operandi. In: USENIX Workshop on Large-Scale Exploits and Emergent Threats (01 2008). https://api.semanticscholar.org/CorpusID:2069430
8. Moore, T., Clayton, R.: Examining the impact of website take-down on phishing. In: Proceedings of the Anti-Phishing Working Groups 2nd Annual eCrime Researchers Summit 2007. vol. 269, pp. 1–13 (10 2007). https://doi.org/10.1145/1299015.1299016
9. Nationwide: September 2023 nationwide cybersecurity survey report. Tech. rep., Nationwide (2023). https://news.nationwide.com/download/bb78aae2-c720-4a7d-a6d8-423354c2cb53/cybersecurity-nationwidesurvey2023.pdf
10. Oest, A., et al.: Sunrise to sunset: Analyzing the end-to-end life cycle and effectiveness of phishing attacks at scale. In: 29th USENIX Security Symposium (USENIX Security 20), pp. 361–377. USENIX Association (08 2020). https://www.usenix.org/conference/usenixsecurity20/presentation/oest-sunrise

11. Sheng, S., Wardman, B., Warner, G., Cranor, L., Hong, J., Zhang, C.: An empirical analysis of phishing blacklists. International Conference on Email and Anti-Spam (07 2009). https://api.semanticscholar.org/CorpusID:1048051
12. Skula, I., Kvet, M.: Domain blacklist efficacy for phishing web-page detection over an extended time period. In: Proceeding of the 33rd Conference of FRUCT Association, pp. 257–263 (05 2023). https://doi.org/10.23919/FRUCT58615.2023.10142999

1st Workshop on AI in Education

A Conceptual Architecture for Building Intelligent Applications for Cognitive Support in Dementia Care

Ana Beatriz Silva[✉] and Vítor Duarte dos Santos

NOVA Information Management School, Lisboa, Portugal
abeatriz.cms@gmail.com, vsantos@novaims.unl.pt

Abstract. In response to the growing global aging population and the expected rise in dementia cases, this research addresses the critical need for innovative technological solutions in elderly dementia care. With a focus on alleviating the imminent strain on healthcare systems, this study advocates for leveraging technology to create tailored solutions, ranging from user-friendly wearable alarms to the integration of cutting-edge service robots. Emphasizing the importance of patient-centered design, this research aims to discern key factors in cognitive aging care, steering the development and implementation of digital solutions. Beyond proposing an intelligent model for brain aging, this work aspires to transcend academic boundaries by fostering a wider community interest in mental health discourse. Furthermore, it seeks to empower and assist caregivers through the integration of advanced technologies, including AI-driven tools, cognitive screening applications, and remote assessments.

Keywords: First Dementia · Elderly · Digital Solutions · Conceptual Architeture · Artificial Intelligence

1 Introduction

In the current landscape, mental health care is a focal point of extensive interest and research, aligning with the United Nations Sustainable Development Goals' emphasis on enhancing healthcare and overall well-being. Globally, the aging population is projected to double by 2050, resulting in an estimated 152 million individuals afflicted by dementia. The impending surge in demand for specialized healthcare services, particularly for those aged over 80, emphasizes the urgency to address the strain on healthcare systems. Projections indicate that by 2030, the costs associated with dementia care will escalate, necessitating adaptable services from urban centers [1, 2].

Despite the escalating demand, there remains a notable dearth of innovation in information and technology tailored to chronic diseases, notably dementia, among the elderly. However, the potential for leveraging technology looms large, promising the development of automated therapeutic and telehealth solutions designed for older adults and individuals with dementia [3, 4].

© The Author(s), under exclusive license to Springer Nature Switzerland AG 2024
Á. Rocha et al. (Eds.): WorldCIST 2024, LNNS 990, pp. 229–238, 2024.
https://doi.org/10.1007/978-3-031-60328-0_22

Recent studies have introduced conceptual architectures for digital solutions specifically targeting dementia patients, advocating for the implementation of successful guidelines. [1, 5]. A spectrum of technologies aims to augment diagnosis and enhance patients' quality of life, spanning from wearable alarms to emerging service robots. Acknowledging the distinct requirements of the elderly demographic, emphasis is placed on enhancing the usability and adoption of digital solutions. This necessitates a focus on specific design features and a patient-centered approach, vital for surmounting engagement barriers [6, 7].

This study centers on identifying pivotal factors within cognitive aging care, driving the imperative to design and deploy digital solutions tailored for elderly individuals grappling with dementia [8]. This encapsulates the central exploration and objective of this research.

The escalating discourse on mental health coincides with an imminent technological revolution. Proposing an intelligent based model that influences brain aging could captivate healthcare, research, and pharmaceutical sectors, and can profoundly impacting society and individual lives. This research aims not only to introduce a conceptual architecture but also to stimulate community interest in these critical topics. Additionally, exploring tailored digital solutions for elderly dementia patients supports caregivers, enabling independent living while maximizing the efficiency of dementia care workers and caregivers through technology like AI, cognitive screening tools and remote assessments.

2 Methodology

This investigation follows Design Science Research (DSR), a methodology that centers on developing and evaluating IT artifacts within socio-technical systems, emphasizing the creation of practical and innovative solutions [9, 10]. This methodology involves a comprehensive process of artifact development, research contribution, design assessment, and disseminating findings [11–14].

This study employs DSR to develop a conceptual architecture for a digital solution catering to elderly individuals dealing with dementia-related health issues. It begins with an extensive review of AI systems in healthcare and dementia care. Key features for an architectural conceptual architecture model serving as an AI solution are established, drawing upon established theories. The model's efficacy is evaluated through interviews with experts, and survey data is analyzed to uncover implications and limitations. The artifact construction aligns with established theories and integrates relevant concepts and existing solutions to foster a robust, innovative approach.

DSR serves as the guiding methodology in crafting a comprehensive AI-driven solution for addressing dementia-related health issues in the elderly. The process involves a meticulous literature review to comprehend the current landscape of health and dementia-related AI systems. Subsequently, defining key features for an architectural conceptual model as an AI solution draws from established theories, ensuring the artifact design is grounded in existing knowledge. This groundwork aims to develop an innovative solution tailored to address cognitive decline associated with aging and dementia. Dementia and Health Information Systems.

2.1 Dementia Overview

Dementia presents a significant challenge for the elderly worldwide, leading to disability and increased dependency. The World Health Organization projects a staggering rise in dementia prevalence globally, estimating that around 50 million people currently live with dementia, a figure expected to reach 152 million by 2050. Low and middle-income countries are particularly affected, hosting nearly two-thirds of these individuals [15]. This surge in dementia poses significant economic strain globally, with an estimated cost of USD 818 billion to the world economy. As the elderly demographic, especially those over 90 years old, continues to be the fastest-growing category, there's an urgent need to address their specific healthcare requirements and improve overall well-being [16].

Defined by cognitive decline and persistently worsening physical and psychological symptoms, dementia necessitates long-term care that strains both healthcare systems and families involved in caregiving. The escalating prevalence of dementia worldwide amplifies the urgency to address its impact on aging populations and the broader healthcare landscape [17, 18].

2.2 Information Systems for Mental Diseases

The rapid advancements in computerized cognitive treatments have transformed traditional therapies and led to the development of new multimedia systems to support health and independent living in older age [19, 20]. Adaptations of traditional cognitive interventions for devices like smartphones, tablets, and PCs offer a cost-effective alternative and potential means for maintaining cognitive function in individuals with mild cognitive impairment and healthy older adults [21–23]. These digitally delivered interventions enhance cognition, memory, attention, and psychosocial functioning, offering tailored exercises that adjust based on performance, provide immediate feedback, and are accessible on portable digital devices [24–26].

As the field of computerized cognitive therapies expands, recent research focuses on their efficacy for dementia patients, exploring virtual reality training as effective rehabilitation techniques, particularly in early cognitive decline stages.

Health information technology (HIT), such as electronic health records and telemedicine, plays a pivotal role in storing, sharing, and assessing health data. It minimizes redundancies in medical testing, decreases errors, enables remote collaboration among experts, and reduces chronic care expenses, potentially enhancing healthcare quality.

Research in digital healthcare for cognitive impairment, especially in apps, games, and virtual/augmented/mixed reality, has surged between 2018 and 2021, also focusing on big data/AI and smart home/telemedicine innovations. Assistive technologies aim to boost patient independence, safety, social interaction, mood, and overall quality of life, facilitating aging in place and reducing medical expenses and caregiver strain through remote patient monitoring. With a focus on accessibility for all, these technologies have the potential to improve health equity.

In the 21st century, deep learning and machine learning-based artificial intelligence (AI) are transforming the medical industry, leveraging data like advanced neuroimaging and genetic testing. Machine learning models demonstrate promise in early neurodegenerative disease diagnosis and planning care for dementia types using clinical data. AI, particularly deep neural networks, plays a significant role in dementia research, predicting outcomes and monitoring brain health via genetic and neuroimaging data. AI-driven technologies like assistive robots, smart sensors, and mHealth applications aid in dementia therapy, improving caregiving, monitoring health, and sustaining social interactions [27]. These technological innovations in healthcare represent potential breakthroughs in enhancing dementia care and patient well-being.

2.3 Brain and Cognitive Games

Brain and cognitive games have emerged as potential strategies to counter cognitive decline in the elderly, aiming to enhance attention, memory, and processing speed. These games offer varied results in reducing dementia risk, with studies like those by Nouchi and Lampit [28, 29] indicating positive cognitive improvements post-training.

When developing applications for dementia mitigation, the design of these games is pivotal. Factors such as game type, frequency, duration, intensity, challenge level, and alignment with affected cognitive domains play crucial roles. User-friendly interfaces and progress feedback are essential to boost motivation and engagement. Research, exemplified by Toril [30] meta-analysis, underscores video game training's potential to enhance attention and memory in older adults, demanding further investigation.

While studies by Hill [31] and Lampit [29] highlight cognitive training's positive impact, limitations in sample sizes and study durations emphasize the urgency for more extensive, well-designed studies to ascertain brain games' specific effects on dementia and cognitive decline in the elderly.

3 Conceptual Architecture for Cognitive Support in Dementia Care

In the realm of developing a digital solution, the role of software architecture cannot be overstated. It serves as the guiding framework that shapes organization, facilitates interaction, and ensures seamless maintenance, thereby establishing essential constraints for the overall design and development process.

Building upon insights gleaned from an extensive literature review and an understanding of the context surrounding mental health disorders, a comprehensive architecture tailored for dementia care has been delineated. This foundational framework, versatile in its instantiation, serves as a conceptual architecture poised to address the intricate needs of this specific population.

Central to this conceptualization is a focus on catering to the unique usage requirements of individuals grappling with dementia. Recognizing the diversity and nuances of this demographic, the architecture is meticulously crafted to encompass the distinctive functional needs of users, underscoring the paramount importance of user-centric design. In the pursuit of developing effective and user-friendly dementia applications, a deep comprehension of the functional requirements specific to this user group is imperative. This understanding forms the bedrock upon which the architecture is erected, ensuring that the resultant digital solution aligns seamlessly with the nuanced needs of those navigating the challenges of dementia.

3.1 Functional Requirements

Functional requirements describe what the application should do, in terms of its intended purpose and features, and focus on the functionality and capabilities of the application.

The application development should prioritize key components, such as the ones listed below:

- Tracking and Monitoring: Including features for activity, sleep, medication, and mood tracking enables caregivers to observe and respond to changes in the patient's health and behavior effectively.
- Cognitive Training and Brain Games: Incorporating activities like memory games and puzzles to stimulate cognitive functions is crucial for improving memory, attention, and slowing down dementia progression.
- Communication Features: Enabling video, voice, and messaging options between patients, caregivers, and healthcare professionals reduces social isolation and ensures better healthcare management.
- Reminders and Alerts: Customizable prompts for medications, appointments, and daily tasks help users adhere to schedules and remember important information, enhancing overall care.
- Personalization Features: Customizable interfaces, font sizes, and color schemes cater to users with varying cognitive abilities, enhancing accessibility and ease of use.
- Access to Support and Resources: Providing links to healthcare professionals, support groups, and educational materials ensures efficient care management and improves the quality of life for both patients and caregivers.

Overall, the requirements should be focused on providing support and assistance to users with cognitive impairment, while also considering the needs and preferences of caregivers. By providing these features, the application can help improve the quality of life for dementia patients and make caregiving more manageable.

3.2 Conceptual Architecture

While design requirements specify software behavior, system architecture outlines the system's structure. The solution's software architecture meticulously incorporates modules, interfaces, and connections, ensuring an efficient system.

The conceptual architecture diagram presented bellow (Fig. 1) serves as a guide for the design and implementation of the digital software, providing a visual representation of the architecture modules.

The User Interface Module is responsible for managing the user interface of the application. It includes the design of the application screens, buttons, icons, and menus to ensure that they are user-friendly, accessible, and intuitive.

Cognitive Training Module provides various exercises and activities to improve cognitive abilities such as memory, attention, and language skills. It includes designing activities that are challenging yet achievable and providing feedback on progress.

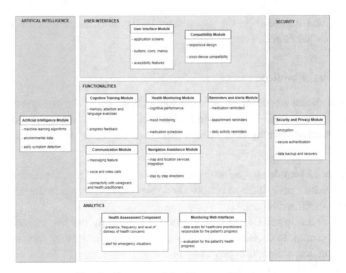

Fig. 1. Conceptual Architecture Diagram

The Health Monitoring Module is responsible for tracking the patient's health data relevant to dementia and mental health, such as cognitive performance, mood, and medication schedules. It includes designing a user interface that allows the patient and caregiver to easily input and view this information.

Reminders and Alerts Module is responsible for providing reminders and alerts for appointments, medication schedules, and daily activities. It includes designing a notification system that is easily understandable and can be customized to the patient's needs.

Navigation Assistance Module is responsible for assisting the patient with navigation and directions, especially when going to unfamiliar places. It includes integrating with map and location services and providing step-by-step instructions.

The Communication Module is responsible for providing communication features to help patients stay connected with their loved ones and caregivers, as well as healthcare practitioners. It includes designing features such as messaging, voice, and video calls that are easy to use and accessible.

Compatibility Module is responsible for ensuring that the application is compatible with different devices and operating systems. It includes designing the application to be responsive to different screen sizes and device capabilities.

Security and Privacy Module is responsible for ensuring the security and privacy of patient data. It includes designing security features such as encryption, secure authentication, and data backup and recovery mechanisms.

Finally, the Artificial Intelligence Module assumes a pivotal role in the conceptual architecture by enabling the system to adapt and learn from diverse data sources. Leveraging advanced machine learning algorithms, including those operating on patient records, environmental data, sensor readings, and user interactions, this module enables the system to provide personalized and context-aware support tailored to the unique needs of individuals affected by dementia. Additionally, this module facilitates intelligent decision-making processes by generating valuable insights and recommendations through continuous analysis of data from multiple sources. These outputs offer valuable guidance to healthcare providers, caregivers, and individuals with dementia, thereby supporting optimized care, early symptom detection, and personalized interventions. The AI module's capacity to dynamically adapt to evolving user needs and preferences enhances the overall effectiveness and efficiency of the digital solution in addressing the challenges posed by dementia in the elderly population.

By implementing this conceptual architecture, the digital solution has the potential to significantly enhance the lives of elderly individuals affected by dementia. Through the seamless integration of these modules within the conceptual architecture, the proposed solution strives to offer a holistic and innovative approach to dementia care, providing crucial support to elderly individuals and their caregivers.

3.3 Prototype

Developing an application prototype requires thorough planning and consideration of key aspects. First, defining the purpose, target audience, and functionalities is crucial. Gathering requirements from stakeholders follows, leading to the design of the software's architecture and user interface. Then, the development, testing, and debugging stages ensure the software meets requirements before deployment.

Afterwards, ongoing maintenance and support become essential to keep the software updated, secure, and efficient. User-centered assessments help refine the application based on users' feedback and interaction satisfaction. To build a prototype targeting dementia in the elderly, understanding user needs and functionality is paramount.

The prototype should feature various modules like login, home, memory games, reminders, messaging, personalized settings, activity tracking, caregiver dashboard, and help and support screens. Each screen should prioritize simplicity, using clear graphics and easily readable fonts for improved accessibility and ease of use.

4   Conclusion

This study has proposed a functional architectural model for a digital solution targeting elderly patients with dementia. The proposed architecture addresses the critical characteristics identified in the literature and provides a conceptual architecture for the development of a reliable, scalable, and user-friendly application.

Despite its comprehensiveness, this study presents several limitations. Firstly, the proposed architecture requires empirical validation and user feedback to ensure its real-world effectiveness and usability. Empirical studies involving a diverse sample of elderly individuals with dementia are necessary to refine the digital solution based on user experiences.

Secondly, technological constraints might hinder implementation. Compatibility issues and the need for continuous adaptation to evolving technologies, such as artificial intelligence or IoT devices, may challenge the digital solution's functionality.

Resource availability, including funding and technical expertise, is crucial for sustainable development, deployment, and maintenance of the solution.

Ethical concerns regarding privacy, data security, and the ethical implications of technology substituting human interaction should be carefully addressed. Lastly, the evolving nature of dementia care necessitates ongoing research and adaptation to incorporate new findings and guidelines.

Acknowledging and addressing these limitations will refine the proposed conceptual architecture, ensuring better care for elderly individuals with dementia.

References

1. Lima, M.R., et al.: Conversational affective social robots for ageing and dementia support. IEEE Trans. Cogn. Dev. Syst. (2021). https://doi.org/10.1109/TCDS.2021.3115228
2. Nasr, M., Islam, M.M., Shehata, S., Karray, F., Quintana, Y.: Smart healthcare in the age of AI: recent advances, challenges, and future prospects. IEEE Access **9**, 145248–145270 (2021). https://doi.org/10.1109/ACCESS.2021.3118960
3. Almeida, A., Mulero, R., Rametta, P., Urošević, V., Andrić, M., Patrono, L.: A critical analysis of an IoT—aware AAL system for elderly monitoring. Futur. Gener. Comput. Syst. **97**, 598–619 (2019). https://doi.org/10.1016/J.FUTURE.2019.03.019
4. Bravo-Torres, J.F., Ordoñez-Ordoñez, J.O., Gallegos-Segovia, P.L., Vintimilla-Tapia, P.E., López-Nores, M., Blanco-Fernández, Y.: A context-aware platform for comprehensive care of elderly people: Proposed architecture. In: 2017 CHILEAN Conference on Electrical, Electronics Engineering, Information and Communication Technologies, CHILECON 2017 - Proceedings, 2017-January, 1–6 (2017). https://doi.org/10.1109/CHILECON.2017.8229507
5. Göransson, C., Wengström, Y., Hälleberg-Nyman, M., Langius-Eklöf, A., Ziegert, K., Blomberg, K.: An app for supporting older people receiving home care - usage, aspects of health and health literacy: a quasi-experimental study. BMC Med. Inform. Decis. Making **20**(1) (2020). https://doi.org/10.1186/S12911-020-01246-3
6. Christie, H.L., Boots, L.M.M., Peetoom, K., Tange, H.J., Verhey, F.R.J., de Vugt, M.E.: Developing a plan for the sustainable implementation of an electronic health intervention (partner in balance) to support caregivers of people with dementia: case study. JMIR Aging **3**(1), e18624 (2020). https://doi.org/10.2196/18624
7. Ye, S., et al.: A computerized cognitive test battery for detection of dementia and mild cognitive impairment: instrument validation study. JMIR Aging **5**(2), e36825 (2022). https://doi.org/10.2196/36825
8. Thordardottir, B., Fänge, A.M., Lethin, C., Rodriguez Gatta, D., Chiatti, C., Magalhães, F.H.: Acceptance and Use of Innovative Assistive Technologies among People with Cognitive Impairment and Their Caregivers: A Systematic Review (2019). https://doi.org/10.1155/2019/9196729
9. Hevner, A., Gregor, S.: Envisioning entrepreneurship and digital innovation through a design science research lens: a matrix approach. Inf. Manag. **59**(3), 103350 (2022). https://doi.org/10.1016/J.IM.2020.103350
10. Peffers, K., Tuunanen, T., Rothenberger, M.A., Chatterjee, S.: A design science research methodology for information systems research. J. Manag. Inf. Syst. **24**(3), 45–77 (2007). https://doi.org/10.2753/MIS0742-1222240302
11. Morschheuser, B., Hassan, L., Werder, K., Hamari, J.: How to design gamification? A method for engineering gamified software. Inf. Softw. Technol. **95**, 219–237 (2018). https://doi.org/10.1016/J.INFSOF.2017.10.015

12. March, S.T., Smith, G.F.: Design and natural science research on information technology. Decis. Support. Syst. **15**(4), 251–266 (1995). https://doi.org/10.1016/0167-9236(94)00041-2
13. Peffers, K., Tuunanen, T., Rothenberger, M.A., Chatterjee, S.: A design science research methodology for information systems research. J. Manag. Inf. Syst. **24**(3), 45–77 (2014). https://doi.org/10.2753/MIS0742-1222240302
14. Hevner, A.R.: The duality of science: knowledge in information systems research. J. Inf. Technol. **36**(1), 72–76 (2021). https://doi.org/10.1177/0268396220945714/FORMAT/EPUB
15. Nichols, E., et al.: Estimation of the global prevalence of dementia in 2019 and forecasted prevalence in 2050: an analysis for the global burden of disease study 2019. Lancet Public Health **7**(2), e105–e125 (2022). https://doi.org/10.1016/S2468-2667(21)00249-8
16. Woods, B., Aguirre, E., Spector, A.E., Orrell, M.: Cognitive stimulation to improve cognitive functioning in people with dementia. Cochrane Database Syst. Rev. **2** (2012). https://doi.org/10.1002/14651858.CD005562.PUB2/INFORMATION/EN
17. Shin, J., Cho, E.: Patterns and risk factors of cognitive decline among community-dwelling older adults in South Korea. Arch. Gerontol. Geriatr. **104** (2023). https://doi.org/10.1016/J.ARCHGER.2022.104809
18. Livingston, G., et al.: The lancet commissions dementia prevention, intervention, and care: 2020 report of the lancet commission the lancet commissions. Lancet **396**, 413–446 (2020). https://doi.org/10.1016/S0140-6736(20)30367-6
19. de Marco, M., et al.: Technologies for cognitive training and cognitive rehabilitation for people with mild cognitive impairment and dementia. A systematic review. Front. Psychol. **1**, 648 (2020). www.frontiersin.org. https://doi.org/10.3389/fpsyg.2020.00648
20. Bail, K.: Using health information technology in residential aged care homes: an integrative review to identify service and quality outcomes. Int. J. Med. Inform. **165**, 104824 (2022). https://doi.org/10.1016/J.IJMEDINF.2022.104824
21. Lee-Cheong, S., Amanullah, S., Jardine, M.: New assistive technologies in dementia and mild cognitive impairment care: a PubMed review. Asian J. Psychiatr. **73**, 103135 (2022). https://doi.org/10.1016/J.AJP.2022.103135
22. Paay, J., Kjeldskov, J., Aaen, I., Bank, M.: User-centred iterative design of a smartwatch system supporting spontaneous reminiscence therapy for people living with dementia. Health Inform. J. **28**(2) (2022). https://doi.org/10.1177/14604582221106002
23. Book, S., Jank, M., Pendergrass, A., Graessel, E.: Individualised computerised cognitive training for community-dwelling people with mild cognitive impairment: study protocol of a completely virtual, randomised, controlled trial. Trials **23**(1) (2022). https://doi.org/10.1186/S13063-022-06152-9
24. Cafferata, R.M.T., Hicks, B., von Bastian, C.C.: Effectiveness of cognitive stimulation for dementia: a systematic review and meta-analysis. Psychol. Bull. **147**(5), 455–476 (2021). https://doi.org/10.1037/BUL0000325
25. Anderberg, P., et al.: The effects of the digital platform support monitoring and reminder technology for mild dementia (SMART4MD) for people with mild cognitive impairment and their informal carers: protocol for a pilot randomized controlled trial. JMIR Res. Protocols **8**(6) (2019). https://doi.org/10.2196/13711
26. Sohn, M., Yang, J., Sohn, J., Lee, J.-H.: Digital healthcare for dementia and cognitive impairment: a scoping review. Int. J. Nurs. Stud. 104413 (2022). https://doi.org/10.1016/J.IJNURSTU.2022.104413
27. Su, Z., et al.: 6G and artificial intelligence technologies for dementia care: literature review and practical analysis. J. Med. Internet Res. **24**(4), E30503 (2022). https://www.jmir.org/2022/4/e30503. https://doi.org/10.2196/30503
28. Nouchi, R., et al.: Brain training game improves executive functions and processing speed in the elderly: a randomized controlled trial. PLoS ONE **7**(1) (2012). https://doi.org/10.1371/JOURNAL.PONE.0029676

29. Lampit, A., et al.: The timecourse of global cognitive gains from supervised computer-assisted cognitive training: a randomised, active-controlled trial in elderly with multiple dementia risk factors. J. Prevention Alzheimer's Disease 1–7 (2014). https://doi.org/10.14283/JPAD.2014.18

30. Toril, P., Reales, J.M., Ballesteros, S.: Video game training enhances cognition of older adults: a meta-analytic study. Psychol. Aging **29**(3), 706–716 (2014). https://doi.org/10.1037/A0037507

31. Hill, N.T.M., et al.: Computerized cognitive training in older adults with mild cognitive impairment or dementia: a systematic review and meta-analysis. Am. J. Psychiatry **174**, 329–340 (2017). https://doi.org/10.1176/appi.ajp.2016.16030360

1st Workshop on Artificial Intelligence Models and Artifacts for Business Intelligence Applications

Improving Customer Service Through the Use of Chatbot at Enma Spa Huancayo, Peru

Elvis Araujo$^{(\boxtimes)}$ ⓘ, Diana Javier ⓘ, and Daniel Gamarra ⓘ

Faculty of Engineering, Universidad Continental, Huancayo, Peru
{71104672,76437971,dgamarra}@continental.edu.pe

Abstract. This paper shows the implementation of EnmitaBot (chatbot) in the Spa and personal care micro-enterprise, based on the identification of problems through interviews, in order to improve customer service. The platform used was ManyChat, which was developed with a ten-step process from identifying the problem to testing with customers. The key dimensions used to measure chatbot performance were knowledge and personalization. The measurement was carried out from frequent customer surveys, obtaining a Net Promoter Score (NPS) result of 71% and on the Likert scale of 4.03, achieving a high correlation between these two indicators. Measurements by questions and dimensions are presented, and the significant correlation between responses on the Likert scale and NPS is highlighted. The dimensions used to measure were; response time, knowledge and database, personalization, intuition and understanding, natural interaction and language; obtaining the highest NPS in knowledge and database with 72.25% and in personalization with 72.33%. Regarding the lowest NPS obtained, Intuition and comprehension were 63.00% and natural interaction and language with 69.00%.

Keywords: Chat implementation · ManyChat Platform · Customer Service Enhancement · NPS (Net Promoter Score) Measurement

1 Introduction

In recent years, the advancement of artificial intelligence has revolutionized numerous business areas globally, especially in terms of optimizing services and improving the customer experience [1, 2]. This progress has led to the widespread adoption of technologies such as chatbots, artificial intelligence systems designed to interact with users in a human-like manner, offering accurate and relevant answers in real-time [3, 4].

In this context, start-ups and small companies, such as Enma Spa in Huancayo, Peru, face challenges in customer service. The growth of this micro-enterprise dedicated to personal care and beauty has been accompanied by problems identified through an interview with the administrator, who reported that the problems were recurrent, affecting approximately 50% of the queries that were not attended to on time and adequately, a significant level of dissatisfaction was also identified among those who did get answers. as the quality of these did not meet their expectations. It was also found that dissatisfied customers were unaware of the scope of work carried out in the company. It should be

Á. Rocha et al. (Eds.): WorldCIST 2024, LNNS 990, pp. 241–250, 2024.
https://doi.org/10.1007/978-3-031-60328-0_23

noted that the management of appointments was carried out manually, contributing to the problems identified, thus reflecting the need to adapt to the current technological environment [5].

This is the scenario focused on by the present work, centered on the successful implementation of EnmitaBot (a chatbot) at Enma Spa, using the ManyChat platform [6, 7]. This project was developed through a ten-step process, from problem identification to testing with clients. Two key dimensions, knowledge and customization, formed the basis for evaluating the chatbot's performance.

The main objective of this study was to improve the user experience when interacting with EnmitaBot, overcoming the limitations of a micro-enterprise, such as limited investment in resources [8]. The results are based on measurements obtained through frequent customer surveys, revealing an NPS of 71% and a score of 4.03 on the Likert scale. In addition, the significant correlation between these two indicators is detailed, highlighting the dimensions that had the greatest impact on customer satisfaction.

This article also reviews previous research related to the use of chatbots in various sectors, from customer service in tourism to their role in query management. The advantages and limitations of these technologies are discussed, as well as the importance of human intervention in customer satisfaction.

The methodology used to carry out this study includes the evaluation of specific dimensions of the chatbot, the collection of data through online surveys, and statistical analysis to interpret the results [9]. The importance of these tools for measuring customer satisfaction is highlighted and a detailed discussion of the findings obtained is presented.

The article offers an analysis of the chatbot implementation at Enma Spa, showing how the adoption of virtual agents can improve the customer experience in small businesses. The results and conclusions provide a pathway to understand the positive impact these technology solutions can have on customer satisfaction and suggest recommendations for future implementations in similar organizations.

2 Related Work

Chatbots based on Artificial Intelligence have the ability to interact with customers in a natural language style in order to satisfy communication needs, answering doubts, queries, providing necessary information and simulating conversations with the customers of a company, communication with the chatbot is as if the customer were talking to another person, That is why a study shows that most people express interest in using chatbots because of the topic of online health consultations. However, another portion of people express concerns about data security and privacy issues when interacting with chatbots [10].

On the other hand, chatbots in the tourism industry took a lot of importance due to the low cost and efficiency it provides in customer satisfaction, however, it is perceived that many researchers did not adequately attend the emotional expressions of chatbots in services, for this reason they are based on the theory of expectations violations and they manage to understand that the emotional expressions of chatbots affect customer satisfaction, through the communication and expression of concern that chatbots have in relation to the user, in the sector of recommendations of tourist attractions and orientation towards the purpose of customers, which resembles human attention [11]. The growing use of text-based chatbots to provide one-on-one support to customers is primarily focused on comparing chatbots to each other or to humans, leaving a gap in research into task-based dialogues. Identifying the Characteristics of Dialogues Lead to Successful Task-Based Conversations [12].

The importance of chatbots as valuable tools to provide instant and accurate answers to user questions, as they act as first-line intermediaries, opening up innovative possibilities to improve services for both customers and businesses. There is enormous potential in the use of AI-assisted chatbots to manage customer service queries, allowing processes to be automated and improve the effectiveness of interactions, however, it is important to take into account the benefits and limitations of the technology, as well as the importance of human intervention in customer satisfaction [4]. Therefore, chatbots are increasingly integrated into interactions with brands and companies, facilitating services such as customer service, reservations and online shopping. Henry Fernández, CEO of Cari AI, defines chatbots as tools that use AI techniques to interact with users, improving customer service by offering 24-h availability, multimedia communication and adaptability to various channels [13].

NPS is presented as a key measure to assess customer satisfaction, offering valuable feedback to improve their experience. An NPS score above 70 is considered world-class, with ratings above 50 being considered excellent and above 0 generally positive. The interpretation of a positive NPS score can vary by industry and geographic location [14]. In a study on the Problem-Based Learning (PBL) strategy, using a questionnaire of 15 questions organized into three dimensions, the overall results exceed 75%. In knowledge, the figures range from 77.5% to 94.4%, while in skills and attitudes they range from 75% to 98.3%. Although the NPS scores of the questions exceed 50%, additional actions are proposed to further improve these indices [15].

Research in Finland on the Finna national service, which involves archives and museums, focused on assessing user satisfaction using NPS, identifying critical points and success factors. This assessment resulted in improvements, raising the NPS from 29.9% in 2016 to 45.6% in 2019 [16]. Measuring U evaluated 11 companies from various industries using different NPS scales. Netflix, on the 5-point scale, achieved an outstanding 61%, achieving an excellent rating [17]. Another study addressed the modality bias between web and telephone surveys, showing an NPS of 54.7% in mixed mode. A decrease was simulated when switching 10% of users from phone to web. The importance of considering the business context and objectives when determining the most appropriate survey channel is highlighted [18]. In the Czech Republic, the study focused on customer-facing communication in retail, using Mystery Shopping and revealing an NPS of 17 in small businesses and 19 in large ones, with no significant differences. This approach made it possible to directly measure service performance against various current standards [19].

3 Methodology

This study adopted a quantitative approach and an ex post-facto retrospective design to assess user satisfaction with customer service at Enma Spa. Data collection was carried out through a questionnaire composed of 15 items formulated in positive terms, using a five-level Likert scale to capture the participants' perception [20].

These items were grouped into five different dimensions and a measure of overall service satisfaction was included, detailed in Table 1 [21].

Table 1. Dimensions to evaluate the chatbot.

Id dim	Dimensions	Questions
D01	Response Time	5
D02	Knowledge & Database	2, 4, 10, 13
D03	Customization	1, 7
D04	Intuition and understanding	6
D05	Natural Interaction and Language	3, 8, 9, 10, 11, 12
D06	Overall satisfaction with service	13, 14, 15

Prior to data collection, the questionnaire was subjected to content validity through the review of three information technology experts, obtaining a content validity coefficient of 0.88 [22]. Subsequently, a reliability analysis was carried out which resulted in a Cronbach's alpha of 0.92.

Data collection was carried out through online questionnaires, administered to 100 clients who interacted with the virtual agent during three consecutive weekends [23–25]. The collected data were processed using SPSS v.23, R Studio 2023.09.1 and Excel version 2016, obtaining the NPS and the percentage of Likert scale scores, both global and for the dimensions [26, 27].

The chatbot was implemented using the ManyChat platform on Facebook Messenger [28]. The 10-step method used in the process of developing and implementing the chatbot on the Enma Spa Facebook page, presented in Fig. 1, is shown below.

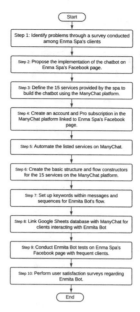

Fig. 1. Development and implementation of the chatbot in Messenger Facebook.

4   Results and Discussion

Results are shown by question and dimension. Likewise, the standard deviation and 95% confidence interval were found so that the results obtained are reproduced more accurately.

Table 2 shows that the item with the highest NPS score is item 13 with 80%, with 86% as promoters and 6% as detractors. Likewise, it is observed that the lowest NPS identified is item 5 with 63%; with 73% of promoters and 10% of detractors. On the other hand, the highest score according to the Likert scale presents items 7 and 12 with 4.11; which differs from the identification of the highest NPS value and the lowest with 3.89, with item 1 being the one that agrees with the lowest NPS value. Finally, the standard deviation of the NPS results is 4.39 and the Likert 0.06; showing an indication of a more consistent trend [29].

At a 95% confidence interval, the NPS results have a margin of error of ± 2.43, with an upper limit of 74.43% and a lower limit of 69.57%. Meanwhile, the margin of error of the Likert scale is ± 0.03 with an upper limit of 4.03 and a lower limit of 3.97.

The correlation is significant at the level 0.01 (bilateral), obtaining Spearman's rho coefficient per question equal to 0.819, which indicates that the relationship between the variables is direct and its degree is high, having a proximity to 1 [30].

Table 3 shows that the item with the highest NPS score is dimension 6 with 75.67%, with 82.67% as promoters and 7.00% as detractors. Likewise, it is observed that the lowest NPS identified is dimension 4 with 63.00%; with 73.00% of promoters and 10.00% of detractors. On the other hand, the highest score according to the Likert scale presents dimensions 3 and 6 with 4.08; which differs from the identification of the highest

Table 2. NPS & Likert Results by Questions

Item	Detractors	Liabilities	Promoters	NPS	Likert
1	8.00%	18.00%	74.00%	66.00%	3.89
2	6.00%	19.00%	75.00%	69.00%	3.97
3	5.00%	19.00%	76.00%	71.00%	4.06
4	10.00%	11.00%	79.00%	69.00%	4.07
5	10.00%	17.00%	73.00%	63.00%	4.00
6	9.00%	13.00%	78.00%	69.00%	3.99
7	6.00%	12.00%	82.00%	76.00%	4.11
8	9.00%	11.00%	80.00%	71.00%	4.06
9	8.00%	14.00%	78.00%	70.00%	4.05
10	7.00%	15.00%	78.00%	71.00%	4.08
11	8.00%	11.00%	81.00%	73.00%	4.06
12	6.00%	10.00%	84.00%	78.00%	4.15
13	6.00%	8.00%	86.00%	80.00%	4.11
14	6.00%	14.00%	80.00%	74.00%	4.09
15	9.00%	9.00%	82.00%	73.00%	4.04

Table 3. NPS and Likert results by Dimensions.

	Dimensions	Detractors	Liabilities	Promoters	NPS	Likert
1	Response Time	7.00%	15.00%	78.00%	71.00%	4.00
2	Knowledge & Database	7.25%	13.25%	79.50%	72.25%	4.06
3	Customization	7.17%	13.33%	79.50%	72.33%	4.08
4	Intuition and understanding	10.00%	17.00%	73.00%	63.00%	4.00
5	Natural Interaction and Language	9.00%	13.00%	78.00%	69.00%	3.99
6	Overall Care Satisfaction	7.00%	10.33%	82.67%	75.67%	4.08

NPS value of dimension 3 and the lowest with 3.99, with dimension 5 being consistent with the lowest NPS value.

The correlation is significant at the level of 0.05 (bilateral), obtaining Spearman's rho coefficient per dimension equal to 0.899, which indicates that the relationship between the variables is direct and its degree is high due to the proximity to 1 Fig. 2.

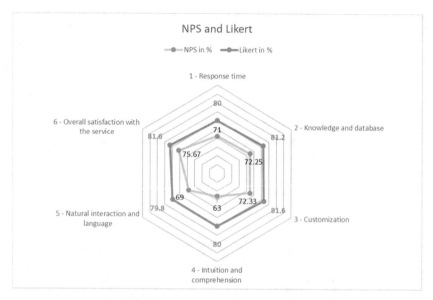

Fig. 2. NPS & Likert Radar Diagram

It can be seen that the results obtained from dimensions 4 and 5; intuition and perception and natural interaction and language; they differ in terms of score, because with the NPS assessment, dimension 5 has a lower score with 63% compared to dimension 4 with 69%. While with the Likert assessment, the highest score is given to dimension 5 with 80%, unlike dimension 4 which has 79.8%. Likewise, dimensions 1, 2, 3 and 6 show direct proportionality in their results.

The research highlights the specific methodology used in the study of chatbots, focusing on the evaluation of specific dimensions and tools such as online surveys and statistical analysis. At the same time, the shared importance of consciously adapting the measurement method to the unique characteristics of each study is highlighted, emphasizing the need for critical reflection and the constant search for improvement, regardless of the specific topic of the research [18].

In the study carried out, the percentage of detractors is less than 10%. If we make a constant improvement in dimensions such as intuition and comprehension, the percentage of detractors can be further reduced, as can be seen in the Finnish research on the national service Finna, who had a progressive improvement from 29.9% to 45.6% in three years because they achieved further development of the service, usability improvements and constant evaluation through surveys of their customers to correct the critical comments detected by the detractors. It is important for each company to determine its own NPS levels based on its experience, it must segment its customers according to the information necessary to ensure a good level of service [16].

The results of the global satisfaction dimension of attention reflect 75.67%, while Netflix, according to the magazine Measuring U, obtained 61% on the same NPS scale, the latter being lower [17]. This is because satisfaction is measured differently in each context, and the scores obtained reflect users' unique expectations and experiences in each situation. However, in both cases they exceed the excellent rating above 50%, which supports the effectiveness of NPS as an evaluation tool in various business contexts [14].

The study conducted in the Czech Republic provides valuable insights into customer-centric communication and its impact on loyalty in the retail space. It highlights the general uniformity of the results and the absence of significant differences between small and large companies [19]. However, the findings we obtained underscore the importance of addressing specific dimensions that present opportunities for improvement. Both the NPS and the Likert scale were used to gain a more complete understanding of the situation. It is relevant to note that the dimensions with the highest NPS coincide with those that obtained the highest scores on the Likert scale, indicating a coherence in positive perception.

5 Conclusions

The successful implementation of the chatbot at Enma Spa has proven to be an effective strategy for improving customer service. The positive results support the idea that artificial intelligence, in the form of chatbots, can be a valuable tool even for small businesses, offering an improved user experience and contributing to customer loyalty.

It contributes to the body of literature by providing a specific methodology focused on the assessment of specific dimensions of chatbots in the context of microenterprises. The findings reinforce the usefulness of the NPS and Likert scale as complementary tools for understanding customer satisfaction.

In future implementations, customer feedback should be included, in order to make continuous improvement, especially in dimensions with opportunities for improvement.

References

1. Akour, I., et al.: Artificial intelligence and financial decisions: empirical evidence from developing economies. Int. J. Data Netw. Sci. 8(1), 101–108 (2024). https://doi.org/10.5267/j.ijdns.2023.10.013
2. Moposita, D., Jordán, J.: Chatbot: a customer service tool in times of COVID-19. Rev. Episteme Units 9(3), 327–350 (2022). https://dialnet.unirioja.es/servlet/articulo?codigo=8591153
3. Casazola Cruz, O.D., Alfaro Mariño, G., Burgos Tejada, J., Ramos More, O.A.: The perceived usability of chatbots on customer service in organizations: a review of the literature. Interfaces (014), 184–204 (2021). https://doi.org/10.26439/esh2021.n014.5401
4. Rojas Ahumada, K.A., López Zavaleta, V., Mendoza de los Santos, A.C.: The impact of artificial intelligence on improving customer service: a systemic review. Innov. Softw. 4(2), 201–222 (2023). https://doi.org/10.48168/innosoft.s12.a90

5. Stoeckli, E., Dremel, C., Uebernickel, F., Brenner, W.: How affordances of chatbots cross the chasm between social and traditional enterprise systems. Electron. Markets **30**(2) (2020). https://doi.org/10.1007/s12525-019-00359-6

6. Sebastian, D., Restyandito, Nugraha, K.A.: Developing of middleware and cross platform chat application study case: telegram, LINE. Int. J. Adv. Comput. Sci. Appl. **12**(11), 79–85 (2021). https://doi.org/10.14569/IJACSA.2021.012110

7. Tamrakar, R., Wani, N.: Design and Development of CHATBOT: A Review, no. April, pp. 369–372 (2018)

8. Mahato, J., Jha, M.K.: Does financial inclusion promote sustainable livelihood development? Mediating effect of microentrepreneurship. J. Financ. Econ. Policy **15**(4/5), 485–499 (2023). https://doi.org/10.1108/JFEP-05-2023-0134

9. Hayes, B.E.: Measuring Customer Satisfaction and Loyalty: Survey Design, Use, and Statistical Analysis Methods, Third edit. ASQ Quality Press (2008)

10. Yin, R., Neyens, D.M.: Examining how information presentation methods and a chatbot impact the use and effectiveness of electronic health record patient portals: an exploratory study. Patient Educ. Couns. **119**, 108055 (2024). https://doi.org/10.1016/j.pec.2023.108055

11. Zhang, J., Chen, Q., Lu, J., Wang, X., Liu, L., Feng, Y.: Emotional expression by artificial intelligence chatbots to improve customer satisfaction: underlying mechanism and boundary conditions. Tower. Manag. **100**, 104835 (2024). https://doi.org/10.1016/j.tourman.2023.104835

12. Rese, A., Tränkner, P.: Perceived conversational ability of task-based chatbots – which conversational elements influence the success of text-based dialogues? Int. J. Inf. Manage. **74**, 102699 (2024). https://doi.org/10.1016/j.ijinfomgt.2023.102699

13. Lorduy, J.: Chatbots, companies' commitment to improve customer service. Portfolio (2022). https://www.portafolio.co/negocios/empresas/chatbots-apuesta-de-empresas-para-mejorar-atencion-al-cliente-567405

14. Hanks, C.: How to Calculate NPS and Rank Customer Loyalty. TextExpander (2023). https://textexpander.com/blog/how-to-calculate-nps

15. Gamarra-Moreno, A., Gamarra-Moreno, D., Gamarra-Moreno, A., Gamarra-Moreno, J.: Assessing problem-based learning satisfaction using net promoter score in a virtual learning environment. In: EDUNINE 2021 - 5th IEEE World Engineering Education Conference Future of Engineering Education: Current Challenges and Opportunities Proceedings (2021). https://doi.org/10.1109/EDUNINE51952.2021.9429104

16. Laitinen, M.A.: Fix, develop, keep! net promoter score as a tool for improving customer experience. Qual. Quant. Methods Libr. **8**(1999), 147–158 (2019)

17. Sauro, J.: Changing the Net Promoter Scale: How much does it matter? measuringu.com (2017). https://measuringu.com/nps-scale-change/. Accessed 14 Jan 2024

18. Van Bennekom, F.C., Klaidman, S.: Survey Mode Impact Upon Responses and Net Promoter Scores (2013)

19. Eger, L., Mičík, M.: Customer-oriented communication in retail and net promoter score. J. Retail. Consum. Serv. **35**, 142–149 (2017). https://doi.org/10.1016/j.jretconser.2016.12.009

20. de Ghent, Á.G.C., González, W.E.S., Ortega, J.B., Castillo, J.E., Fernández, A.S.: Likert scale: an Alternative to elaborate and interpret an instrument of social perception. Rev. la Alta Tecnol. Soc. **12**(1), 38–45 (2020). http://eds.a.ebscohost.com/eds/pdfviewer/pdfviewer?vid=1&sid=9dada2ee-bbb8-4496-bf61-0b421d6368e4%40sdc-v-sessmgr03

21. Maroengsit, W., Piyakulpinyo, T., Phonyiam, K., Pongnumkul, S., Chaovalit, P., Theeramunkong, T.: A survey on evaluation methods for chatbots. In: Proceedings of the 2019 7th International Conference on Information and Education Technology, pp. 111–119 (2019). https://doi.org/10.1145/3323771.3323824

22. Galicia Alarcón, L.A., Balderrama Trápaga, J.A., Edel Navarro, R.: Content validity by experts judgment: proposal for a virtual tool. Aperture **9**(2), 42–53 (2017). https://doi.org/10.32870/ap.v9n2.993
23. Islam, M.N., Islam, M.S.: Data collection and analysis. In: Islam and Democracy in South Asia, pp. 49–65. Springer, Cham (2020). https://doi.org/10.1007/978-3-030-42909-6_3
24. Hox, J.J., Boeije, H.R.: Data collection, primary versus secondary. Encycl. Soc. Meas. **1**, 593–599 (2005)
25. Otzen, T., Manterola, C.: Sampling techniques on a study population. Int. J. Morphol. **35**(1), 227–232 (2017). https://doi.org/10.4067/S0717-95022017000100037
26. InSitu Advisors & Researches: Satisfaction Index and Net Promoter Score, Universidad Continental. Huancayo
27. Fire, D.: Measuring Customer Experience with Net Promoter Score. Datadecisionsgroup (2014). https://www.datadecisionsgroup.com/blog/bid/194635/measuring-customer-experience-with-net-promoter-score
28. Illescas-Manzano, M.D., Vicente López, N., Afonso González, N., Cristofol Rodríguez, C.: Implementation of chatbot in online commerce, and open innovation. J. Open Innov. Technol. Mark. Complex. **7**(2), 125 (2021). https://doi.org/10.3390/joitmc7020125
29. Wan, X., Wang, W., Liu, J., Tong, T.: Estimating the sample mean and standard deviation from the sample size, median, range and/or interquartile range. BMC Med. Res. Methodol. **14**(1), 1–13 (2014). https://doi.org/10.1186/1471-2288-14-135
30. Roy-García, I., Rivas-Ruiz, R., Pérez-Rodríguez, M., Palacios-Cruz, L.: Correlation: not all correlation implies causation. Rev. Alerg. Mexico **66**(3), 354–360 (2019). https://doi.org/10.29262/ram.v66i3.651

NLP in Requirements Processing: A Content Analysis Based Systematic Literature Mapping

Bell Manrique-Losada[1]([✉]) [iD], Fernando Moreira[2] [iD], and Eidher Julián Cadavid[1] [iD]

[1] University of Medellín, Medellín, ANT, Colombia
bmanrique@udemedellin.edu.co, ecadavid641@soyudemedellin.edu.co
[2] Portucalense University, Porto, PT, Portugal
fmoreira@upt.pt

Abstract. As a result of the evolution of agile methodologies in the software development industry, there are currently various applications of Natural Language Processing techniques, models, and tools to classify, extract, and analyze documents within the stages of the software development process. However, their utility has been relatively unexplored in relation to the processing of user stories, such as the most widely used technique for capturing and specifying requirements in the last decade. This article presents a content analysis based on a systematic literature mapping on the application of natural language processing in user stories, following Petersen's methodological proposal. The search methodology is based on obtaining relevant articles from Dimensions, ScienceDirect, IEEE, and Scopus. Initially, 483 articles published between 2018 and 2022 were identified, and inclusion and exclusion criteria were applied, filtering down to 125 articles for review. Finally, a quality assessment was conducted, resulting in 57 articles relevant. Analyzing these primary studies, findings are identified, and current/future lines of work are proposed as contributions to this field of knowledge.

Keywords: User stories · Natural Language Processing · Software requirements

1 Introduction

1.1 Context

Natural Language Processing (NLP), as a field of knowledge focused on the study and analysis of linguistic aspects of text through computer programs [1], has been applied in recent years in various disciplines where the management of oral and written language is sought conversing with a machine, conducting internet searches, categorizing products for online shopping, extracting relevant data from texts, among others. According to the IEEE and ACM in the Computing Curricula [2] there is a broad spectrum of NLP development, due to advances in digital transformation processes. Within the framework of software development processes, the literature presents a wide variety of studies applying NLP to various tasks or specific phases, yielding contributions, as asserted by McCarthy [3], primarily in automation and the reduction of efforts in times of analysis and human intervention.

Á. Rocha et al. (Eds.): WorldCIST 2024, LNNS 990, pp. 251–260, 2024.
https://doi.org/10.1007/978-3-031-60328-0_24

1.2 Problem and Justification of Study

With the rise of agile methodologies such as Scrum [6], the concept of user story [7] appears as a work unit that speeds up the administration of requirements, reduces the number of formal documents and time needed, and is normally written by a product owner. This story, although written in natural language, can be the object of analysis from the NLP to facilitate its management and subsequent understanding by the members of the team developing a software product.

Currently, there are several applications of NLP techniques, models, and tools to classify, extract, and analyze documents [11] within the framework of a software development process. However, its usefulness has been little explored about the user stories, as the most used requirements specification technique in the last decade [8].

Some authors present automated approaches, such as Abualhaija et al. [4] to demarcate requirements using machine learning, as Friedrich et al. [5] to generate notation and business process models from natural language text. On the other hand, Elallaoui et al. [8] plan a process to transform user stories into use cases, to facilitate the work of the development team, reduce ambiguity in requirement specifications and automatically generate design models. Summarizing this, Robeer et al. [9], suggest automatically deriving conceptual models from user stories with the purpose of facilitating communication between stakeholders, and Nasiri et al. [10], opens the possibility of automatically transforming user histories into class diagrams to facilitate the design of analytical tasks of the project team, minimizing time and costs in agile projects.

Using tools applied to the user history review process will allow us to identify entities, syntaxes and categories that make it easier for the development team to understand and identify possible artefacts of the development and quality process. Which, later, would result in a better product.

This article presents content analysis, following a systematic literature mapping methodology, focused on requirements analysis and processing, mainly based on user stories for the development of software products. Additionally, an analysis of finding is presented comprising approaches, trends, gaps, and important aspects based on analysis that envisions needs and emerging areas of work in this context.

The rest of the paper is organized as follows: Sect. 2, explain the related studies and Requirements Elicitation and NLP respectively. Section 3 includes details of the applied methodology. The results and discussion are shown in Sect. 4, followed by the conclusions and future work in Sect. 5.

2 Background and Related Work

This section introduces the background on RE, NLP techniques, and agile techniques, and the synergy among these fields. In addition, we describe related work to our study.

2.1 Requirements Engineering

Requirement engineering (RE) is one of the first stages of the software engineering process. Given its complexity, defects such as ambiguities, inconsistencies and incomplete requirements may arise. These defects have been reported by practitioners to be causing problems in software projects, such as poor product quality and time and budget overruns [23]. Moreover, the costs for correcting these RE-related problems increase throughout the software development life cycle [9]. These additional costs reinforce the importance of identifying such defects at early stages.

2.2 Requirements Specification with User Stories

Requirements are specified in a user story format The software industry has gradually increased the use of agile and hybrid methods in its projects [33]. In this context, user story is the most frequently used artifact for requirement specification [48]. Therefore, this study is focused on user stories. The stories are often analyzed independently and structured in a sentence as follows: As a [role], I want to [feature], so that [reason].

User stories define the requirements of the software to be developed and are often written by analysts, requirements engineers, and even product owners, usually making them more ambiguous and abstract than traditional requirements. US specify software requirements by describing how actors or end-users will interact with the system under construction [63].

Most sources of requirements elicitation are available in natural language form. However, the approach to extracting user stories from natural language is still limited [74]. User stories' popularity among practitioners and their simple yet strict structure make them ideal candidates for automatic reasoning based on NLP.

2.3 Related Work

Extracting Information from User Stories. In the previous literature review, heuristics and rule-based approaches has been developed to parse and extract information from user stories to be used for several purposes. In many approaches was reported that most of the NLP techniques applied over the user stories as a source, have goals associated to extract information to define test cases, to generate database scripts, to design UML diagrams or conceptual models [70]. Moreover, as a result of several studies, there are several datasets and templates of user stories in many contexts, but most of them were not made available as they were proprietary data from industry partners [63], or in the only public dataset we found almost half the user stories did not conform to defined templates. We additionally observed that such approaches were limited in how much information they could extract from user stories. In all cases, such approaches were not able to identify specific data objects or operations.

Related Literature Reviews. As the base knowledge for this systematic mapping, the following related works are recognized, which serve as a reference for the current review process: Raharjana et al. [12] present a systematic review exercise of user stories and natural language processing; Bakar et al. [13] develop a systematic review of feature extraction approaches from natural language requirements for reuse in software product lines; Nazir et al. [14] present systematic review on the applications of NLP to software requirements engineering.

3   Methodology

3.1 Systematic Mapping

We conduct a systematic mapping study to provide an overview of RE research from the perspectives of NLP as well as to structure this research area, and mainly based on user stories processing. In this study we applied a systematic mapping methodology for conducting an evidence-based literature mapping of research publications addressing NLP applications to requirements processing techniques, methods, and approaches.

By following the guidelines proposed by Petersen *et al.* [34],[71], we adjusted a systematic mapping defining the following procedures: (i) Research Questions; (ii) Search strategy; (iii) Papers filtering; and (iv) Quality Assessment.

Research Questions. The following questions guided the discovery and characterization of studies under research. We define:

General Question. How have NLP techniques been used to optimize the Requirements Engineering stage from software development process?

Focused Questions. Comprises questions detailing existing solutions of 'general question' to structure models, identify patterns, limitations, and gaps for future research [72]. The focused questions are as follows:

- Which NLP techniques are used to process, manage, or define requirements in software development processes?
- Which NLP models are used in requirements engineering applications?
- Which development tools (*i.e.,* programming language, frameworks, or software) are used for implementing the proposed solutions?
- What are the domain areas of user stories-based applications with NLP techniques?

Search Strategy. We defined three steps for this strategy, as follows: (1) specify search string; (2) select databases; and (3) collect results, as we present next:

Search Strings. The first step was identifying the major terms and their most relevant synonyms—based on preliminary research and related works. The mayor terms were 'requirements', 'user stories', and 'natural language processing'. Subsequently, the search string merged the major terms with Boolean operators. The search string was conformed including the Spanish terms, thus: ("Historias de usuario" OR "User stories" OR "User story") AND ("Procesamiento de Lenguaje Natural" OR "NLP" OR "Natural Language Processing" OR "PLN") AND ("requirements" OR "requisitos").

Select Databases. The search strategy encompassed 4 digital libraries: Dimensions (linking data from publications, grants, datasets, and policy documents, among other), IEEE Xplore, Science Direct, and Scopus. They were selected, as suggested Silva and Braga [73], because they're prioritized well-known research sources, which is essential to find applications in this area of knowledge.

Collect Results. In addition, in the study we applied filters for language (English and Spanish), type of venue according to the filtering process (journals, conference papers, book chapters), when available in the digital library (full access), temporary window with articles published since 2018 to February 2023. Then, inclusion and exclusion criteria were applied, and duplicates were eliminated.

Papers Filtering. Papers filtering started at the initial search from each digital library, removing studies unrelated to main subject of the research questions. The filtering was supported on the following exclusion criteria:

- The study is not written in English or Spanish.
- The study has a publication date before 2018.
- The study venue is neither conference, research book, nor journal.
- The study context is different from the research purposes.
- The study is not accessible in full text.
- The study is derived from a poster-paper or a short paper.
- The study is not cited by other papers.

Quality Assessment. We evaluate the selected papers' quality applying the focused questions previously defined. The quality assessment of the results is based on the verification of such question, highlighting the user stories specification. Each one was evaluated based on the response options YES or NO.

4 Results and Discussion on Findings

4.1 Data Extraction Results

The set of initial results in all digital libraries summarized 4729 studies. Then the first filters, a set of 483 studies were defined and grouped by data source, as shown in Fig. 1, thus: Dimensions (178 studies), IEEE (19 studies), Science Direct (41 studies), and Scopus (245 studies). After the whole filtering process, the selection of the main works resulted in the 57 papers. In Fig. 2 we show the prioritized studies by digital library.

According to the accepted studies, we identify that the most studies were published in 2021 and 2020, as we show in Fig. 3.

The selected studies completely answer the focused questions, so in Table 1 we present the list of them which accomplish the quality criteria: all works have their YES response. This result indicates good quality papers—detailing application of NLP techniques in RE processes.

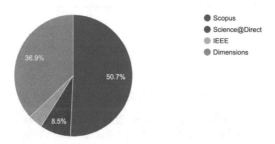

Fig. 1. Filtered studies by library.

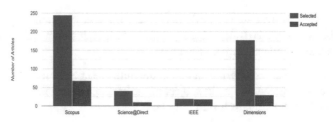

Fig. 2. Accepted vs. selected studies by library.

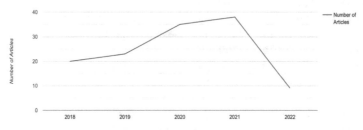

Fig. 3. Research networks.

A total of 40 empirical studies, 5 non-empirical studies, and 5 literature review were selected, as shown in Table 1.

Table 1. Classification of studies.

Classification	User stories from other artifacts	User stories as input
Empirical	[18], [19], [10], [20], [23], [24], [25], [26] [30], [31], [33], [34], [35], [36], [37], [38], [40], [41], [42], [45], [52], [53], [22], [70]	[46], [47], [48], [54], [55], [56], [57], [59], [60], [61], [62], [63], [64], [65], [68], [69]
Non-empirical	[21], [22], [44], [67]	[44]
Literature review	[17], [12], [32], [66], [75]	

4.2 Findings Analysis

General Analysis. Papers filtering process resulted in a group of 125 studies accepted, as we show in Table 2.

It is also identified, that over the years the number of articles on the topic has increased. Examining the results obtained in Dimensions by means of a bibliometric analysis in VOSviewer, research networks have been identified highlighting an interesting relationship found between Dalpiaz and Ferrari, as we show in Fig. 4.

Relevant Findings. The following sub-paragraphs present the research trend conducted by primary studies, organized in categories by most relevant findings:

Table 2. Analysis of accepted papers.

Binomial Test					
Variable	Level	Counts	Total	Proportion	p
Source	Dimensions	178	483	0.369	<.001
	Science Direct	41	483	0.085	<.001
	IEEE	19	483	0.039	<.001
	Scopus	245	483	0.507	0.785
Year	2018	54	463	0.117	<.001
	2019	86	463	0.186	<.001
	2020	138	463	0.298	<.001
	2021	146	463	0.315	<.001
	2022	39	463	0.084	<.001
State	Accepted	125	483	0.259	<.001
	Duplicates	114	483	0.236	<.001
	Rejected	244	483	0.505	0.856

Note. Proportions tested against value: 0.5.

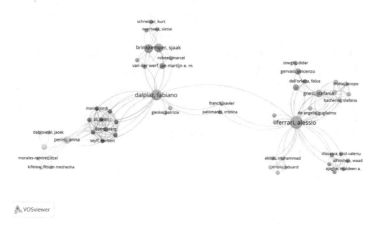

Fig. 4. Published studies by year.

Contextualizing. The studies prioritized on natural language processing (NLP) techniques are trying to help to improve the process of user stories specification, adding more flexibility to write scenarios that could be semantically interpreted to meet the behaviors described in initial requirements.

Although these studies continue reporting several challenges in the literature, the number of solution proposals to address these concerns is broad, diverse, but limited. Existing approaches employ inspection techniques to verifying security or identifying security goals from textual documents. These proposals are not completely focused on agile but have reached approaches with these kinds of requirements. Given the contemporary state of reported evidence, we identify an existing literature gap concerning security requirement verification in agile contexts.

Basic Application of NLP Techniques for Specifying User Stories. The main approaches commonly used for processing information sources to specify user stories comprise preprocessing (includes preparing the data, tokenization, filtering, and stop-word removal), part-of-speech tagging (POS tag), syntactic parsing (representing lexical categories and grammatical relationships) and semantic analysis [6].

Several methods have been used to obtain and understand the semantic connections in user stories, such as similarity function [25] [70] [50] [59] and clustering for semantic similarity [29] [65] [59]. Some methods were used to identify topics (*e.g.*, security requirements, non-functional requirements, privacy-related entities, and quality attributes) inside user stories for heuristic analysis using LDA [50], Word Vectors [Younas et al.] [59] and word embeddings [45], [52], [59]. Also, some approaches using NLP to perform lexical, syntax and semantic analysis of user stories [29], [65], [30],[33], [74], and automatically identify concepts and relationships to facilitate model generation [55] [47], [68], [69].

For generating user stories from technical documents, we found the following approaches: extracting types of events from an app review [40]; identifying user story from online news [43]; extraction and classification of requirements from software engineering contracts [76] and from components of domain documents [77].

NLP for Generating Model from User Stories. This aspect has become popular with the increase of agile software development, such as generating goal models [5], conceptual models [8] or UML models [9] from user stories. In this analysis category the purpose is generating software artifacts from a user story (as input of the generated artifact). Several studies reported algorithms, techniques, or methods for generating software model/artifacts from user stories, such as the following: conceptual models [35], [36]; goal models [37], [41]; test case [39], [46]; UML diagrams [40], [41], [48]; BPMN diagrams [47], [68], [69].

Improving the Quality of User Stories with NLP. The main purpose of this category is improving the requirements quality represented in user stories and finding defects using NLP techniques. The main studies are related to identify ambiguity [33], [58], [63], [70]; identifying quality attributes [65]; supporting assisted requirements elicitation (*e.g.*, via chatbots) [66], [78]; detecting privacy-related entities [50]; predicting story popularity and detecting untrue stories [62]; and identifying security requirements [30], [31], [51], [79].

Research Support Related to NLP in User Stories. This category comprises NLP techniques, connection with other techniques/methods, and tools used as a support of the approaches available in research related to NLP in user stories, as we present as follows:

- NLP techniques: semantic similarity and clustering [25], [70], [50], [59]; Text analytics [29], [30], [33], [65]; named entity recognition (NER) [50], [63], [74]; word embedding methods [45], [52], [59]; dependency parsing [74].

- Connection with other: convolutional neural network [33]; Transfer Learning technique [33]; recurrent neural network models [38] and Recurrent Neural Network with Convolutional Neural Network [60]; Deep learning [63].
- Tools: Stanford CoreNLP [30], [39], [57]; NFR Locator [80]; Natural text processor spaCy [36]; ArTu for generating goal models [41]; DizReq engine for digitalizing requirements [53]; NLP4IP recommendation for story or bug prioritization [54]; FRED and NLQuery: existing knowledge processing engines [64].

5 Conclusions and Future Work

This study presented a systematic mapping of literature. The main purposes highlighted from this study of NLP in user story research still majorly focus on identifying defects, improve the requirements quality, and generating models. Most of the approaches in the current research still concentrates on user story processing, applying the NLP basic techniques related to text analytics (preprocessing, part-of-speech tagging, syntactic parsing, and semantic analysis.

Generating artifacts from user stories is widely worked in the reviewed research. Most of them processed them to generate software artifacts, such as a UML models (e. g., class diagram or sequence diagram), conceptual models or business process models).

Natural language processing is a discipline that is increasingly attracting the interest of academic and research communities. This is evidenced by the growth indicators of the reference publications prioritized in this mapping over the last 6 years.

To ensure obtain practical adoption models from user stories, first, user stories should follow good writing specifications, and secondly, the information that are automatically identified from user stories need to be confirmed manually. A corresponding research goal is how to reduce manual tasks and produce useful results in the practice.

Among the main findings of the systematic mapping, we can summarize the following: (i) About 40% of articles are using the user stories from agile methodologies as the source for applying NLP techniques, to generate any kind of model, aspect, or element regarding software development process. (ii) About 32% of articles are focused on the generation of models or diagrams; (iii).

The findings of the study allow us to affirm that delving deeper into NLP-user stories research would contribute both to improving the understanding of their content for development teams, but also help to create models, diagrams, and other software artifacts (e.g., interface, including code, tests, or functional prototypes) from the user stories.

As future work, progress is being made in proposing a method for extracting information from user stories of agile software projects in a Latin America context, which makes it easier the comprehension, acceptance criteria compliance, and disambiguation task for the development teams.

It is thus necessary to generate knowledge and opportunities for the optimization of requirements management processes in agile contexts, for contributing to reduce the manual or human intervention. The application of NLP techniques can contribute to the intervention of this scenario and the RE process automation. Potential beneficiaries may be agile development teams that have the user stories as the core of the whole agile development process. Thus, it will also be possible to have useful methodological and knowledge products as a reference for related research and software industry efforts.

Appendix

All references are available in: https://drive.google.com/file/d/1ux0fmO7aINR9UpPal P0id7P8zIH2Wkx9/view

References

1. Gil Leiva, I., Rodríguez Muñoz, J.: El procesamiento del lenguaje natural aplicado al análisis del contenido de los documentos. Rev. Gen. Inf. y Doc. **6**(2), 205–218 (1996)
2. C2020 Task Force. Computing Curricula 2020 (2020)
3. McCarthy, J.: What is artificial intelligence? Eng. Mater. Des. **32**(3), 1–14 (2004)
4. Abualhaija, S., Arora, C., Sabetzadeh, M., Briand, L.C., Traynor, M.: Automated demarcation of requirements in textual specifications: a machine learning-based approach. Empir. Softw. Eng. **25**(6), 5454–5497 (2020)
5. Friedrich, F.P.F., Mendling, J., Robeer, M.: Process model generation from natural language text. ARPN J. Eng. Appl. Sci. **12**(8), 2581–2587 (2017)
6. Schwaber, K.: SCRUM Development Process. In: Business Object Design and Implementation, pp. 117–134. Springer, London (1997)
7. Cohn, M.: User Stories Applied for Agile Software Development, vol. 53, no. 9 (2004)
8. Elallaoui, M., Nafil, K., Touahni, R.: Automatic transformation of user stories into UML use case diagrams using NLP techniques. Procedia Comput. Sci. **130**, 42–49 (2018)
9. Robeer, M., Lucassen, G., Martijn, J., Van Der Werf, E.M., Dalpiaz, F., Brinkkemper, S.: Automated Extraction of Conceptual Models from User Stories via NLP; Automated Extraction of Conceptual Models from User Stories via NLP (2016)
10. Nasiri, S., Rhazali, Y., Lahmer, M., Chenfour, N.: Towards a generation of class diagram from user stories in agile methods. Procedia Comput. Sci. **170**, 831–837 (2020)
11. Mukhamedyev, F.N.A.R.I., Symagulov, A., Kuchin, Y.I., Abdullaeva, S.: Cloud Services for Natural Language Processing
12. Raharjana, I.K., Siahaan, D., Fatichah, C.: User stories and natural language processing: a systematic literature review. IEEE Access **9**, 53811–53826 (2021)
13. Bakar, N.H., Kasirun, Z.M., Salleh, N.: Feature extraction approaches from NL req. for reuse in soft. product lines: a systematic lit. review. J. Syst. Softw. **106**, 132–149 (2015)
14. Nazir, F., Butt, W.H., Anwar, M.W., Khan Khattak, M.A.: The applications of natural language processing (NLP) for software requirement engineering - a systematic literature review. In: Kim, K., Joukov, N. (eds.) ICISA 2017. LNEE, vol. 424, pp. 485–493. Springer, Singapore (2017). https://doi.org/10.1007/978-981-10-4154-9_56

1st Workshop on The Role of the Technologies in the Research of the Migrations

"From Letters and Phone Calls to WhatsApp and Social Media: The Evolution of Immigration Communication"

Jessica Ordóñez Cuenca and Analy Poleth Guamán Carrión[(✉)]

Universidad Técnica Particular de Loja, San Cayetano Alto S/N, Loja, Ecuador
{jaordonezx,apguaman6}@utpl.edu.ec

Abstract. This document centers on communication as a pivotal transnational practice traditionally engaged in by migrants, underscoring its significant role in strengthening social networks. It highlights how communication aids in preserving family ties, cultural habits, and values, thereby easing the integration of family members into the migrant's everyday life. Over the years, communication strategies and mediums have undergone a notable evolution. Initially, migrants relied on letters and telephone calls. With the advent of the Internet, "cyber cafes" emerged as the new hubs for transnational interactions. Presently, platforms like WhatsApp and various technological applications have revolutionized and democratized transnational communication, making it more accessible and efficient.

Keywords: international migration · transnational practices · social networks

1 Transnationalism and Migratory Networks

Migrant transnationalism, as described by [7], fosters a heightened level of connectivity among individuals, societies, and communities (p.11). These transnational practices are multifaceted, ranging from the sending of economic remittances to more intangible elements like affection, longing, and love for one's homeland. This phenomenon underscores the significance of communication within the realms of globalization and intercultural relations, particularly in the migration context. This document elucidates how communication is an integral aspect of the migration process and a vital tool for strengthening family and social networks. It has been propelled by technological advancements, the widespread availability of mobile devices, social media, and mobile applications.

According to [5], transnational practices are established to preserve familial bonds, especially now that the traditional family model, where all members reside in the same location, has given way to a decentralized model with family members spread across various global locations. In this scenario, individuals tend to choose communication methods that are most accessible, user-friendly, and cost-effective. However, the utilization of emails, as noted by this author, is contingent upon access to a computer and the ability to use it. Consequently, the widespread use of telephone calls and the transmission of various messages has become more prevalent. [3] mentions that the intensity of transnational remittances is assessed by considering the frequency and extent of connections, which hinge on the migrant's socioeconomic status. For instance, those with limited resources may not engage in certain transnational practices as frequently as those with higher and more stable incomes.

1.1 Communication Through Letters and Telephone Calls: An Overview

In their ethnographic study, [6] evaluates the transnational practices of Ecuadorian migrants in Spain, the United States, and Italy from 1996 to 2006. These practices include various actions, notably the sending of clothes, toys, letters, and greeting cards. The latter is termed as 'social remittances', reflecting the intangible aspects of migrant life such as care, love, concern, and longing.

The study highlights the case of "Vilma", an Ecuadorian migrant in the United States. Like many migrants, Vilma relied on letters and phone calls to stay connected with her family. Over a decade, she consistently communicated with her children weekly. She used conventional telephones, as cell phones were not yet prevalent, and the cost of international calls was high.

Letters provided a way for migrants like Vilma to be a part of their family's daily lives, bridging the geographical gap. Through various forms of transnational communication, they could extend and relocate aspects of motherhood, aiding in their children's education and maintaining familial bonds. Despite the long separations necessitated by financial constraints, regular phone calls and other means (like letters and gifts) allowed migrant mothers to preserve emotional ties. This phenomenon is described by [2] as "being a mother from a distance", where emotional connections with children are maintained through various communication methods.

The evolution and intensification of media and communication means have made responses to messages more immediate. An example is a letter from Silvia to Vilma dated May 31, 1996, expressing affection, sharing personal news, and requesting a "pupera" blouse. This letter illustrates how communication methods served as a way to share information, ideas, feelings, future plans, and even local colloquialisms from the migrant's new residence.

For Vilma, letters were not just a means of interaction with her family but also a way to share and exchange ideas across countries. This led to transnational learning and adapting new concepts to their living situations. [11] observed that family members, though not in the same physical space, were part of a transnational social space. This space, maintained through strong ties, supported family unity and social cohesion across different regions.

1.2 Emerging Trends in Communication: Exploring Modern Methods

Observing transnational life from a perspective devoid of rapid communication methods, it would undoubtedly be more complex. The intensification of transnational exchanges can be attributed in part to technological innovations (Sorensen y Guarnizo, 2007). The advent of advanced communication and transportation means has significantly bridged the distance between migrants and their countries of origin, creating a new form of presence and participation in their homeland's daily reality [3].

While earlier emotional connections were maintained through letters and occasional phone calls, today information is relayed instantly and at a lower cost via applications like WhatsApp, Messenger, and Facebook. These platforms enable migrants to recreate moments that strengthen emotional bonds, with the telephone serving as the primary mode of interaction. They facilitate immediate interaction networks, with advantages such as reduced call costs and enhanced audiovisual quality [8]. What once were weekly phone calls made from call centers, have now transformed into daily calls or messages in our current era of globalization and immediacy.

This technological evolution has also enabled the documentation of various stages of migration, from departure to settlement or return. It provides a means of virtually accompanying the migrant during their journey and documenting events, challenges faced, and available support mechanisms. This documentation often takes place on social media platforms like Instagram and Facebook, sometimes in real time, especially during undocumented travels and illegal crossings.

According to C. Diego (personal communication, November 17, 2022), technology has simplified communication. Having a WhatsApp group, for example, allows family members to feel present at gatherings by sharing photos and updates, thus bridging distances.

From the perspective of migration management, a report about migrants and refugees in transit for Medellín-Antioquia by [9] indicates a preference for communication via WhatsApp (31%) and social networks (26%), followed by flyers and posters (18%) and verbal information (14%). Non-governmental organizations have harnessed these applications to provide information and prevent abuse or violence. [1] offers an information service through WhatsApp and the Turn.io tool, enabling refugee migrants to learn about regularization processes, rights, obligations, and access to services in the country. This tool generates automated queries and connects users with [1] staff for reporting violence or abuse.

In 2023, the platform [10] released a country-specific information document for refugees, highlighting the use of mobile devices to access information pertinent to their needs. The document underscores the prominence of WhatsApp as the primary communication channel, and the accessibility of information via emails and icons from digital platforms like Facebook, Twitter, YouTube, and Instagram, and quick barcode scanning for information retrieval.

1.3   Final Considerations

In summary, transnational communication plays a crucial role in reinforcing migratory networks and reducing the costs associated with migration, particularly the psychological toll and the burden of the adaptation process. It enables the extension of the concept of 'home' beyond physical borders through various media. Traditional forms of communication such as letters and phone calls have evolved into emails, instant messaging, and the use of various technological applications.

Moreover, communication and technology have been instrumental in highlighting the 'journey' of migrants and the challenges they face until they reach their destination. The speed and immediacy of technological platforms not only facilitate widespread dissemination of these experiences but also contribute to the creation of a collective consciousness. This consciousness fosters greater empathy among non-migrants towards these groups, thereby promoting a deeper understanding of the migrant experience.

References

1. ACNUR. Nuevo canal informativo de Whatsapp en Ecuador permite a los refugiados comunicarse con ACNUR https://news.un.org/es/story/2020/11/1483742. Accessed 12 Dec 2023
2. Carbajal, M., Ljuslin, N.: Jeunes sans-papiers d'Amérique latine en Suisse ou devenir adulte sur fond de recomposition de rôles. *Lien social et Politiques*. https://doi.org/10.7202/100 1404ar (2011)
3. Cavalcanti, L., Parella, S.: Una aproximación cualitativa a las remesas de los inmigrantes peruanos y ecuatorianos en España ya su impacto en los hogares transnacionales. Revista Española de Investigaciones Sociológicas (REIS) **116**(1), 241–257 (2006)
4. Hondagneu-Sotelo, P., Avila, E.: "I'm here, but I'm there" the meanings of Latina transnational motherhood. Gend. Soc. **11**(5), 548–571 (1997)
5. Carbajal, M.: Ser madre en la distancia1: análisis de una práctica transnacional: El caso de mujeres latinoamericanas en Suiza. L'Ordinaire des Amériques (208–209), 163–181 (2008). https://doi.org/10.4000/orda.3304
6. Guamán Carrión, A.: Prácticas transnacionales de migrantes ecuatorianos con relación a las remesas sociales [Tesis para optar por el título de economista, Universidad Técnica Particular de Loja] Repositorio Institucional UTPL. (2022)
7. Oroza Busutil, R., Puente Márquez, Y.: Migración y comunicación: su relación en el actual mundo globalizado. Revista Novedades en Población **13**(25), 10–16 (2017). http://scielo.sld.cu/scielo.php?script=sci_arttext&pid=S1817-40782017000100002&lng=es&tlng=es. Accessed 22 Nov 2023
8. Ramírez Martínez, J.P.,: Uso de tecnologías de la información y la comunicación en familias caleñas con migrantes en España. Revista de Estudios Sociales (48), 110–123 (2014). https://doi.org/10.7440/res48.2014.09
9. R4Va. SITUACIÓN DE PERSONAS REFUGIADAS Y MIGRANTES EN TRÁNSITO MEDELLIN- ANTIOQUIA (TERMINAL DEL NORTE). https://reliefweb.int/report/col ombia/reporte-de-situacion-situacion-de-personas-refugiadas-y-migrantes-en-transito-med ellin-antioquia-terminal-del-norte-1-mayo-31-agosto-2023. Accessed 31 May 2023
10. R4Vb. INFORMACIÓN POR PAÍS PARA PERSONAS REFUGIADAS Y MIGRANTES. https://www.r4v.info/sites/default/files/2021-06/Directorios_de_contactos.pdf. Accessed 31 May 2023
11. Sorensen, N.N., Guarnizo, L.E.: La vida de la familia transnacional a través del Atlántico: la experiencia de la población colombiana y dominicana migrante en Europa. (2007)
12. Puntos de Vista: Cuadernos del Observatorio de las Migraciones y la Convivencia Intercultural de la Ciudad de Madrid (oMci), (9), 7–28

Visual Ethnographic Analysis of the Transit Migration of Venezuelans in Huaquillas, Ecuador

Pascual Gerardo García-Macías[1](✉), Marcel Angel Esquivel-Serrano[2](✉), and Edison Javier Castillo-Pinta[1](✉)

[1] Universidad Técnica Particular de Loja, Loja 110111, Ecuador
pasgegar84@gmail.com
[2] Universidad Autónoma de Sinaloa, Culiacán, Sinaloa, Mexico
marcelesquivel73@gmail.com

Abstract. Culture has been the subject of extensive scholarly inquiry, approached through diverse methodologies including textual analysis, iconographic studies, and field research. Central to this multifaceted exploration is ethnography, a specialized branch of science dedicated to the systematic study and detailed description of cultures. Ethnography encompasses not only the academic study of cultural phenomena but also serves as a robust methodology for gathering insights into the nature of social reality, the intricacies of interpersonal relationships within groups, and other sociocultural dynamics. One particularly compelling approach within this field is the utilization of visual mediums such as photography and imagery. This visual perspective offers a unique vantage point for extracting and interpreting cultural elements, enabling researchers to delve into the meanings, symbology, interpretations, and narratives inherent within a community's visual testimony. Visual ethnography, therefore, stands as a powerful tool in the exploration of cultural landscapes.

The document presented herein aims to employ visual ethnography as a lens to examine and elucidate the culture of Venezuelan migrants in transit through the community of Huaquillas, Ecuador, which lies adjacent to Peru. This analysis is predicated on the collection and interpretation of images and photographs obtained during fieldwork. The objective is to glean insights from these visual materials, offering a nuanced understanding of the migrants' cultural milieu and experiences. Through this approach, the study seeks to contribute to the broader understanding of the dynamics at play within transient migrant communities and the socio-cultural tapestry they inhabit.

Keywords: Ethnography · Visual ethnography · Transit migration · Huaquillas · Ecuador

1 Introduction

In the present article, an endeavor has been made to systematically collate and elucidate the evolution and fortification of visual ethnography as a scholarly discipline. This methodological framework, which predominantly leverages visual elements such

© The Author(s), under exclusive license to Springer Nature Switzerland AG 2024
Á. Rocha et al. (Eds.): WorldCIST 2024, LNNS 990, pp. 267–272, 2024.
https://doi.org/10.1007/978-3-031-60328-0_26

as images and photographs, serves as a potent instrument for comprehending and inter-preting the multifaceted dimensions of reality. Notably, visual ethnography has emerged as a particularly valuable tool in the study of diverse thematic areas, with a special emphasis on migration studies. Furthermore, this document advocates for the applica-tion of visual ethnography as an enriched methodological approach in exploring the nuances of Venezuelan transit migrations between Ecuador and Peru. This approach is grounded in participant observation, a technique that offers profound insights into the experiential realities of migrant populations.

To facilitate a comprehensive understanding, the structure of this document is metic-ulously organized into several distinct sections. It commences with a theoretical section, which lays the foundational concepts and academic discourse surrounding visual ethnog-raphy. This is followed by a methodological section, delineating the specific procedures and approaches employed in the study. Subsequently, the results section presents the empirical findings derived from the visual ethnographic analysis. The document culmi-nates with a conclusions section, which synthesizes the insights garnered and discusses their implications within the broader context of migration studies and ethnographic research. This systematic organization ensures a coherent and in-depth exploration of the subject matter, underpinning the scholarly rigor of the study.

2 Theoretical Framework

Ethnography, as a discipline, manifests in two distinct but interrelated forms: firstly, as an offshoot of anthropology, where its inception is intertwined with the early development of the latter; and secondly, as a methodological paradigm. In its methodological guise, ethnography is inherently qualitative, striving to furnish comprehensive descriptions of peoples, ethnic groups, and their encompassing cultures. This includes an examination of their lifestyles and linguistic developments, as emphasized by [1]. Naranjo [2] further elucidates this point, stating that "culture can be read and interpreted as a text, where ethnography serves as a research tool […] in the quest for profound cultural meanings" [2, p. 21]. A pivotal tool within the ethnographic repertoire has been the use of images. These visual elements contain rich, layered information awaiting interpretation and com-prehension. Far from being superficial representations, to the trained and experienced eye, images encapsulate narratives, life stories, experiences, cultural symbols, and world-views. They contribute significantly to our understanding of populations, ethnic groups, and identities [2]. By the late 19th century, ethnography began gravitating towards a more visual methodology. The works of Kroehle and Huebner in the Amazonian region of Peru exemplify this shift, contributing to the development of disciplines like phys-iology and physical anthropology. Their focus was on physical textures and material elements reflective of identity and culture [3]. However, figures like Malinowski and Radcliffe-Brown sought to reclaim the origins of anthropology, particularly in visual and audiovisual ethnography, challenging the contributions of Boas and Haddon [4].

Franz Boas and Alfred Haddon (1898) are acknowledged as pioneers in employ-ing audiovisual media for research. Boas's investigations in Canada and Haddon's in Australia, particularly his scientifically methodical expedition in the Torres Strait, were groundbreaking. Haddon's aim extended beyond mere research to the documentation and preservation of native cultures imperiled by Western colonialism [4, 5]. During the early 20th century, audiovisual media gained traction in various studies. In the 1930s, Boas's disciples (like Margaret Mead) and Haddon's (like George Bateson) produced significant audiovisual content. Their goal, however, was not documentary filmmaking but rather the use of these tools for ethnographic recording, capturing images for deeper analysis, particularly focusing on "types of non-verbal behavior" [5, p. 33].

As Expósito [5] explicates, anthropology has extensively utilized imagery to understand cultural conceptions, acknowledging that culture often becomes palpable through visual representations. Hence, the relevance of adopting a visual ethnography approach, which encompasses not just evident photographs but all forms of visual media, including videos and films. This approach, termed 'visual ethnography', is a qualitative methodology that facilitates the understanding of social phenomena and processes through visual information [2]. The primary objectives of visual anthropology, therefore, include producing audiovisual elements to support research and innovatively generate new knowledge; and utilizing existing audiovisual materials to enhance investigations, enabling the discovery of novel elements and knowledge creation [5].

Rodríguez [6] posits that images, as interpretable documents, offer explanatory elements vital to ethnography. Visual ethnography should not replace traditional methods or written texts but rather complement them in a symbiotic relationship. It acknowledges that sometimes images can convey what words cannot, and certain processes or ideas might elude textual capture. This understanding operates on two levels. Firstly, it revisits the historical experience, drawing from it to produce new knowledge. This is evident in the continued relevance of Amazonian studies, which yield insights into the formation and nature of societies, identities, and cultures. Secondly, it recognizes the contemporary significance of the internet and social media, where images and audiovisual materials play a crucial role. These platforms not only capture material reality but also contribute to the formation of social imaginaries, showcasing the pervasive influence of visual media in contemporary discourse [7, 8].

Thus, visual ethnography emerges as a critical interdisciplinary tool, bridging the gap between traditional anthropological methods and the dynamic, image-driven landscape of modern society. It embodies the evolution of ethnographic practice, adapting to and incorporating the technological and cultural shifts of the 21st century.

3 Methodology

In this scholarly investigation, we have elected to employ visual ethnography, specifically harnessing the power of photography. This decision is anchored in the recognition of photographs as pivotal representations of reality, a concept elucidated by [9]. Photographs, as Iturra notes, have the unique capacity to illuminate details and aspects of life that might otherwise remain obscured, thereby "making those details and characteristics of life that are visible and that could have gone unnoticed 'pop out'" [9, p. 26]. Furthermore, as tangible representations rather than abstract constructs, photographs serve as catalysts for discussion and interpretation, providing a rich tapestry of information accessible to a variety of disciplinary perspectives. The methodological approach to engaging with the subjects of study is a critical element in ethnography, as emphasized by [10]. To this end, the participant observer technique has been adopted. This technique involves the researcher immersing themselves in the lives of the study population, maintaining an appropriate balance of emotional and affective detachment, to ensure objectivity and analytical rigor.

The focus of this study is on transit migration at the Ecuador-Peru border, with a specific emphasis on the community of Huaquillas. This locale serves as a nexus for transit migrants originating from Venezuela and aiming for Chile as their final destination. Fieldwork was conducted within this migrant community, enabling direct engagement and observation. Subsequently, a visual ethnographic analysis was undertaken, utilizing a selection of photographs collected during the research period. This analysis aims to provide a nuanced understanding of the migrant experience, contextualized within the broader framework of transit migration and its socio-cultural dynamics.

4 Results

This research prominently features the application of visual ethnography, a method that integrates elements like photography and imagery to delve into cultural studies. Far surpassing the realm of simple graphic representation, visual ethnography engages in a profound interpretation of the meanings, symbols, and cultural narratives that manifest visually. This approach is instrumental in capturing facets of culture that are often elusive in verbal expression, thus furnishing a rich and detailed perspective. Photographs, in this context, provide an invaluable window into the social dynamics of migrants, illuminating their interactions, relationships, and the manner in which they engage with their surroundings. These interactions may encompass the formation of temporary communities, the dynamics of mutual support, and the intricate process of identity negotiation within a transient milieu. The images also document the physical and environmental facets of migration, including landscapes, transit spaces, and makeshift settlements. These visual records are pivotal in conveying the challenges, risks, and adaptive strategies and resources employed by migrants.

Accordingly, the document sets forth a proposal to apply visual ethnography in analyzing the culture of a specific demographic: Venezuelan migrants in transit in Huaquillas, Ecuador. This methodological approach is poised to uncover intimate details and perspectives regarding the migration experience, including the challenges faced and cultural adaptations undertaken by this group. The empirical collection of images and photographs during fieldwork will serve as a foundational basis for a deeper, more nuanced comprehension of the migrants' experiences and viewpoints. Visual ethnography is particularly adept at revealing the emotional and psychological states of migrants. Through the analysis of facial expressions, body language, and gestures captured in photographs, it is possible to infer a range of emotions such as hope, fear, anxiety, and resilience.

Thus, the text posits that visual ethnography, when applied to the study of specific groups like Venezuelan migrants, presents a unique and potent means to grasp the complexities of cultural experiences in contexts characterized by mobility and transition. Incorporating this methodology into ethnographic research significantly broadens our understanding of culture, paving new paths for the exploration and documentation of the dynamically evolving social landscapes (Fig. 1).

Fig. 1. Photograph by the author, taken in 2023 in Huaquillas, Ecuador.

5 Conclusions

In concluding this discourse, it is imperative to underscore the ethical considerations inherent in employing visual ethnography, particularly in delicate contexts such as migration studies. The researcher bears a profound responsibility to uphold the dignity and privacy of subjects, ensuring that their narratives and experiences are portrayed with the utmost integrity and sensitivity. This ethical framework is not merely an adjunct to the research process but a fundamental aspect that underpins the entire methodology. To encapsulate, the text delineates a comprehensive framework for the application of ethnography, with an emphasis on visual ethnography, as an efficacious instrument for the profound exploration of cultures. It specifically addresses the study of a demographic group within a defined geographical and socio-cultural milieu. The document presents a visual ethnographic analysis that sheds light on transit migration at the Ecuador-Peru border. The interpretative examination of photographs, gathered during fieldwork, aims to elucidate the experiences and realities of Venezuelan migrants en route to Chile, thereby enriching our understanding of the migratory phenomenon within this region.

Central to this research is the conscientious navigation of ethical considerations related to the visual representation of vulnerable populations, such as migrants. This involves the judicious handling of images, respecting the privacy and dignity of the individuals depicted, and critically reflecting on the potential impact of these visual representations on public perceptions of migration. Such ethical vigilance is essential to ensure that the research not only contributes to academic discourse but also adheres to the highest standards of moral and ethical research practices.

References

1. Peralta, C.: Ethnography and ethnographic methods. Anal. Colomb. J. Humanit. **74**, 33–52 (2009)
2. Naranjo, M.J.: Migratory networks of women and decolonizing spaces, study of visual ethnography: collective "women with a voice-Valencia". [Doctoral Thesis] University Polytechnic of Valencia, Valencia (2021)
3. Flores, R.: Visual ethnography and rubber colonization. In: Cánepa, G. (ed.) Visual Imagination and Culture in Peru, pp. 197–219. Editorial Fund of the Pontifical Catholic University of Peru (2011)
4. Lema, D.: Visual anthropology from a phenomenological perspective. Corporeality in observational ethnofilm methodology. [Doctoral thesis] University of Zaragoza (2022)
5. Expósito, J.: Visual anthropology: from the ethnographic record to shared cinema. Bull. Chil. Mus. Pre-Columbian Art **25**(2), 31–47 (2020)
6. Rodríguez, F.: Metamorphosis and strengths of visual anthropology: methodological contributions from photoethnography and ethnographic documentary. [Doctoral thesis] University of Granada, Granada (2015)
7. Limpo, A.: Images in code. A visual ethnography essay on <software> development. Cadernos de Arte e Antropologia **9**(1), 125–138 (2020)
8. Ortega, M.: Methodology of visual sociology and its ethnological correlate. New Era **22**(59), 165–184 (2009)
9. Iturra, L.: The use of photography to problematize the urban. Methodological transfers, visual ethnography in the teaching of architecture. Of Architecture (28–29), 22–30 (2013)
10. Romero, R.: The use of the image as a primary source in social research. Methodological experience of a visual ethnography in the case study: territorialities of daily life in the Zócalo plateau of Mexico City. Sequence **82**, 175–194 (2012)

Evaluation of the Benefit of Artificial Intelligence for the Development of Microeconomics Competencies

Luís Rojas[✉] and Álvaro Méndez

Universidad Técnica Particular de Loja (UTPL), Loja, Ecuador
{larojas,aemendez3}@utpl.edu.ec

Abstract. The objective of the research is to quantitatively and qualitatively determine the academic benefit obtained by students of the Open and Distance Learning Modality and the Presential Modality of the Universidad Técnica Particular de Loja, enrolled in the subject of principles of microeconomics during seven consecutive academic periods and who participated in the university connectivity project to the virtual learning environment. In order to fulfill the objective, the Propensity Score Matching (PSM) technique was applied, maintaining the control groups according to the results obtained. It was possible to determine that there are statistically significant differences in the academic performance of the students to whom the technique was applied and the control groups, but this does not allow us to conclude that there is an important qualitative change in terms of academic performance since the absolute value of the differences in average is not significant.

Keywords: Communication · Education · Artificial Education

1 Introduction

Education is one of the elements that has a significant influence on the development of peoples; it has become a necessary factor in all senses since it allows the inhabitants of nations to reduce the asymmetries that exist in societies with respect to employment, health, access to food, in short, better levels of social welfare.

According to studies by the Organization for Economic Cooperation and Development [1], there is a close relationship between the level of development of countries and the strength of their educational and scientific and technological research systems; an additional year of schooling increases a country's GDP per capita by between 4% and 7%. The World Bank states that education is a human right, an important driver of development and one of the most efficient instruments for eradicating poverty, increasing hourly income by 9% for each additional year of schooling.

Naturally, formal and non-formal education that takes place through educational institutions at all levels, as well as in the family and society, is strongly influenced by economic, political and cultural factors, which in some way generate differences in the

Á. Rocha et al. (Eds.): WorldCIST 2024, LNNS 990, pp. 273–279, 2024.
https://doi.org/10.1007/978-3-031-60328-0_27

quality of education received and therefore lead to differences in the standard of living of each of the peoples of our society. In view of the above, it is vitally important to have an educational policy aimed at closing the large gaps generated by the current educational system, specifically in the supply of public and private institutions with respect to the differences that exist in the management of the availability of human and support potential: infrastructure, classrooms, laboratories, properly equipped libraries and technology that state educational institutions scarcely offer.

Based on a bibliographic review of scientific articles, determines that there is a great difference between the academic average of students enrolled in a private and a state institution and that the difference is more acute when comparing institutions in the city with institutions in the rural sector. In his study, he attributes the socioeconomic composition of the student as one of the reasons for these results. In view of this, higher education institutions play a fundamental role in the formation of citizens with critical thinking, applying methodologies and practices in the classroom that encourage analysis and reflection and that are connected and focused on providing solutions to real problems.

The amount of information that is available today is one of the indicators that has allowed some theorists to call it the knowledge society, others relate it to a digital society or the information society, whose basis is based on the media and information and communication technologies (ICT), and in recent years Artificial Intelligence (AI) has been added. These conceptions allow describing that its origins and development are due to innovation and changes in technology, strictly related to ICT and AI in the field of educational planning and training, in the organizational field (knowledge management) and knowledge work [2].

The objective of this research is to quantitatively determine the academic benefit obtained by the students of the Open and Distance and Presential Modality of the Universidad Técnica Particular de Loja, enrolled in the subject of principles of microeconomics during seven consecutive academic periods and who participated in the university connectivity project to the virtual learning environment (EVA), using the technique of Propensity Score Matching (PSM).

2 State of the Art

All the theories analyzed agree that students can construct their learning autonomously or in collaboration and interaction with other people and this can be done through ICTs and not depend solely on the educator in the classroom [3].

At the present time, there are researchers who speak of a neo-constructivism [4], highlights a new constructivism, also known as techno-educational constructivism, which provides an effective and positive meeting space between research and pedagogical practice and technological advances, based on five major processes: awareness, elaboration, personalization, application and evaluation.

Computational theory or also known as information processing theory, this theory proposes a metaphor between the functioning of the human mind and a computer [5], is framed in cognitive psychology, which is responsible for the study of the functioning of human cognition; this refers to the way in which people elaborate, transform, encode, store, retrieve and use the information they receive from their environment. Robert Gagné is considered to be its manager. The formal bases of the computational theory of mind are based, on the one hand, on the mathematical formalism that conceived a discipline such as mathematics as the art of manipulating symbols based on formal rules and, on the other hand, on the experiments of Alan Turing, who implemented a mathematical model consisting of an automaton capable of constructing any mathematical problem expressed through algorithms. One of the criticisms of the theory is that they consider it empirically implausible, because a brain that manipulates computational symbols does not seem entirely biological.

The theory of communicative action is based on the linguistic communicational relationship. The structure of this model has been used to describe the model of communication that is established through web pages, based on rigor, rationality and criticism, promoting a certain capacity to express oneself, make oneself understood and act coherently, it is also congruent with the edges of telematics and its logical resources [6]. Today more than ever it is a theory that is widely accepted since communication has no time barriers, it can occur synchronously and asynchronously, so that the student can always be in communication with the teacher or classmates. To achieve the desired success it is essential the interaction of two or more people capable of language and interaction, with an action oriented on the basis of a mutual understanding to achieve an agreement that is to commit, such action must be given with the only force that gives the power of the argumentation. The interaction, which is considered as the communicative act, must establish an interpersonal relationship with a mutual understanding that is distinguished from the others. For this to be possible in the objective of the communicative action, shared knowledge must be involved, with an agreement based on the validity and mutual trust in the learner and educators [7].

The theory of situated knowledge, proposed by Young [8] this learning is based on the main value attributed to work experience (practice) and active practice in learning and understanding, privileging social learning over individual learning, i.e. social learning and the application of this learning in the student's daily life is very important. In other words, it is a teaching method based on the social construction of learning based on real situations, solving real and everyday problems, the main ideas for its application are: knowledge is contextual (we learn better immersed in reality and in relation to others), teaching should focus on real or everyday practices, learning through interaction between the teacher and peers (group work). Currently, according to this theory, the Internet is a means of learning, because it fosters innovative environments, conceives knowledge as an active relationship between a subject and the environment, so that learning occurs when the student is actively involved in a complex and realistic context.

Collaborative conversational model, proposed by García y Ramírez [9]. It arises from the new paradigm of e-learning, but includes a new actor and conversationalist this action facilitates that the internet participates in this type of learning because through this tool, conversation between groups is allowed, through networks or specific groups, where a forum for discussion or the traditional brainstorming and exchange of documentation can take place, among others. Some authors compare it to the model of the theory of communicative action, since groups can communicate through the Internet for collaborative work.

Jhon MCCarthy defined the term Artificial Intelligence in the 1950s as the science and ingenuity behind making intelligent machines, especially intelligent computer programs. Although AI in particular has been developed largely in the technology, robotics and financial sectors, it has also been proliferating in the field of education. In recent years, the branch that studies the implementation of AI in the field of education has been called AiEd (Artificial Intelligence in Education) [10], AiEd is nothing more than the result of the application of computer science, statistics, artificial intelligence and education; the interaction between these fields also leads to learning analysis, educational data mining and ICT-based education.

According to [11], IES (Intelligent Education Systems) refers to educational processes in which AI plays a fundamental role, such as intelligent tutors, game-based learning, among others. He also points out that AI will make it possible to eliminate barriers to access to education by reducing its economic costs and making education more accessible through the network.

[12] point out that the implementation of AI in education eliminates tedious and repetitive tasks that teachers used to deal with in educational processes, allowing them to focus on responding in a timely manner to students, focusing on personalized student learning. In turn, indicates that the main objective of AiEd should be to combine the best of the machine (AI) and the best of the human being for the benefit of the student. He argues that one of the main advantages of AiEd is how innovative and entertaining the use of this tool can be for the student [13], the main benefit of AiEd is the personalization of education according to the needs of the student, allowing the teacher to know firsthand even the mood of the students in the impact of the results of the subject.

Finally [14] highlight the increased presence of AI in education and the consequences it may have, because although it is true that the benefits of AI (AiEd) are known, the consequences of malpractice in teaching combined with AI have not yet been studied in depth. Similarly, it should be noted that no consensus has been reached, nor have guidelines been implemented for the use and implementation of AiEd in the classroom.

3 Methodology

One of the problems faced by students enrolled in an open and distance education program is the availability of time to achieve greater connectivity in the tutoring schedules proposed by their tutor, with the purpose of encouraging student participation in the tutoring schedules, the university connectivity to the virtual learning environment (VLE) project is proposed, which provides academic incentives to students who connect and actively participate in the spaces. Something similar happens with students enrolled in the face-to-face modality with respect to attendance and participation in activities in contact with the teacher (tutorials). To calculate academic improvement, the Propensity Score Matching (PSM) technique was applied. Propensity Score Matching (PSM) is intended to strengthen arguments about the cause-likelihood of relationships between variables in quasi-experimental studies. This is because PSM reduces selection bias, i.e., the bias that is generated because individuals cannot be randomly assigned to an intervention or a control group, to balance all those characteristics that can affect the estimation of the effect [15].

One of the advantages attributed to this technique is that the results obtained are very similar to the results obtained in real experiments, that is to say, the results obtained when we try to determine the influence of one or more independent variables against a dependent variable, but with the application of the PSM we can avoid the different difficulties (bias) that occur in a real experimental situation, In this particular case, we are talking about the fact that the students enrolled in the subject of principles of economics in the Open and Distance Modality and in the Presential Modality of the Universidad Particular de Loja present a series of previous characteristics such as the minimum knowledge required by the subject, internet coverage in the area of residence, knowledge and use of the AI, time dedicated to their studies, etc., which may affect the academic results of the students who are enrolled in the course. Which may affect the academic results that are of interest to the researcher.

The PSM has been used very successfully in the social sciences, specifically to determine the effect that social programs have had, with this technique research questions could be solved such as: what would the performance of students from low quality educational institutions be like if they were placed in high quality institutions? What would the performance of students from public institutions be like if they were placed in private institutions? How would academic performance be affected if students were placed in a pre-placement program?

The group to which the project was applied was the students of the Universidad Técnica Particular de Loja, Open and Distance and Presential Modality enrolled in the subject of principles of microeconomics during seven consecutive academic periods. Each parallel consisted of 50 students and they registered 14 parallels giving a total of 700 students, of which 200 students were those who remained as a control group.

The first step was to calculate the PS (propensity score) in this case the independent variables are of binary type, after having the propensity scores, these were used to make pairings between the treatment group and between the control groups, for the pairing the statistical software (R or SPSS) was used since they allow applying an exact pairing to the entire selected sample as applied in the research of [16].

The matching will be performed with the closest score. After completing the matching process, it is necessary to verify that there is no matching bias by using a histogram of the propensity score distribution, The same is done with the cases of the control group, it is also necessary to verify that there is a reduction in terms of the mean and standard differences (Cohen's d) in the propensity scores and covariates (pre- and post-matching comparison), this will allow us to be fully confident that the matching was successful.

Cases that are not matched are discarded and those that are matched are used as a new database and a statistical analysis, such as a regression model, is applied to them.

4 Results

In the present work, the role played by (AiEd) in education was put into practice in order to determine the academic performance of a group of students who participated in the University Connectivity to the Virtual Learning Environment (VLE) project. Specifically, the impact of the project was analyzed from a quantitative point of view.

With regard to the results using the PSM technique with data from students enrolled in the subject of principles of microeconomics, a statistically significant difference is obtained in the average academic performance between the group of students who will participate in the project and the students in the control group, but this does not allow us to conclude that there is an important qualitative change in terms of academic performance, since the absolute value of the differences in average are not significant.

There is little socialization on the part of teachers towards students regarding the use of the (AiEd), and the way it can be applied to the contents of the subjects. 76% of the students participating in the project state that they are unaware of the benefit they can have in their learning process and that it was on their own that they initiated its use and that it is limited to specific queries.

63% of the participants in the project consider that the AiEd, contributed significantly to their learning process, considering it an innovative learning process and making it attractive and fun. Likewise, they consider that it promotes learning based on the acquisition of skills that are related to the resolution of real problems, allowing them to relate the contents of the subject to their daily decision making.

One of the characteristics of the students who participated in the project is that, in addition to their student responsibilities, they have work responsibilities, which means that they have very little time to dedicate to the study of the subject (less than 1 h a day). Another characteristic is that the tutoring schedules proposed by the teachers are carried out during working hours, which makes it difficult to maintain communication with their tutor.

The poor connectivity available to students in rural areas is another factor that influences their decision not to participate in the project, since the lack of computer devices prevents them from having continuous access to the information available and shared by their tutor.

5   Conclusion

The implementation of AiEd in the subject of microeconomics in the open and distance modality of the Universidad Técnica Particular de Loja has shown that there is a statistically significant difference in the average academic performance between the group of students in which AI resources have been implemented within the teaching compared to the group of students who did not use it. Although this does not allow us to conclude that there is a qualitative change in academic terms; the absolute values of the average differences are not significant.

On the other hand, 76% of students indicate that they personally implemented the use of (AiEd), within their studies since there is little socialization of teachers towards students on these issues. This is related to Holmes, Bialik, Fadel (2019), i.e., there is no consensus or guidelines regarding the use and implementation of AI in the education of students.

63% of the project participants consider that AI has a positive impact within their learning processes. Which supports what is stated by Becker (2018) who refers to AIEd as a novel and Innovative process that gives the best of human and machine for the benefit of the learner.

Finally, in rural areas, due to poor connectivity, students were not able to take part in the project, which differs from what Sánchez and Lama (2007) mentioned, since the gap in access to connectivity in rural areas has not yet been closed, especially in developing countries.

References

1. Organización de las Naciones Unidas para la Educación, la Ciencia y la Cultura (Unesco). Estándares de Competencias en TIC para Docentes. (2009). Disponible en: http://www.unesco.org/en/competencystandards-teachers
2. Krüger, K.: El concepto de sociedad del conocimiento. Revista Bibliográfica de Geografía y Ciencias Sociales **9** (683) (2006). Disponible en: http://www.ub.edu/geocrit/b3w-683.htm
3. Castillo G.M.Y., Jiménez Puello, J.de J.: Las teorías de aprendizaje, bajo la lupa TIC. Acción Y Reflexión Educativa, (44), 144–158 (2019). Disponible en: https://revistas.up.ac.pa/index.php/accion_reflexion_educativa/article/view/693
4. Díaz-Barriga, F.: TIC en el trabajo del aula. Impacto en la planeación didáctica. Revista Iberoamericana de Educación Superior **4**(10), 3–21 (2013)
5. Cristobal, R., Sebastian, V.: Data mining in education (2012). Disponible en: https://doi.org/10.1002/widm.1075
6. Jordi, A.: Tendencias en educación en la sociedad de las tecnologías de la información. EDUTEC. Revista Electrónica de Tecnología Educativa **7** (2007). Disponible en; http://www.uib.es/depart/gte/revelec4.html
7. Grundy, S.: Producto o praxis del currículum. Ediciones Maroto Tercera edición Madrid (1991)
8. Brett, B.: Artificial Intelligence in Education; What is it, where is it now, where is it going? 43–46 (2018). Disponible en: https://www.brettbecker.com/wp-content/uploads/2019/04/Becker-AI-in-Education-with-cover-sheet.pdf
9. Ovalle Ramírez, C.: Sobre la técnica de Puntajes de Propensión (Propensity Score Matching) y sus usos en la investigación en Educación. Educación y Ciencia, **4**(43), 80–89 (2015)
10. Randolph, J.J., Falbe, K., Kureethara-Manuel, A., Balloun, J.L.: A step by step guide to Propensity Score Matching (2014). Disponible en: https://scholarworks.umass.edu/cgi/viewcontent.cgi?article=1330&context=pare
11. Sánchez Vila, E.M., Lama Penín, M.: Monografía: Técnicas de la Inteligencia Artificial Aplicadas a la Educación. Inteligencia Artificial. Revista Iberoamericana de Inteligencia Artificial, **11**(33), 7–12 (2007)
12. Chan, K.S., Zary, N.: Applications and challenges of implementing artificial intelligence in medical education: integrative review. JMIR Med Educ; **5**(1), e13930 (2019). Disponible en: https://mededu.jmir.org/2019/1/e13930
13. Zovko, V, Gudlin, M.: Artificial Intelligence as a Disruptive Technology in Education 2–6 (2019). Disponible en: https://conference.pixel-online.net/files/foe/ed0009/FP/5803-ENT3951-FP-FOE9.pdf
14. Ojeda, D., Gómez, R.B.: Álvaro. ¿Qué son las puntuaciones de propensión? **144**(3), 364–370 (2016). Disponible en: http://www.scielo.cl/scielo.php?script=sci_arttext&pid=S0034-98872016000300012&lng=es&nrm=iso>. ISSN 0034-9887. https://doi.org/10.4067/S0034-98872016000300012
15. Ramírez Valbuena, W.Á.: La inclusión: una historia de exclusión en el proceso de enseñanza aprendizaje. Cuadernos de Lingüística Hispánica, (30), 211–230 (2017). Disponible en: https://doi.org/10.19053/0121053X.n30.0.6195
16. Baldassarri, S.: EDUCAUSE Horizont Report. Teaching and Learning Edition, TE y ET, vol. 32, p. e14 (2022)

Ethnography of Tourism in Saraguro: Exploring the Dynamic Legacy of Sumak Kawsay in Local Culture and Heritage

Edison Javier Castillo-Pinta, Ochoa Jiménez Diego, and Pascual García-Macías[✉]

Universidad Técnica Particular de Loja (UTPL), Loja, Ecuador
daochoa@utpl.edu.ec, pasgegar84@gmail.com

Abstract. Exploring the Dynamic Legacy of *Sumak Kawsay* in Local Culture and Heritage investigates the intricate cultural fabric of Saraguro, a locale increasingly reliant on tourism for economic and cultural vitality. It centers on the *Sumak Kawsay* principle, an Andean ethos of 'good living', and its endurance in the face of burgeoning tourism. Employing participant observation and semi-structured interviews with locals and tourists, the research scrutinizes the interplay between tourism and indigenous cultural practices, focusing on their influence on cultural heritage perception and conservation.

The findings indicate that while tourism bolsters Saraguro's economy, it also serves as a conduit for promoting and revitalizing *Sumak Kawsay*. Nonetheless, the study identifies challenges, notably the risk of cultural commodification and the imperative for sustainable tourism practices that honor and maintain local traditions. The research underscores the necessity of a synergistic approach to tourism development, advocating for community involvement in decision-making to ensure tourism fortifies Saraguro's cultural essence. This study significantly advances the understanding of tourism as a nuanced, dynamic entity, deeply intertwined with and reflective of a community's cultural practices and perspectives.

1 Introduction

In the early 21st century, the concept of Buen-Vivir (Good Living) emerged as a transformative project within modern civilization, aiming to address the historical colonization and oppression of indigenous peoples. Manifesting under various nomenclatures such as Buen-Vivir or Vivir Bien, Sumak Kawsay in Quechua, and Suma Gamaña in Aymara, this concept is deeply rooted in indigenous worldviews. It represents a critical challenge to Western ideological constructs, offering a deconstruction of the philosophical underpinnings traditionally associated with the notion of 'good living' [1]. According to Hidalgo-Capitán and Cubillo-Guevara [2], contemporary debates around Sumak Kawsay span at least six thematic realms, encompassing its meaning, translation, origin, cultural references, its relation to development, and its historical context. Lalander and Cuestas-Caza [3] posit that both Sumak Kawsay and Buen-Vivir hold significant potential as foundational concepts for envisaging alternative societal structures and new paradigms of human-nature relationships.

Á. Rocha et al. (Eds.): WorldCIST 2024, LNNS 990, pp. 280–284, 2024.
https://doi.org/10.1007/978-3-031-60328-0_28

Over the past decade, ongoing academic discourse and historiographical examination have elevated these concepts as viable alternatives to conventional narratives, extending even beyond traditional development paradigms. Vanhulst [4] emphasizes that discourses surrounding Good Living particularly underscore the interdependence between society and the natural environment, symbolized by Pachamama or Mother Earth, and a conception of culture as a multifaceted reality. This perspective marks a fundamental departure from prevailing Western ideologies, notably the society-nature dualism and Eurocentric notions of unending progress toward well-being. Buen-Vivir's narratives challenge the concept of cultural and social homogeneity, recognizing the impracticality of such uniformity in a diversely characterized world. Instead, they advocate for harmony within the ethos of 'unity in diversity.' These discourses vary from radical, particularist views to more moderate, pluralistic approaches.

2 Methodology

The primary methodological framework of this strategically positioned ethnographic study was centered around participant observation, a deeply immersive and reflective technique. This approach enabled the systematic observation and analysis of practices, interactions, and discourses within the everyday lives of the subjects. By transcending mere superficial observation, this methodology facilitated a comprehensive and intricate comprehension of the participants' perceptions, emotions, and thoughts concerning contemporary events and circumstances. Moreover, the ethnographic research was augmented through the deployment of semi-structured interviews. These interviews, conducted concurrently with participant observation, were designed to investigate both the recent past experiences and current scenarios of the subjects in a reflective manner. The study engaged nine individuals from the gastronomic service sector, encompassing both employees and proprietors, to gather insights. Concurrently, from January to July 2023, fifteen residents of Saraguro were interviewed to assess and comprehend the feasibility and potential of Saraguro as a rural tourism destination.

The selection of interview participants was guided by the "snowball" sampling technique, which progressively enabled access to pivotal informants as fieldwork and interactions advanced. The determination of the number of interviews was grounded in the principle of qualitative data saturation, ensuring the collection of adequate information for a comprehensive and meaningful analysis. This approach was further strengthened by a triangulation process, encompassing not just the gathered data but also the diverse methodological techniques employed. This triangulation ensured enhanced validity and rigor in the research outcomes.

3 Qualitative Ethnographic Methodology with Semi-structured Interviews: A Holistic Approach in Social Research

Within the diverse array of research methodologies, the qualitative paradigm, particularly ethnographic methodology employing semi-structured interviews, is distinguished for its profound capability to elucidate human experiences, perceptions, and cultural dynamics.

Distinct from quantitative methodologies, which prioritize numerical data and statistical analysis, qualitative approaches emphasize exploring and interpreting the nuanced qualities, meanings, and subjective experiences inherent in social phenomena. This focus on qualitative analysis facilitates an in-depth exploration of the 'how' and 'why' behind social interactions, yielding a richly contextualized and dynamic understanding of the studied phenomena.

Ethnography, a technique with roots in anthropology yet extensively utilized across various social science disciplines, entails an intensive examination of groups and communities through participant observation and immersive engagement within the study context. This method enables researchers to authentically apprehend cultural practices, social interactions, and belief systems from an 'insider' perspective, thereby providing a nuanced and comprehensive comprehension of cultural and social environments.

In the specific context of Saraguro, semi-structured interviews played a pivotal role in the qualitative research methodology. Contrasting with the rigidity of structured interviews, semi-structured interviews offer a level of flexibility that is imperative for in-depth exploration of emergent themes, particularly relevant to understanding Saraguro as a rural tourist destination. This adaptability is crucial for investigating the intricate and multifaceted dimensions of human experience.

The deployment of ethnographic qualitative methodology, augmented by semi-structured interviews, was instrumental in this research focused on Saraguro. This approach facilitated a holistic investigation, extending beyond the scope of tourism to encompass local self-perceptions of their community and their relationship with the Sumak Kawsay concept. The methodology afforded a profound and intricate appreciation of the complexities embedded within human experiences and cultural contexts. Its application effectively surpasses the confines of quantitative methodologies, paving the way for a more inclusive and empathetic comprehension of human societies.

4 Rural Tourism

Rural tourism in Ecuador plays a crucial role in the country's tourism sector, offering authentic experiences in rural communities and natural environments. Defined as tourism in non-urban areas, it connects visitors with rural life and culture, including agriculture, crafts, and traditions. Ecuador's diverse regions, from the Andes to the Amazon, offer varied rural tourism experiences, showcasing the country's rich biodiversity and cultural mosaic of indigenous, Afro-Ecuadorian, and mestizo communities. This form of tourism is sustainable and community-oriented, generating alternative income for rural areas while preserving natural and cultural heritage. It contributes to diversifying local economies beyond traditional agriculture, enhancing quality of life, and supporting community development. Mediano and Domínguez & Lennartz suggest that rural tourism should focus on inland agricultural areas, offering traditional lifestyles and activities for tourists.

Nations focusing on rural quality of life often promote rural tourism, with policies supporting multifunctionality in agriculture. This approach includes remunerating farmers for various agricultural outputs and addressing land concentration due to technological advances. Moral-Moral et al. analyze rural tourism's impacts. Economically, it brings job creation, income improvement, and investment appeal, but also faces challenges like inflation and seasonal employment. Environmentally, it aids in conserving natural areas, but risks include environmental degradation and unplanned urban development. Socioculturally, it enhances life quality and preserves local customs, yet risks cultural dilution and conflicts between locals and tourists.

In summary, rural tourism in Ecuador is a sustainable, community-focused model that enriches the local economy and cultural landscape, while facing challenges in economic, environmental, and sociocultural dimensions.

5 Marketing and Digital Tourism

In Saraguro, Ecuador, the strategic application of digital tools, particularly the internet and digital platforms, is critical for enhancing tourism promotion and improving the tourist experience. This includes promoting Saraguro's attractions through specialized websites, social media, and online platforms, along with implementing online booking systems and interactive content like virtual tours. These innovations allow preliminary virtual engagement with destinations. Effective social media engagement is key for enabling tourists to share experiences and reviews, contributing to an organic promotional strategy. Maintaining current, accurate online information about local attractions and events is essential for providing relevant visitor information. Digital tourism marketing in Saraguro focuses on its rich cultural heritage, especially its indigenous population and their traditions. The marketing emphasizes unique cultural aspects, such as festivals, handicrafts, music, and cuisine, using various digital media platforms like social networks, blogs, and mobile apps. These platforms showcase Saraguro's attractions, narrate traveler experiences, and provide practical information, thus broadening the reach and enhancing understanding of the region's cultural identity.

The digital approach highlights experiences ranging from nature trails to participation in local traditions. Involving the local community in these marketing strategies is crucial to ensure benefits to residents and respect for cultural and environmental integrity. However, there is a need for more sophisticated digital marketing strategies in Saraguro. The current gap in robust approaches underscores the necessity for well-developed strategies that effectively promote tourism while aligning with community values and preserving cultural and ecological heritage.

6 Conclusions

In Ecuador, rural tourism emerges as a pivotal sector with multifaceted benefits, yet it confronts a spectrum of challenges that necessitate careful navigation. Among these challenges are the imperative to enhance infrastructural facilities, the commitment to adopting sustainable practices, and the responsibility to manage tourism's impact on local communities and the environment. Despite these hurdles, rural tourism presents substantial opportunities for fostering rural development, preserving cultural and environmental heritage, and crafting authentic, enriching tourism experiences.

A specific case in point is Saraguro, where digital tourism marketing plays a crucial role. This approach involves leveraging digital tools to position Saraguro as a premier tourist destination, one that values and safeguards its unique cultural and natural heritage. Central to this strategy is the emphasis on sustainable tourism practices, underscoring the importance of conserving Saraguro's environmental and cultural assets.

In conclusion, rural tourism in Ecuador stands at the crossroads of culture, nature, and sustainable development. It offers a valuable alternative to conventional tourism models, contributing significantly to the socioeconomic upliftment of rural communities. Furthermore, it enriches the visitor experience, providing a deeper insight into the diverse and rich tapestry of the country. This synergy of local development, cultural preservation, and environmental stewardship establishes rural tourism in Ecuador as a paradigm of sustainable and responsible travel.

References

1. Tavares, M.: A Filosofia Andina: uma interpelação ao pensamento Ocidental. Colonialismo, colonialidade e descolonização para uma interdiversidade de saberes (J. Estermann). EccoS – Rev Cient [Internet]. (32), 197–235 (2014). Disponible en: http://www.redalyc.org/articulo.oa?id=71530929012
2. Hidalgo-Capitán, A.L., Cubillo-Guevara, A.P.: Seis debates abiertos sobre el Sumak Kawsay. Íconos Revista de Ciencias Sociales [Internet] (1997). [citado el 10 de diciembre de 2023]. Disponible en: https://iconos.flacsoandes.edu.ec/index.php/iconos/article/view/1204/1103
3. Lalander, R., Cuestas-Caza, J.. Kawsay, S., Ecuador, B.-V.: Digitala Vetenskapliga Arkivet [Internet] (2017). [citado el 10 de diciembre de 2023]. Disponible en: https://www.diva-portal.org/smash/record.jsf?pid=diva2%3A1166918&dswid=-2255
4. Vanhulst, J.: El laberinto de los discursos del Buen vivir: entre Sumak Kawsay y Socialismo del siglo XXI. Polis [Internet] [citado el 10 de diciembre de 2023], **14**(40), 233–261 (2015). Disponible en: https://journals.openedition.org/polis/10727
5. Mediano Serrano, L., Vicente Molina, A.: Análisis del concepto de turismo rural e implicaciones de marketing. Boletín Económico de ICE [Internet]. [citado el 10 de diciembre de 2023] (2002). Disponible en: https://www.researchgate.net/profile/Maria-Vicente-Molina/publication/28120928_Analisis_del_concepto_de_turismo_rural_e_implicaciones_de_marketing/links/09e4150b8a03a8fa2d000000/Analisis-del-concepto-de-turismo-rural-e-implicaciones-de-marketing.pdf
6. Domínguez Gómez, J.A., Lennartz, T.: Vista de Turismo rural y expansión urbanística en áreas de interior. Análisis socioespacial de riesgos. Revista Internacional de Sociología [Internet]. [citado el 10 de diciembre de 2023] (2015). Disponible en: https://revintsociologia.revistas.csic.es/index.php/revintsociologia/article/view/615/658
7. Barrera, E.: Política agropecuaria: Multifuncionalidad y Turismo Rural van de la mano. Ecopuerto Buenos Aires [Internet]. [citado el 10 de diciembre de 2023] (2006). Disponible en: https://argentinambiental.com/wp-content/uploads/pdf/AA49-28-Turismo_Rural_Politica_Agropecuaria_Multifuncionalidad_Y_Turismo_Rural_Van_De_La_Mano.pdf
8. Moral-Moral, M., Fernandez-Alles, M.T., Sanchez-Franco, M.J.: Análisis del turismo rural y de la sostenibilidad de los alojamientos rurales. Revista ESPACIOS [Internet]. [citado el 10 de diciembre de 2023], **40**(01) (2019). Disponible en: https://www.revistaespacios.com/a19v40n01/19400103.html

12nd Workshop on Special Interest Group on ICT for Auditing and Accounting

A Guide to Identifying Artificial Intelligence in ERP Systems in Accounting Functions

Célia Rocha Santos[1]([⊠]) [iD], Graça Azevedo[2,3] [iD], and Rui Pedro Marques[2] [iD]

[1] University of Aveiro, Aveiro, Portugal
`celia.rsantos@ua.pt`
[2] Higher Institute of Accounting and Administration (ISCA-UA), University of Aveiro, Aveiro, Portugal
[3] CICF, IPCA, Barcelos, Portugal

Abstract. Artificial Intelligence (AI) is a topic that has been heard by many and fantasized about for decades. This study proposes a guide for semi-structured interviews to identifying AI in ERP systems in accounting functions. The guide presented in this study was tested by three different techniques, (1) internal testing, (ii) expert assessment and (iii) field-testing. With this guide we can evaluate the existence, the challenges and opportunities and the future trends of AI in the accounting functions of ERP systems.

Keywords: artificial intelligence · ERP · semi-structured interviews · interviews guide

1 Introduction

Artificial Intelligence (AI) has been a topic we have been hearing about for decades [1] and fantasizing about in the future [2]. Indeed, it is a fact that AI is fascinating [3]. It combines software and hardware to mimic human intelligence [4, 5] and enables the machine to learn from experience [5, 6]. In today's world companies need to combinate the forces of AI and human intelligence [7], which can lead to greater success [8]. Discussing organizational success necessitates examining Enterprise Resource Planning (ERP) [9].

ERP systems have been implemented in many organization [10, 11] to provide timely and consistent information across different functional areas of an organization [11]. The ERP has evolved into an essential component of any organization [9] and has seen significant advancements and modernization in recent years [12].

ERP is a unified, adaptable, and customizable information system that optimizes business processes by centralizing organization-wide data [13, 14]. Today, for ERP to remain a tool for creating value in companies, it needs to embrace digital innovations, allowing it to integrate technologies such as AI [15]. For the accounting profession, ERP systems and the technologies integrated within them represent a transformational influence [11]. It is clear that new technologies will impact accounting profession [16].

Á. Rocha et al. (Eds.): WorldCIST 2024, LNNS 990, pp. 287–295, 2024.
https://doi.org/10.1007/978-3-031-60328-0_29

Accountants play a crucial role in operating ERP systems to deliver timely and adaptable accounting information for decision-making [17]. Based on this, this study aims to propose a guide for semi-structured interviews to identify AI in ERP systems in accounting functions. The choice of semi-structured interviews is due to the value and richness they contribute to the data [18]. The use of semi-structured interviews also allows researchers to expand their horizons [19].

The literature on research interviews in general, and semi-structured interviews, in particular, is considerable and growing [20]. However, as qualitative research gains greater recognition and importance, it is essential to conduct it in a rigorous and systematically to produce meaningful and valuable outcomes [21]. Interviews constitute a valuable method in the scope of software engineering research [22] and accounting [23], because they help to understand practical work and processes [24].

Therefore, it is evident that there is a necessity to present guidelines and tools to support researchers in conducting reliable qualitative research [25]. The development of the semi-structured interview guide is a process that is rarely described in scientific articles, which makes it difficult to assess the success of the study methodology [21].

This work is divided into five sections. The next section refers to literature review, followed by the presentation of the study's methodology and results. Finally, some conclusions are presented.

2   Literature Review

2.1   Semi-structured Interview

Since the 1990s, that semi-structured interview has proliferated as an independent research method [26]. Now, they are widely used in qualitative research [24]. The semi-structured interview is a frequently used data collection approach [21], to obtain insights of the human experience [27]. It is a versatile and potent instrument for capturing people's perspectives and the way they derive meaning from their experiences [24, 28, 29]. The semi-structured interview enables focused interviews while granting the researcher the freedom to delve into pertinent ideas that may emerge during the interview [30, 31], this represents the semi-structured part of this technique [26]. Logically, a high-quality qualitative study grounded in semi-structured interviews depends on the researcher's expertise, abilities, insight, and ethical conduct during the analysis [28].

In semi-structured interviews the emphasis is in the interview guide, which encompasses a range of overarching themes to address during the interview, guiding the conversation towards the subjects and concerns of interest to the interviewees [29]. The same questions are made to participants in the same order, to ensure the data collected are comparable [26]; logically, before that stage, it is crucial to define the general purpose of the study [32]. A duration of approximately one hour is deemed a reasonable maximum length for semi-structured interviews to minimize fatigue for both the interviewer and the respondent [33].

2.2 Semi-structured Interview – Guide

The interview guide is a list with high-level topics that you plan to cover, along with questions that we want to answer in each topic [22]. This guide, crafted by the responsible researcher, determines the structure of the interview [34]. It is important to prepare a guide for semi-structured interview with primary questions or question stems, followed by sub-questions [26]. This stage is crucial for enhancing the trustworthiness of qualitative research in various ways, as it allows for the development of a robust semi-structured interview guide [21]. The guide usually is limited to one page so that it is easy to refer to [22] and to ensure that we are not losing the control of the interview [22, 32]. It is essential that these questions are phrased in an open-ended manner, designed to prompt unstructured responses, and foster meaningful discussions [26].

The process of creating the semi-structured interview guide help to focus and organize the line of thinking [22] and allow to identify the crucial information they aim to collect during the interview process [32]. It is important that questions are formulated to ensure that they are communicated correctly and clearly [26] based on literature about the theme [18].

How Elaborate a Guide?

To conduct semi-structured interviews is necessary to prepare the guide, write the questions items and test the guide [26]. In Table 1, we present the principles presented by McIntosh and Morse [26] for elaborating a semi-structured interviews guide. Nothing surpasses a solid grasp of the subject matter, so it is important verify the prior literature and work [28]; this will allow to identify the topic's domain, categories, and items [26].

Table 1. Principles to elaborate semi-structured interviews guide.

Prepare the interview guide	
Identifying the domain of the topic	Before constructing the interview, interviewers are already aware of the topics they intend to cover and may have some of the questions they plan to pose in mind
Identifying the categories	The domain is divided into categories based on their specific shared characteristics
Identifying the items	Items constitute the primary framework or question stems developed for the interview schedule
Writing the question items	Questions must be formulated to guarantee they are conveyed effectively and with clarity
Piloting the interview guide	
Testing	Testing enables the interview guide to be practiced in simulated conditions that closely resemble the actual process, facilitating adjustments before the primary data collection

There are important components to include in a guide: (i) establish how you are presented to the interviewed, and (ii) develop the questions and follow-up strategies

[28]. To incorporate the principles of an interview guide [26], there are some things that need to be considered: (i) avoid incorporating too many topics into the agenda; (ii) remember that closed-ended questions can serve as effective entry points for subsequent open-ended probing; (iii) use the language commonly used by the target groups, ensuring not to talk down to them in the process; and (iv) when composing the initial question order, attempt to anticipate the most probable and seamless sequence [33].

How Validate a Guide?

Designing and refining the guide is the most important key to successful interviews [19]. Once we have a guide, it is evident that testing it is extremely important [21, 26, 28, 33, 35], as this could help to improve it [28]. Thus, if possible, pilot tests should be conducted with a few intended respondents to help refine the guide [33].

The pilot test of the interview guide could be conducted using three different techniques: (1) internal testing, (ii) expert assessment and (iii) field-testing [21].

Internal Testing

Internal testing refers to the assessment of the initial interview guide by researchers from the research team [21]. After completing the draft interview guide, it is essential to try it out with a colleague to see if the questions are coherent and transition smoothly from one to the next [18].

Expert Assessment

Expert assessment involves soliciting feedback to enhance the interview guide [33]. It includes subjecting the initial interview guide to scrutiny and evaluation by external experts not affiliated with the research team [21]. Consulting with experts in the field, either from the study area or those specializing in qualitative studies, will help you gain confidence in the prepared guide [28].

Field-Testing

Field testing is a method where the initial interview guide is evaluated by potential study participants [21]. In this case, a few intended respondents assist the interviewer in improving their guide [33] and identifying any necessary final changes [36]. This step is recommended to ensure the guide's quality [18].

After the interview guide has been field-tested and adjusted as necessary, the interview process can begin [37]. Although thorough preparation for piloting might seem excessive, it will ultimately pay off at the study's conclusion [18].

3   Methods

This study has the main goal to present a definitive guide for semi-structured interviews. Through this, the authors intend to present a guide capable of identifying where AI is present in the accounting functions of ERP systems. This will allow for the identification of challenges, opportunities, and future directions for AI in ERP systems.

Based on the literature reviewed previously, the methodology of this study followed the activities presented on Table 2. First, we need to identify the domain topics, the categories, and the items [26]. To do this, we followed the steps presented by McIntosh and Morse [26]. We began with a literature review on the terms AI and ERP, which allowed us to determine what would be included in the semi-structured interview guide.

After identifying the content of the interview, it is necessary to write the draft guidelines and then test them. The guide testing involved three phases, as shown in Table 2. Firstly, we conducted an internal test with the research team. The internal test results in some alterations because some of the questions were found to be repetitive. It also helped to make the questions clearer.

Secondly, we contact two experts in ERP, AI, accounting, or qualitative studies to obtain external feedback, which provided more assurance regarding the guide's perception.

Finally, we conducted a field test with a potential study participant. This step validated the guide's comprehensibility by someone who works in the field.

Table 2. Methodology of the study

Activities	Based on Authors
Identifying the domain of the topic of interviews	[26]
Identifying the categories of the interviews	[26]
Identifying the items of the interviews	[26]
Writing the draft of semi-structured interviews guide	[26, 28, 32]
Testing the interview guide Internal testing Expert assessment Field-testing	[21]

All these activities combined have led to the results presented in the following section.

4 Results

The guide that will be presented in this section was constructed after a literature review on the terms ERP and AI [28]. In accordance with McIntosh and Morse [26], we agree that this is the only way to define the domain topic. We have identified three categories/sections to divide our guide and then the items that constitute each one.

We aimed to incorporate only the essential topics into the guide [33], and after test the guide [21] we present the guide bellow.

4.1 Interview Guide Proposal

About the interviewee:
  What is your role associated with ERP?
  On a scale of 1–10, how would you rate your knowledge related to accounting?
  On a scale of 1–10, how would you rate your knowledge related to AI?
  What training do those who provide ERP services undergo?
  Section 1: AI in ERP Systems

1. *What are the main recent advances related to AI in your accounting ERP system?*
2. *Which AI technologies are used, and in what features?*
3. *How do these technologies benefit accountants?*

Section 2: Challenges and Opportunities

4. *How do you convince users who are resistant to change? How do you show the opportunities and benefits of using AI in ERP?*
5. *What is requested to be implemented in the ERP (IA) to improve processes?*
6. *What challenges have been identified when implementing AI in ERP systems for accountants?*
7. *How are you addressing these challenges to ensure a smooth and effective transition?*
8. *Are there limitations or areas where AI is not yet as effective as desired?*

Section 3: Future Trends

9. *Where is AI taking the ERP evolution?*
10. *Which AI capabilities are believed to have a major impact on accounting in the future?*

Figure 1 represents the steps and the estimated time for each interview. Initially, it is necessary to contextualize the main goal of the interview and provide a summary to the interviewee. In step two, we aim to obtain more information about the interviewee, such as technical knowledge and academic background. We estimate that these two steps will take about 10 min. Following the guide, we move on to section 1 of the guide to learn about the existing AI in ERP systems. Then, in section 2, we address the challenges and opportunities of AI in ERP. It is estimated that each of these two sections will require 15 min. Section 3, which should take about 10 min, allows us to discuss the future trends of AI in ERP systems. To conclude the interview, it is important make a brief summary, thank the interviewee for sharing their knowledge and appreciate their time.

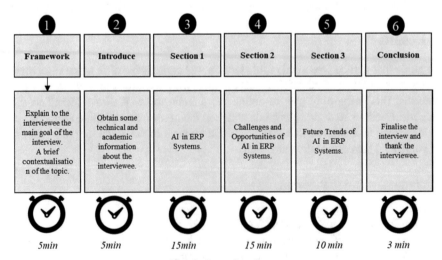

Fig. 1. Interview Steps

5   Conclusion

In this work, our purpose was to highlight how to elaborate and validate a guide for semi-structured interviews. The intention is that this guide will enable us to evaluate the use of AI in the accounting functions of the ERP.

The use of semi-structured interviews in qualitative research is very common. However, the importance of constructing an objective and executable guide make the difference in the outcomes of the interviews. Thus, it is extremely important to test the guide to ensure that it fits the main goal of the interviews.

The guide presented in this study was tested through three tests (Internal testing, Expert assessment, and Field-testing) [21]. These tests helped to refine the guide and allowed to present a comprehensive guide for evaluating the existence, challenges, opportunities and future trends of AI in ERP accounting functions.

This study could serve as a reference for other researchers in constructing and validating a guide for semi-structured interviews.

"This work is financed by national funds through FCT - Foundation for Science and Technology, I.P., within the scope of multi-annual funding UIDB/04043/2020."

References

1. Dwivedi, Y.K., et al.: Artificial Intelligence (AI): multidisciplinary perspectives on emerging challenges, opportunities, and agenda for research, practice and policy. Int. J. Inf. Manage. **57**, 101994 (2021). https://doi.org/10.1016/j.ijinfomgt.2019.08.002
2. Hautala, J., Heino, H.: Spectrum of AI futures imaginaries by AI practitioners in Finland and Singapore: the unimagined speed of AI progress. Futures **153**, 103247 (2023). https://doi.org/10.1016/j.futures.2023.103247
3. Knox, A., Bass, N., Khakoo, Y.: You can run but you can't hide: artificial intelligence is here. Pediatr. Neurol. **147**, 163–164 (2023). https://doi.org/10.1016/j.pediatrneurol.2023.03.010
4. Askary, S., Abu-Ghazaleh, N., Tahat, Y.A.: Artificial intelligence and reliability of accounting information. In: Al-Sharhan, S.A., et al. (eds.) I3E 2018. LNCS, vol. 11195, pp. 315–324. Springer, Cham (2018). https://doi.org/10.1007/978-3-030-02131-3_28
5. Lee, C.S., Tajudeen, F.P.: Usage and impact of artificial intelligence on accounting: evidence from Malaysian organisations. Asian J. Bus. Account. **13**(1), 213–239 (2020). https://doi.org/10.22452/ajba.vol13no1.8
6. Chapinal-Heras, D., Díaz-Sánchez, C.: A review of AI applications in Human Sciences research. Digit. Appl. Archaeol. Cult. Herit. **30**, 00288 (2023). https://doi.org/10.1016/j.daach.2023.e00288
7. Lichtenthaler, U.: Beyond artificial intelligence: why companies need to go the extra step. J. Bus. Strateg. **41**(1), 19–26 (2020). https://doi.org/10.1108/JBS-05-2018-0086
8. Makridakis, S.: The forthcoming Artificial Intelligence (AI) revolution: its impact on society and firms. Futures **90**, 46–60 (2017). https://doi.org/10.1016/j.futures.2017.03.006
9. Goundar, S., Nayyar, A., Maharaj, M., Ratnam, K., Prasad, S.: How artificial intelligence is transforming the ERP systems. In: Goundar, S. (ed.) Enterprise Systems and Technological Convergence: Research and Practice, pp. 85–98. Information Age Publisher (2021)

10. Sarno, R., Sinaga, F., Sungkono, K.R.: Anomaly detection in business processes using process mining and fuzzy association rule learning. J. Big Data **7**(1), 1–19 (2020). https://doi.org/10. 1186/s40537-019-0277-1

11. Grabski, S.V., Leech, S.A., Schmidt, P.J.: Review of ERP research: a future agenda for accounting information systems. J. Inf. Syst. **25**(1), 37–78 (2011). https://doi.org/10.2308/jis.2011. 25.1.37

12. El Hairech, O.E., Lyhyaoui, A.: The new generation of ERP in the era of artificial intelligence and Industry 4.0. In: Kacprzyk, J., Balas, V.E., Ezziyyani, M. (eds.) AI2SD 2020, vol. 1417, pp. 1086–1094. Springer, Cham (2022). https://doi.org/10.1007/978-3-030-90633-7_96

13. Malik, M.O., Khan, N.: Analysis of ERP implementation to develop a strategy for its success in developing countries. Prod. Plan. Control **32**(12), 1–16 (2021). https://doi.org/10.1080/095 37287.2020.1784481

14. Kanellou, A., Spathis, C.: Accounting benefits and satisfaction in an ERP environment. Int. J. Account. Inf. Syst. **14**(3), 209–234 (2013). https://doi.org/10.1016/j.accinf.2012.12.002

15. Katuu, S.: Enterprise resource planning: past, present, and future. New Rev. Inf. Netw. **25**(1), 37–46 (2020). https://doi.org/10.1080/13614576.2020.1742770

16. Santos, C., Azevedo, G., Marques, R.P.: The impact of new technologies on the accounting profession. Int. J. Bus. Innov. **2**(2), e32718 (2023). https://doi.org/10.34624/IJBI.V2I2.32718

17. Nguyen, T.H., Nguyen, Q.T., Vu, L.D.H.: The effects of accounting benefit, ERP system quality and management commitment on accountants' satisfaction. Accounting **7**(1), 127–136 (2021). https://doi.org/10.5267/J.AC.2020.10.005

18. Sankar, P., Jones, N.L.: Semi-structured interviews in bioethics research. in empirical methods for bioethics: a primer. In: Jacoby, L., Siminoff, L.A. (eds.), pp. 117–136. Emerald Group Publishing Limited, Bingle (2007). https://doi.org/10.1016/S1479-3709(07)11006-2

19. Horton, J., Macve, R., Struyven, G.: Chapter 20 - qualitative research: experiences in using semi-structured interviews. In: Humphrey, C., Lee, B. (eds.) The Real Life Guide to Accounting Research, pp. 339–357. Elsevier, Oxford (2004). https://doi.org/10.1016/B978-008043 972-3/50022-0

20. Brown, A., Danaher, P.A.: CHE Principles: facilitating authentic and dialogical semi-structured interviews in educational research. Int. J. Res. Method Educ. **42**(1), 76–90 (2019). https://doi.org/10.1080/1743727X.2017.1379987

21. Kallio, H., Pietilä, A.M., Johnson, M., Kangasniemi, M.: Systematic methodological review: developing a framework for a qualitative semi-structured interview guide. Wiley Online Library **72**(12), 2954–2965 (2016). https://doi.org/10.1111/jan.13031

22. Bird, C.: Interviews. In: Menzies, T., Williams, L., Zimmermann, T. (eds.) Perspectives on Data Science for Software Engineering, pp. 125–131. Morgan Kaufmann, Boston (2016). https://doi.org/10.1016/B978-0-12-804206-9.00025-8

23. Dai, N.T., Free, C., Gendron, Y.: Interview-based research in accounting 2000–2014: informal norms, translation and vibrancy. Manag. Account. Res. **42**, 26–38 (2019). https://doi.org/10. 1016/j.mar.2018.06.002

24. Rowley, J., Jones, R., Vassiliou, M., Hanna, S.: Using card-based games to enhance the value of semi-structured interviews. Int. J. Mark. Res. **54**(1), 93–110 (2012). https://doi.org/10. 2501/IJMR-54-1-093-110

25. Nowell, L.S., Norris, J.M., White, D.E., Moules, N.J.: Thematic analysis: striving to meet the trustworthiness criteria. Int. J. Qual. Methods **16**(1), 1–13 (2017). https://doi.org/10.1177/ 1609406917733847

26. McIntosh, M.J., Morse, J.M.: Situating and constructing diversity in semi-structured interviews. Global Qual. Nurs. Res. **2**, 1–12 (2015). https://doi.org/10.1177/233339361559 7674

27. Bearman, M.: Focus on Methodology: eliciting rich data: a practical approach to writing semi-structured interview schedules. Focus Health Prof. Educ. Multi-Prof. J. **20**(3), 1–11 (2019). https://doi.org/10.11157/fohpe.v20i3.387

28. Rabionet, S.E.: How i learned to design and conduct semi-structured interviews: an ongoing and continuous journey. Qual. Rep. **16**(2), 563–566 (2011)

29. Qu, S.Q., Dumay, J.: The qualitative research interview. Qual. Res. Account. Manag. **8**(3), 238–264 (2011). https://doi.org/10.1108/11766091111162070

30. Adeoye-Olatunde, O.A., Olenik, N.L.: Research and scholarly methods: semi-structured interviews. J. Am. College Clin. Pharm. **4**(10), 1358–1367 (2021). https://doi.org/10.1002/jac5.1441

31. Brinkmann, S.: The interview. In: Denzin, N.K., Lincoln, Y.S. (eds.) The SAGE Handbook of Qualitative Research, 5th edn., pp. 997–1038. Sage Publications Inc., Los Angeles (2018)

32. Mahat-Shamir, M., Neimeyer, R.A., Pitcho-Prelorentzos, S.: Designing in-depth semi-structured interviews for revealing meaning reconstruction after loss. Death Stud. **45**(2), 83–90 (2021). https://doi.org/10.1080/07481187.2019.1617388

33. Adams, W.C.: Conducting semi-structured interviews. In: Newcomer, K.E., Hatry, H.P., Wholey, J.S. (eds.) Handbook of Practical Program Evaluation, 4th edn, pp. 492–505. Jossey-Bass (2015). https://doi.org/10.1002/9781119171386.ch19

34. Corbin, J., Morse, J.M.: The unstructured interactive interview: issues of reciprocity and risks when dealing with sensitive topics. Qual. Inq. **9**(3), 335–354 (2003). https://doi.org/10.1177/1077800403009003001

35. Aung, K.T., Razak, R.A., Nazry, N.N.M.: Establishing validity and reliability of semi-structured interview questionnaire in developing risk communication module: a pilot study. Jurnal Ilmiah Pendidikan **2**(3), 600–606 (2021). https://doi.org/10.51276/edu.v2i3.177

36. Naz, N., Gulab, F., Aslam, M.: Development of qualitative semi-structured interview guide for case study research. Compet. Soc. Sci. Res. J. **3**(2), 42–52 (2022)

37. Magaldi, D., Berler, M.: Semi-structured interviews. In: Zeigler-Hill, V., Shackelford, T.K. (eds.) Encyclopedia of Personality and Individual Differences, pp. 4825–4830. Springer, Cham (2020). https://doi.org/10.1007/978-3-319-24612-3_857

Reshaping the Accountant's Future in the Era of Emerging Technologies

Ana Ferreira[1] (iD) and Isabel Pedrosa[1,2(✉)] (iD)

[1] Polytechnic Institute of Coimbra, Coimbra Business School, Quinta Agrícola – Bencanta, 3045-231 Coimbra, Portugal
ipedrosa@iscac.pt
[2] Polytechnic Institute of Porto, CEOS.PP, Porto, Portugal

Abstract. With the advent of the digital age, the accounting profession has undergone a transformation. Adapting to these changes represents an opportunity to showcase accountants' flexibility and perseverance, rather than a mere new challenge. Approaching the new landscape with eagerness and self-assurance will empower accountants to flourish and achieve their professional goals. Emerging Technologies refer to Robotic Process Automation, Artificial intelligence, Big Data and analytics, Cloud, and Blockchain. This study aims to present and analyze the impacts of using emerging technologies on the accounting profession through an exploratory study and interviews with accounting professionals. The results of the interviews revealed that the relationship between emerging technologies and the role and skills of accountants has not yet been established. This work identifies the skills that accountants must detain to be successful, the new professions that will emerge associated with accounting, and activities and processes that will change. The insights presented in this study are highly valuable for those involved in the accounting profession, including professional bodies, regulators, and educational institutions. Researching the use of emerging technologies in the context of accounting may yield valuable knowledge that could inform future modifications to university curricula and instructional approaches for accounting educators.

Keywords: Emerging Technologies · Robotic Process Automation · Accountant · Accounting Education · Skills · Artificial intelligence · Automation · Big Data · Blockchain

1 Introduction

Emerging technologies (ET) are not only reshaping the daily lives of people worldwide but also redefining the roles of accountants. The rapid pace of technological progress is making this transformation increasingly evident. Several factors have played a part in these transformations, such as swift technological advancements, growing globalization, enhanced internet communication, and legal and regulatory reforms [1]. According to [2] in the future, the accountancy profession will continue to be important for companies and organizations in all sectors. With the advancement of technology, it is possible that some routine and repetitive tasks, such as coding accounting entries and compliance

Á. Rocha et al. (Eds.): WorldCIST 2024, LNNS 990, pp. 296–305, 2024.
https://doi.org/10.1007/978-3-031-60328-0_30

with financial reporting standards, will be automated through artificial intelligence (AI) and machine learning (ML) to provide predictive analytics for compliance assisting specialists.

Accounting has evolved significantly in recent years, with technology playing a crucial role in shaping the new perspective and advancement of accountants. Therefore, it is essential to have both technical and Information Technology (IT) skills to become a better-prepared Chartered Accountant. Reviewing the literature on the framework of accounting profession, overview of ET and understanding the profile and skills of accountants in the digital age are crucial for this research.

An important topic in the current conjuncture, this study aims to: (SO1) identify the ET on accounting, (SO2) identify accountants' familiarity with ET, (SO3) identify the skills and roles of future accountant, (SO4) identify the role of accounting education in the future of the profession, (SO5) identify the advantages and disadvantages of ET in the profession, (SO6) identify expectations/concerns in the adoption of ET and (SO7) identify tasks/processes that will change by technology adoption. For this purpose, an exploratory study of a qualitative nature will be carried out, which consists of the analysis of a literature review and semi-structured interviews. In this study, are present the point of view of different accounting professional profiles.

Following this introduction, Sect. 2 provides an overview of the relevant literature. Section 3 outlines the chosen methodological approach, including the procedures used for data collection and analysis. The results of the study are presented and discussed in Sect. 4. Finally, the main conclusions are summarized in the concluding section along with a discussion of the limitations of the study and future research opportunities on the topic.

2 Literature Review

Disruptive technologies in accounting represent a new phase of accounting affected by ET that are part of the industrial revolution 4.0. New accounting technologies have potential to disrupt work but require a revolution in thought to fully realize their potential [1]. In recent years, accounting has undergone substantial evolution. It is so important to review the literature on the framework of the accounting profession, identify ET and understand the profile and skills of the accountant in the digital age.

2.1 Framework of the Accounting Profession

A technological evolution has transformed your day-to-day life, simplifying your work, streamlining processes, and automating procedures, and requiring them to radical change in the exercise of their functions, as well as a quick ability to adapt [3]. The accounting profession is changing from less emphasis on recording operations to more involvement in management support and decision-making [4].

The accountancy profession in Portugal has enjoyed a great boost and recognition, especially since the 90s. However, the road to this point has not been easy. The regulator body in Portugal is the *Ordem dos Contabilistas Certificados* (OCC), which places responsibility for the initial training of prospective accountants on higher education institutions. Referring to the standards of the International Federation of Accountants (IFAC), the OCC defines areas in which accountants should develop during their apprenticeship, based on which higher education institutions create their Curriculum. These IFAC standards were recently revised, and technical components were added. However, in the newly updated Profile for Accountants, the OCC defines dominant areas, such as financial accounting, taxation, management accounting and ethics and deontology that enter the profession, without reference to IT [3].

Most of Portuguese companies are Small and Midsize Enterprises (SMEs), which has an impact on the work of the accountant, on the diversity of clients, on the rapid changes and tax changes, on the role in the financial literacy of SMEs owners, many of them family members or small entrepreneurs. And in turn, the profile of the accountant in Portugal is essentially the Chartered Accountant who performs the compliance from A to Z, from the person who picks up the papers from the client to the person responsible for tax reporting.

2.2 Overview of Emerging Technologies in Accounting

Looking to the future, technology is impacting both competencies and mindsets. In the finance function of the future, the technical capabilities of robotics and algorithms combine with the creativity and empathy of human accountants [5]. By using technology, companies can access real-time data and make better decisions. This means that finance professionals must have a wide range of skills, not just crunching numbers, but also interpreting and examining data.

Among the technological innovations driving change, highlight the disruptive potential of cloud accounting, Big Data and data analytics, AI and ML, and Blockchain, describing the accounting industry as one of the few that is constantly innovating and adapting [6]. [7] point out Big Data and advanced analytics tools, cloud computing, and AI as technologies offering new opportunities and solutions to enhance and adapt accountant's ability to work through the complex and rapid journey of this digital age.

Digitalization is a major and lasting change in society, affecting many areas of life. Cloud-based accounting software is changing traditional bookkeeping, optimizing document flow and processing time. It offers opportunities for expanding services, efficiency gains, and extra profits [8] for small and medium-sized accounting practices.

Accountants could use visualization tools and social media to communicate insights from complex data sets to the public and co-create responses, potentially elevating their role in organizational communication [9]. Expanding accounting data sets enables detailed transaction analysis, integration of internal/external data, soft integration of environmental Big Data, and transformation of accounting, business, and audit processes [10].

Blockchain can revolutionize accounting by securely storing and sharing data, increasing verifiability, and enabling real-time reporting to interested parties [11]. Blockchain technology simplifies secure and cost-effective transfer of value, making it useful for transactions requiring multiple validations and record validation. It can also detect fraud and errors and facilitates direct trading in the private sector [12]. Despite Blockchain can improve accounting accuracy and speed, but also poses a threat to the traditional role of accountants and auditors [13].

Automation technology is simple, fast, and accurate. It reduces working cycles and is widely used in accounting [14]. Automation will replace manual accounting tasks with technology and robots, freeing up accountants to focus on strategy and analysis [15]. RPA is ideal for repetitive digital tasks across multiple systems with minimal human interaction, but not for paper-based or externally changing processes [16]. As per [24] RPA has the potential to revolutionize the way boring and repetitive tasks are performed. Accountants now focus on value-added tasks like financial management [12]. Robots and AI assistants are ubiquitous and changing work and life rapidly [17]. New technologies, particularly AI, will transform accounting tasks and roles [17]. For example, companies can automate activities like accounts payable and receivable, financial reporting and analysis, regulatory compliance, and audit procedures by utilising RPA technology [15]. IT is changing financial reporting. Accounting professionals must possess advanced IT skills in today's fast-paced digital world[18].

2.3 Shaping the Accountant for the Future

There is no doubt that the accounting professions will change because of the digital revolution [18]. Accountants have undergone a significant transformation in recent years, moving from manual accounting to computerized accounting in business [19].

[6] predicted that almost a quarter of jobs (23%) are expected to change in the next five years and that the fastest-declining roles relative to their size today are driven by technology and digitalization, with data entry clerks, accounting, bookkeeping and payroll clerks. The first impact on the accounting profession will be changing tasks or the way they are performed. These changes may impact accountants' functions, they may have to complete existing tasks differently or even perform new tasks. As the accountant's role is closely related to the tasks performed and the functions assumed, all these changes may ultimately affect the accountant's role [1].

The digital age has brought new challenges and opportunities for accountants. It has transformed the way accountants approach their work, shifting their focus from looking backward to looking forward, and encouraging them to calculate and forecast the future. Accountants are no longer solely focused on resolving issues related to back-end accounting processes. They now can propose solutions and drive strategic decision-making [20]. Developing digital skills is crucial for accountant [8]. Future accountant roles will require a unique combination of technical and interpersonal skills. Those with the right skills and mindset are well positioned to seize the opportunities that technology presents and lead their organizations into the future. To equip graduates with the essential skills needed to thrive in the industry, it is imperative to restructure accounting education in accordance with industry requirements [14]. Educators can confidently instil a philosophy of lifelong learning and empower students to adapt to any situation, ensuring they are prepared to tackle the challenges and opportunities of the future [21].

3 Methodology

Formulating the research question represents the initial and crucial phase in the research process and must be important, specific and without obvious or routine answers [22]. Thus, the research question for this study is: What is the impact of ET on the work of accounting professionals?

To provide an answer to the research question, the following specific research questions were defined: RQ1: Which ET is most often studied in accounting field? RQ2: What are the roles and necessary skills for accountants in the organization? RQ3: How can accountants improve the quality of their work and leverage the benefits of technology? A qualitative exploratory study was conducted using semi-structured interviews to gain deeper insights into the subject. Based on the literature review, the interview questions were created and organized into a guiding script.

Data Collection

The data in this research study were collected through individual interviews with actively practicing accounting professionals. These professionals were deliberately selected from diverse backgrounds, industries, and positions. The responses were analyzed and categorized based on similarities, and the final section of this report organizes these findings within the sample population. In this section, were synthesized key themes from the results to elucidate the respondents' perspectives on current ET and their implications for the accounting profession.

The interviews were held from March to April 2023 and invitations were made via phone/WhatsApp/email, as well as scheduling. The invited professionals are from different generations, genders, and areas of residence, as shown in Table 1. The interviews were conducted via zoom, with an average duration of approximately 45 min. All interviews were only attended by the researcher and the interviewee. The interviews were audio-recorded through the zoom platform, and the researcher indicated their confidential nature and only for data processing use, with no objection from the interviewees. The interviews were semi-structured and the questions were developed to research, as summarize in Table 2. After all the interviews, conducted in Portuguese language, the process of transcription, translation, analysis and interpretation of the data was initiated.

4 Analysis and Discussion of Results

In this chapter, the results of ten interviews conducted at the time of data collection of this study are presented. The analysis of the interviews implies the synthesis and categorization of the answers given by the interviewees. This analysis, supported by the MAXQDA software, was considered useful in the coding and categorization of the interviews and facilitated the integration of the interviewees' perceptions, and makes it possible, with greater reliability, to conduct the matrix analyses that are presented below for each research question.

According to the interviewees, technology is increasingly present in the day-to-day of companies, with the robotization and automation of processes. Visualizing data in dashboards allows to predict the future with more certainty. The integration of robots

Table 1. Summary Overview of Participants

Participant Code	Age	Gender	Residence	Training Area	Profile	Business Sector	Interview Date
I1	41	M	Coimbra	Accounting	Chartered Accountant	SME Accounting firm	2023-03-01
I2	34	F	Barreiro	Accounting	Teacher and trainer	Higher Education Institution	2023-03-02
I3	33	F	Alverca do Ribatejo	Accounting and Administration	Lead accountant	Software Company	2023-03-02
I4	46	M	Luanda	IT management	Software consultant	Software Company	2023-03-04
I5	48	M	Santa Maria da Feira	Economics	Chartered Accountant and economist	Financial and Tax Consultancy	2023-03-09
I6	43	M	Lisboa	Management	Manager	Consulting Big Four	2023-03-24
I7	53	F	Coimbra	Accounting and Audit	Chartered Accountant	SME Accounting firm	2023-03-28
I8	21	M	Marinha Grande	Accounting and Audit	Accounting Student	Higher Education Institution	2023-04-06
I9	23	F	Évora	Accounting	Accountant	SME Accounting firm	2023-04-11
I10	46	M	Leiria	Business Sciences (Accounting)	Teacher	Higher Education Institution	2023-04-13

Table 2. Summary of interview scripts

Section	Objective
Ethical and Legal Considerations	Presentation of the objectives of the interview and issues related to confidentiality and data disclosure
	Identifying the interviewee
Auto-Perceptions and Personal Opinions	Familiarity with emerging technologies
	Motivations for the adoption of technology
Experience	Challenges and adjustments faced during the accounting practices
	Identify incorporating technology in accounting work
	Limitations/difficulties and advantages and disadvantages of implementing ET
Training and Skills	Skills for accountants in the digital era
	Training and experience in digital transformation
Perceived Contributions	Incorporating technology for improved efficiency given the full picture of Portuguese companies
	Continuous learning and accounting education
Expectations	Impact of technology on the accounting profession, taking into account factors such as company size
	Expectations of the profession in the face of digital transformation

can assist in streamlining processes, but human oversight remains necessary for ensuring data accuracy. In general, interviewees suggested adapting the undergraduate education system to meet the demands of the digital age and the workforce. One of the interviewees said that research in education is one step behind the economy when it comes to knowing and using cutting-edge technology. Other interviewee mentioned that Universities should produce professionals with broader perspectives and stronger competencies, rather than narrow specialists with little practical knowledge.

Even other interviewees mentioned that if nothing is changed in the teaching of undergraduate degrees, the student will become technologically illiterate. In addition, the interviewees mentioned the challenge of introducing subjects such as programming, data security, big data, blockchain and AI in core degrees and cutting less relevant skills is discussed. These statements are aligned with the literature, since the gap is not new, and it is time to accounting programs must change to keep up with evolving entry-level positions. Failure to do so risks graduates being unable to find jobs or perform as expected [23]. Regarding the personal skills, it is important to have an open mindset and not be limited by rules, and to maintain values of compassion and empathy even in a technological age.

In the interviews, it was possible to verify that there are several perspectives on the profession. Some interviewees enjoy their work, while others view accounting as a steppingstone to more engaging professions, such as management or auditing. Accountants must be attentive to the needs of clients and businesses and offer services in addition to accountability to the tax authority. It is important to sell the importance of information to customers and show the medium-term potential of the paradigm shift. Cooperation between accountants and technology consultants is key to adapting to change, and specialization in business management and tax consulting can offer differentiated services. Through the interviews, it was noted that interviewees acknowledge the potential for technology to elevate the role of accountants. No longer solely tasked with tax delivery, accountants can utilize cutting-edge technology to streamline data processing, automate tasks, and take on new responsibilities previously unattainable.

The main results of the interviews are summarized in Table 3. Overall, the interviews results are in line with the literature, there are certainly potential downsides of the adoption of technology in accounting [9], however the human-machine cooperation is crucial for a promising future. Those with the right skills will thrive [18].

Table 3. Interviews resume

Research Question	Specific Objective	Extracts from Interviews	Resume
RQ1	(SO1) (SO2)	"...I know the concepts, what impacts they're going to have on the profession, but if you ask me specifically, I understand everything about all of these topics, I don't. And I need training, which I think is the big challenge" I10	Knowledge in Robotics, Cloud computing, and OCR. Digitalization and automation are used with the ERP systems. Lack of knowledge in blockchain, Big data and on the potential of the ERP system
RQ2	(SO3)	"... the accountant has to be very adaptable, in the sense that any problem that may arise, he can always find a solution, even if sometimes it is a bit complicated." Code: Soft Skills I8 "it would be the obligation of the faculties to produce future professionals with broader visions and with more reinforced skills in certain aspects. When entry-level workers go out into the market (...) are not aware of the development or use of computer tools that they may have on a daily basis, therefore, I think that there is still very little interaction between what universities are and what companies need." I6	Digital Skills: Excel, Dashboards, programming, data security, Big data, blockchain RPA AI; Soft Skills: communication, open mindset, emotional intelligence, problem-solving. The accountant would be more advisor, more IT accounting specialist
	(SO4)	"I absolutely agree. Teaching must be changed (...) at this moment we are suffering from a paradigm that the university, academic study, was always at the forefront, it is not what happened in practice today. Today we have the Big 4 at the forefront of academic study, then the academic study what the Big 4 are doing and then pass it on to the others, I mean, there is a triangle here, a bit weird." I5	Lack of interaction between universities and companies. Accounting Curricula should include digital skills but also soft skills. Entry-level will have more difficulty adapting and finding a job. Ongoing learning and training solutions
RQ3	(SO5)	"It's going to make the role of the accountant more attractive (...) it's going to take away some of that label of not being the boring people on the financial side, the boring ones, the ones who always have the pencil behind their ear (...) But for the people who want it." I3 "I have clients who need me more to sit down with them and think about the company in the medium and long term than complying with accounting and tax obligations, which I think is the focus of most of my colleagues' work." I1	Advantages: provide more value to their clients; specialization in business management and tax consulting Disadvantages: Lack of digital skills; cost and time consuming of implementation in the SMEs; dependence on the ERP or RPA consultant in improving and solving problems
	(SO6)	"The accountant has to turn to a more management area, taxation and support companies in managing clients' businesses. That is, you have to change this chip, if they continue to do the same, they will be out of work in a while, in my opinion." I4	Expectations: decrease of boring and repetitive tasks, opportunity to turn the profession more interesting; embrace technology, and improve processes. Concerns about losing jobs, accountants over at age of 50 may face some resistance to change; Cybersecurity, and safeguard sensitive financial data
	(SO7)	"If the accountant is able to use digital tools, the time that will be left will be useful for planning, so that clients have more tax optimization, can better parameterize their systems and have enough data for later at the end of the year, when they make compliance have a more accurate compliance." I2	Bookkeeping will be automated, coding accounting entries and compliance with financial reporting standards, will be automated through AI and ML to provide predictive analytics for compliance. Automatically filling in forms by Robots.

5 Conclusions

It arises from the analysis of the results that the link between ET and accountants' role and skills has not yet been made. Moreover, considering the literature review and the context of Portuguese companies, the Chartered Accountants in Portugal appear to be a step backward.

From the analysis of this study the impact of ET on the work of accounting professionals is that the accounting profession is changing its role and functions with the rise of ET as AI, Big Data, Cloud Computing and RPA. The relationship between accountants and technology is seen to make the accountant's job easier and provide more value to clients. Supporters see it as progress, while opponents fear many accountants will struggle to adapt.

Automating routine tasks allows accountants to devote more time to data analysis and tax planning, as well as offering consulting services in financial planning and business management. The standardization of digital tools is essential to facilitate the work of accountants, but it is important to have strong technical support for externally acquired applications and technologies. Familiarity with technology is essential to the evolution of the accounting professional.

Digital transformation and automation have improved the quality of life of accountants, as it allows the reduction of repetitive, non-value-added tasks, reducing hours and hours of work. The robots require organizations to rethink work, support employees, and invest in ongoing skills for success.

To excel in accounting, you need technical skills, clear communication, high ethical standards, up-to-date knowledge of legislation, and strong analytical skills. Build a foundation of technical knowledge and keep developing new skills to meet the evolving market demands for success. It is crucial to acknowledge that entry-level positions have been changing rapidly for the past decade. Therefore, it is essential to update the accounting programs to prevent them from becoming obsolete. It is needed to make these changes before accounting graduates face the challenge of not being able to find entry-level positions or struggling to meet their employer's expectations after being hired.

The insights presented in this study are highly valuable for those involved in the accounting profession, including professional bodies, regulators, and educational institutions. The findings emphasize the importance of adapting standards to the current digital landscape, as failure to do so may lead to underutilization of existing tools by professionals.

Although this study has made valuable contributions, it is important to acknowledge its limitations. The current coverage of accounting technologies is limited in scope as it only caters to a few types, while other types remain unaddressed. As a result, it could not provide detailed descriptions for each one of them. This suggests that further research is required to develop a more thorough comprehension of accounting technologies. Despite the diverse range of interviewees, it is important to note that this study has a limitation in that the sample selected by the researcher is a convenience sample as opposed to a random sample. Although the sample includes a diverse range of participants, the non-random selection process could affect the study's findings' generalizability.

The research explores the increasing interest in ET and proposes that these technologies be studied through collaborative research involving both academia and industry practitioners. Collaborative research endeavors have the potential to advance the evolution of professional accountants' roles and set the foundation for the next generation of professional engagement standards.

Acknowledgements. This work received financial support from the Polytechnic Institute of Coimbra within the scope of Regulamento de Apoio à Publicação Científica dos Estudantes do Instituto Politécnico de Coimbra (Despacho n.° 5545/2020).

References

1. Kroon, N., Do Céu Alves, M., Martins, I.: The impacts of emerging technologies on accountants' role and skills: connecting to open innovation-a systematic literature review. J. Open Innov. Technol. Mark. Complex. **7**(3) (2021). https://doi.org/10.3390/JOITMC7030163
2. Atanasovski, A., Tocev, T.: Research trends in disruptive technologies for accounting of the future – a bibliometric analysis. J. Account. Manag. Inf. Syst. **21**(2), 270–288 (2022). https://doi.org/10.24818/jamis.2022.02006
3. Cristina, A., Covas, D., Carlos, J., Ribeiro, S., Jorge, V., Soares, S.: V. Contabilidade 5.0: o perfil do profissional de contabilidade. In: XX Congress International AECA, pp. 22–23 (2019)
4. Almeida, A.C., Carvalho, C.: As futuras competências essenciais do contabilista: um estudo empírico. In: CICA XVIII–Congresso Internacional de Contabilidade e Auditoria, pp. 1–32 (2020)
5. Chartered Global Management Accountant [CGMA]. Changing competencies and mindsets, pp. 1–12 (2018). https://www.cgma.org/content/dam/cgma/resources/reports/downloadable documents/changing-competencies-mindsets-cgma.pdf
6. Kokina, J., Davenport, T.H.: The emergence of artificial intelligence: how automation is changing auditing. J. Emerg. Technol. Account. **14**(1), 115–122 (2017). https://doi.org/10.2308/jeta-51730
7. Ibrahim, S., Yusoff, W.S., Rashid, I.M.A.: A systematic review of disruptive technology within accounting and accounting sector. AIP Conf. Proc. **2339** (2021). https://doi.org/10.1063/5.0044297
8. Stancheva-Todorova, A.: Are accounting educators ready to embrace the challenges of Industry 4.0. Int. Sci. J. "Industry 4.0" **4**(6), 309–312 (2019). http://www.accaglobal.com/big data
9. Marrone, M., Hazelton, J.: The disruptive and transformative potential of new technologies for accounting, accountants and accountability: a review of current literature and call for further research. Meditari Account. Res. **27**(5), 677–694 (2019). https://doi.org/10.1108/MEDAR-06-2019-0508
10. Vasarhelyi, M.A., Kogan, A., Tuttle, B.M.: Big data in accounting: an overview. Account. Horizons **29**(2), 381–396 (2015). https://doi.org/10.2308/ACCH-51071
11. Dai, J., Vasarhelyi, M.A.: Toward blockchain-based accounting and assurance. J. Inf. Syst. **31**(3), 5–21 (2017). https://doi.org/10.2308/isys-51804
12. Zhang, Y., Xiong, F., Xie, Y., Fan, X., Gu, H.: The impact of artificial intelligence and blockchain on the accounting profession. IEEE Access **8**, 110461–110477 (2020). https://doi.org/10.1109/ACCESS.2020.3000505

13. Schmitz, J., Leoni, G.: Accounting and auditing at the time of blockchain technology: a research agenda. Aust. Account. Rev. **29**(2), 331–342 (2019). https://doi.org/10.1111/auar. 12286
14. Fernandez, D., Aman, A.: Impacts of robotic process automation on global accounting services. Asian J. Account. Gov. **9**, 123–132 (2018). https://doi.org/10.17576/ajag-2018-09-11
15. Kaya, B., Türkyılmaz, C.T., Birol, M.: Impact of RPA technologies on accounting systems. Muhasebe ve Finans. Derg. **82** (2019). https://doi.org/10.25095/mufad.536083
16. Kokina, J., Blanchette, S.: Early evidence of digital labor in accounting: innovation with robotic process automation. Int. J. Account. Inf. Syst. **35**, 100431 (2019). https://doi.org/10. 1016/J.ACCINF.2019.100431
17. Leitner-Hanetseder, S., Lehner, O.M., Eisl, C., Forstenlechner, C.: A profession in transition: actors, tasks and roles in AI-based accounting. J. Appl. Account. Res. **22**(3), 539–556 (2021). https://doi.org/10.1108/JAAR-10-2020-0201
18. Kruskopf, S., Lobbas, C., Meinander, H., Söderling, K., Martikainen, M., Lehner, O.: Digital accounting and the human factor: theory and practice. ACRN J. Financ. Risk Perspect. **9**(1), 78–89 (2020). https://doi.org/10.35944/JOFRP.2020.9.1.006
19. Damasiotis, V., Trivellas, P., Santouridis, I., Nikolopoulos, S., Tsifora, E.: IT competences for professional accountants. a review. Procedia - Soc. Behav. Sci. **175**, 537–545 (2015). https:// doi.org/10.1016/J.SBSPRO.2015.01.1234
20. Pan, P.S., Seow, G.: Preparing accounting graduates for digital revolution: a critical review of information technology competencies and skills development. J. Educ. Bus. **91**(3), 166–175 (2016). https://doi.org/10.1080/08832323.2016.1145622
21. Zhan, A., Dai, J., Vasarhelyi, A.: The impact of diruptive technologies on accounting and auditing education. Eos, Transactions American Geophysical Union. Accessed 30 Jan 2023. https://www.cpajournal.com/2018/09/13/the-impact-of-disruptive-technolog ies-on-accounting-and-auditing-education/
22. F.N., de Souza, S.D.N.: Formular questões de investigação no contexto do corpus latentena internet. Internet Latent Corpus J. **2**(1). https://proa.ua.pt/index.php/ilcj/article/view/14887/ 10234. Accessed 08 Oct 2023
23. Sangster, A.: Revolutionising the accounting curriculum in higher education: a vision of the future. Acc. Manag. Rev. Rev. Contab. Gestão. **26**(1), 49–75 (2022). https://doi.org/10.55486/ amrrcg.v26i.3a

Factors Influencing Statutory Auditors' Perception of the Role of Artificial Intelligence in Auditing

Joana Nogueira, Davide Ribeiro⬡, and Rui Pedro Marques^(✉)⬡

Higher Institute of Accounting and Administration, University of Aveiro, Aveiro, Portugal
ruimarques@ua.pt

Abstract. Artificial Intelligence (AI) is a rapidly advancing field that aims to create intelligent entities capable of performing tasks that would typically require human intelligence. It has become an integral part of our lives, impacting various sectors and industries, offering immense potential and posing challenges, such as on its responsible and ethical use. The integration of AI in auditing has the potential to enhance the effectiveness and efficiency of audit processes. Thus, this study examines the familiarity, utilization, and perceptions of AI technologies among Portuguese Statutory Auditors through a questionnaire survey and contributes to the understanding of how these factors influence the auditing profession. The survey design is aligned with the research questions aimed at capturing a demographic profile while assessing AI technologies' usage and perception. Descriptive and inferential statistics, including hypothesis testing, analyze the survey data to discern trends and influencers on auditors' engagement with AI. The results reveal disparities between auditors' awareness and actual application of AI in audit practices, identifying a prevalent underutilization, notably with exceptions such as Robotic Process Automation which exhibited the highest occasional use. Further statistical analysis reveals that gender, age, and educational qualifications significantly influence both the frequency of use and the importance ascribed to AI technologies in audit tasks. Age emerges as the primary influencing factor, although gender and educational qualifications were significant for some technologies.

Keywords: Artificial Intelligence · Auditing · Influencing Factors · Statutory Auditors

1 Introduction

AI encompasses machines or computers demonstrating behaviors akin to human intelligence, such as cognitive automation, machine learning, natural language processing (NLP), and the intentional alteration of algorithms [1]. These technologies perform tasks that typically require human cognition, and the field aims to build and comprehend intelligent agents. The Institute of Internal Auditors acknowledges AI as a significant technological progression, prompting organizations to invest in systems that automate and enhance, or even mimic, human intelligence, producing insights and analyses potentially exceeding human capability [2].

© The Author(s), under exclusive license to Springer Nature Switzerland AG 2024
Á. Rocha et al. (Eds.): WorldCIST 2024, LNNS 990, pp. 306–316, 2024.
https://doi.org/10.1007/978-3-031-60328-0_31

Despite the potential benefits, the current use of AI in auditing is limited, leaving many auditors unprepared to integrate these technologies into their work or to identify AI-specific risks. The auditing profession is set to evolve, with significant changes anticipated in the roles, experiences, and skills required of auditors. To stay relevant and serve their clients effectively, auditors must embrace continuous learning and adapt to technological advancements [3].

This empirical study is designed to ascertain the extent of auditors' knowledge of AI technologies, the frequency with which they utilize these tools in their audit practices, and their views on the significance of AI in enhancing audit processes. Data will be collected through a structured questionnaire survey, aiming to shed light on the integration of AI in the auditing domain.

The paper is structured into five sections, beginning with this introduction. The next section offers a literature review with a focus on AI's impact on the auditing field. This is followed by a detailed description of the methodology employed in the study. The subsequent section presents and discusses the findings. The article concludes with a section that provides final reflections and considerations.

2 Theoretical Framework

2.1 Artificial Intelligence

AI is a multidisciplinary field aimed at creating systems that perform tasks requiring human intelligence, such as decision-making, problem-solving, language comprehension, and pattern recognition. The essence of AI is the development of machines that can learn, reason, and adapt to their environment. Russell and Norvig [4] outline four main approaches to AI: thinking humanly, thinking rationally, acting humanly, and acting rationally, emphasizing the significance of autonomous reasoning and learning in machine development.

Deep learning has propelled AI forward by enabling machines to learn from experience using layered neural networks. This approach allows systems to process data through multiple levels of abstraction, resulting in improvements in computer vision, speech recognition, and NLP [5]. AI systems can achieve problem-solving and learning without replicating human intelligence, as shown by reinforcement learning, where systems learn and refine their behavior through trial and error [6].

Floridi et al. [7] stress the importance of an ethical framework for AI, suggesting that AI development should prioritize the common good and adhere to human values and ethics, especially as AI becomes more prevalent in sectors like healthcare, transportation, and education. Marcus and Davis [8] advocate for AI systems that are reliable, safe, and truly understand the world, critiquing current AI for its errors and advocating for AI that can generalize knowledge across various contexts.

Combining these perspectives, AI is the pursuit of creating machines that exhibit human-like intelligence, capable of learning, reasoning, and autonomously making decisions. It involves crafting algorithms and models for tasks like facial recognition or language translation and extends to general-purpose technologies adaptable to various tasks. AI spans technologies and methodologies that enable intelligent machine behavior, driven by goals such as optimizing reward signals, processing extensive data, and adhering to ethical standards. It stands at the cutting edge of science and technology, with profound implications for the future.

2.2   Artificial Intelligence in Auditing

Since its introduction to auditing in the early 1980s, AI has evolved from rudimentary systems to sophisticated tools integral to the field. AI now aids in risk assessment, transaction testing, work paper preparation, identifying financial statement distortions, ensuring tax compliance, detecting fraud, and supporting decision-making. Auditors are increasingly trusting AI, leveraging advanced algorithms to enhance the efficiency and effectiveness of their work (Munoko et al., 2020). Despite AI's origins in the 1940s, its classification as an emerging technology is due to its swift advancement and expansion in application. Stahl et al. [9, 10] suggest that ongoing AI developments are poised to have significant social impacts. Manita et al. [11] argue for a paradigm shift in audit firms towards embracing digital technologies. These advancements not only transform audit processes but also free up auditors from repetitive tasks, enabling them to focus on more complex, judgment-intensive activities.

Puthukulam et al. [12] acknowledge the integral role of technology in business and daily life, with AI being a transformative force in auditing. AI's capacity to precisely process large datasets allows for more efficient identification of patterns, anomalies, and risks, optimizing audit effectiveness. Automation of routine tasks by AI tools enables auditors to allocate more time to complex, judgment-reliant tasks. Additionally, AI aids in fraud detection by scrutinizing vast amounts of data for irregularities, thus bolstering the audit's capability to uncover fraudulent activities that might be missed manually. This contributes to more robust and efficient fraud detection mechanisms within the auditing process [13].

According to the AICPA [14], the role, experience, and even the skills of auditors will inevitably undergo reformulations. This does not imply that auditors will be replaced by AI technologies; on the contrary, auditors will not be replaced by automation, but rather there will be changes in how they conduct audits.

PwC [15] identifies three mechanisms through which AI can drive gains: (1) Assisted AI Systems: These systems help in decision-making, forecasting, and process automation, tackling routine and repetitive tasks to bolster audit efficiency and reduce errors. Assisted AI allows auditors to minimize prediction errors and enhance decision-making accuracy [15, 16]; (2) Augmented AI Systems: these aim to boost the cognitive capabilities of auditors, refining the decision-making process and utilizing high-quality data to detect relevant information for problem-solving. They encourage the automation of existing tasks and support auditors in more complex, previously not automated work. (3) Autonomous AI Systems: These operate with a high degree of independence, making decisions and performing tasks without human intervention. In auditing, such systems could potentially conduct entire audits or significant parts thereof autonomously, though this application is more prospective and involves complex considerations regarding accountability and professional judgment [13, 16].

The adoption of AI in auditing presents ethical challenges, particularly regarding data privacy and security. Auditors deal with sensitive information, making it crucial to adhere to data protection regulations and maintain strong cybersecurity measures.

Upholding confidentiality and public trust is essential. It is important to consider how AI systems are trained and the biases they may inherit from data. Transparency in AI-driven decisions is necessary to ensure accountability in the auditing process. Overall, addressing these challenges is crucial for ethical AI adoption in auditing.

3 Methodology

This study seeks to enhance our understanding of artificial intelligence in the field of auditing. To meet its goals, it addresses three key research questions:

Q1: What are the key factors that influence the level of familiarity Statutory Auditors have with AI?
Q2: What factors contribute to the extent of Statutory Auditors' utilization of AI?
Q3: What are the factors that shape Statutory Auditors' perception regarding the significance of AI in performing the audit function?

The study surveyed Portuguese Statutory Auditors to assess the extent of AI technology integration within their firms and to examine their perceptions of its impact on the auditing field. The data were collected in the first semester of 2022 and were analyzed using SPSS software. The questionnaire, with ensured data protection, includes three sections: 1) Profile Characterization of the respondent auditors, 2) Characterization of their Auditing Firms, and 3) Adoption of AI Technologies in Auditing, to gather data on respondents' backgrounds, firms' operations, and attitudes towards AI in audit practices. The study used descriptive and inferential statistics for data analysis, including hypothesis testing with Pearson's Chi-Square test.

4 Presentation and Analysis of Results

This section analyzes the survey results, starting with the demographic profiles of respondents using descriptive statistics, then addressing the research questions.

4.1 Sample Characterization

The study's sample was drawn from the 2022 list of active Statutory Auditors in Portugal, totaling 1372. The survey was emailed to 1365 auditors after removing invalid email contacts. Out of 169 responses received, 100 complete responses were used for the final sample after excluding 69 incomplete ones.

The sample's age distribution is as follows: 40–49 years (36%), 50–59 years (33%), 60+ years (21%), and under 30 years (10%). The gender split is 70% male. Educational background shows 59% with an undergraduate degree, 30% with a Master's, 7% with a Doctorate, and 4% with other qualifications. Management (49%), Accounting (34%), and Economics (31%) are the main fields of study, with the remaining 10% in Finance, Auditing, Engineering, Public Administration, and Law.

The tenure of Statutory Auditors in the sample is as follows: 47% have over 20 years of service, 35% have 10–19 years, 7% have 5–9 years, and 11% have less than 5 years.

Most respondents work in companies with a single auditor, up to a maximum of 300 auditors, averaging 4 auditors per company. Employment distribution is 53% in nationwide firms, 39% in local/regional firms, and 28% in multinational firms.

The client company sizes primarily served by Statutory Auditors in the sample are as follows: 68% work mostly with Medium-sized Enterprises, 59% with Small companies, 23% with Micro-enterprises, and 20% serve Large Companies.

4.2 Analysis and Discussion of Results

This section is dedicated to a detailed analysis of the results with the objective of providing specific answers to each of the posed research questions.

Q1 - What Are the Key Factors that Influence the Level of Familiarity Statutory Auditors have with AI?
Table 1 reveals that the four AI technologies most recognized by auditors are Voice Recognition (69%), Image Recognition (63%), Robotic Process Automation (RPA) (57%), and Machine Learning (ML) (51%). Notably, a minority (14%) of respondents are not familiar with any AI technologies.

Table 1. AI Technologies that respondents are familiar with.

AI Technologies	Absolute Frequency	Relative Frequency
Voice Recognition	69	69%
Image Recognition	63	63%
Robotic Process Automation	57	57%
Machine Learning	51	51%
Process Mining	40	40%
Computer Vision	23	23%
Artificial Neural Networks	22	22%
Deep Learning	21	21%
Natural Language Processing	19	19%
Virtual Agents	16	16%
None	14	14%
Others	1	1%

To answer Q1 on the factors affecting auditors' familiarity with AI technologies, a Chi-Square test was used, focusing on gender, age, qualifications, and auditing experience. The analysis is limited to the AI technologies respondents reported familiarity with, as shown in Table 1.

The hypotheses considered for the Chi-Square test application are as follows:

H_0: The familiarity with AI technologies among Statutory Auditors is not influenced by gender, age, educational qualifications, and the number of years of experience as a Statutory Auditor.

H_1: The familiarity with AI technologies among Chartered Accountants (CAs) is influenced by gender, age, educational qualifications, and the number of years of experience as a Statutory Auditor.

The Chi-Square test's assumptions were not met; therefore, the results presented in Table 2 are based on Fisher's exact test. The rejection of the null hypothesis (H_0) hinges on the significance (Sig.) values, which require careful consideration for each variable analyzed.

Table 2. Chi-Square test statistic value (Sig.) on factors that can influence auditors' familiarity with AI technologies.

AI Technologies	Gender	Age	Qualifications	Experience as Auditor
Machine Learning	0,081	0,032	0,013	0,503
Deep Learning	0,003	0,262	0,160	0,676
Artificial Neural Networks	0,003	0,876	0,009	0,379
Process Mining	0,504	0,550	0,010	0,460
Natural Language Processing	0,051	0,703	1,000	0,548
Computer Vision	0,068	0,106	0,816	0,363
Voice Recognition	0,482	0,305	0,950	0,165
Image Recognition	0,498	0,482	0,761	0,308
Virtual Agents	0,138	0,342	0,588	0,558
Robotic Process Automation	0,385	0,359	0,469	0,813
None	0,345	0,790	0,427	0,399
Others	1,000	0,310	0,040	1,000

Based on Table 2, since most significance (Sig.) values are above 0.05, the null hypothesis (H_0) is generally not rejected. However, in instances where the Sig. value is less than or equal to 0.05, the null hypothesis (H_0) is rejected for those specific cases:

1. Gender significantly impacts the familiarity with Deep Learning, Artificial Neural Networks and NLP.
2. Age is a factor affecting familiarity with Machine Learning.
3. The familiarity with Machine Learning, Artificial Neural Networks, and Process Mining, is influenced by educational qualifications of the Statutory Auditor.

Summarizing the findings in response to research question 1, it is clear that certain AI technologies' familiarity among Statutory Auditors is influenced by independent variables including gender, age, and qualifications.

Q2 - What Factors Contribute to the Extent of Statutory Auditors' Utilization of AI?

To determine the frequency of use of the AI technologies that auditors identified as being familiar to them, the Statutory Auditors were queried, asking them to choose one of the following response options, on a frequency scale, for each AI technology they indicated they were familiar with: Never, Rarely, Occasionally, Frequently, and Very Frequently.

Table 3 indicates that although auditors are familiar with various AI technologies, the most common frequency of use reported for nearly all technologies (except "Other") is "Never." This suggests a gap between awareness and practical application in audit work. Notably, Robotic Process Automation (RPA) stands out as the technology with the highest percentage (22%) indicating at least occasional use.

Table 3. Frequency of use of AI Technologies in audit work

AI Technologies	Never	Rarely	Occasionally	Frequently	Very Frequently
Voice Recognition	54,0%	4,0%	9,0%	2,0%	0,0%
Image Recognition	39,0%	10,0%	6,0%	6,0%	2,0%
Robotic Process Automation	28,0%	7,0%	11,0%	10,0%	1,0%
Machine Learning	34,0%	8,0%	5,0%	4,0%	0,0%
Process Mining	17,0%	11,0%	5,0%	5,0%	2,0%
Computer Vision	18,0%	3,0%	2,0%	0,0%	0,0%
Artificial Neural Networks	21,0%	1,0%	0,0%	0,0%	0,0%
Deep Learning	16,0%	1,0%	0,0%	4,0%	0,0%
Natural Language Processing	13,0%	2,0%	3,0%	1,0%	0,0%
Virtual Agents	12,0%	1,0%	3,0%	0,0%	0,0%
Others	0,0%	1,0%	0,0%	0,0%	0,0%

To analyze the factors that may influence the frequency with which Statutory Auditors use AI technologies in their audit work, the hypotheses considered when applying the Chi-Square test are:

H_0: The frequency with which Statutory Auditors use AI technologies in their audit work is not influenced by gender, age, educational qualifications, and number of years of experience as a Statutory Auditor.

H_1: The frequency with which Statutory Auditor use AI technologies in their audit work is influenced by gender, age, educational qualifications and number of years of experience as a Statutory Auditor.

In this scenario, since the Chi-Square test conditions were not met, Fisher's exact test results are presented in Table 4. The analysis of Table 4 shows that in most instances, the significance (Sig.) value is above the 0.05 threshold. Consequently, the null hypothesis (H_0) is retained for these variables.

Table 4. Chi-Square test statistic value (Sig.) on factors that can influence the frequency of use of AI technologies by auditors.

AI Technologies	Gender	Age	Qualifications	Experience as Auditor
Machine Learning	0,148	0,036	0,549	0,992
Deep Learning	0,464	0,008	0,378	0,177
Artificial Neural Networks	1,000	0,158	0,561	0,096
Process Mining	0,501	0,757	0,710	0,683
Natural Language Processing	0,278	0,025	0,662	0,081
Computer Vision	1,000	0,224	0,899	0,139
Voice Recognition	0,153	0,028	0,050	0,615
Image Recognition	0,035	0,072	0,451	0,681
Virtual Agents	0,085	0,051	0,687	0,188
Robotic Process Automation	0,843	0,396	0,723	0,887

However, in eight instances where the Sig. value is less than or equal to 0.05, H_0 is rejected. This suggests that the corresponding independent variables have a statistically significant impact, which can be characterized as follows:

4. Age is a significant factor affecting the frequency of use of machine learning, deep learning, voice recognition, image recognition and virtual agents in audit work.
5. The adoption of voice recognition technology in audit processes is influenced by both the age and educational qualifications of the Statutory Auditor.
6. Gender significantly impacts the frequency of use of image recognition AI technology in audit tasks.

Considering the preceding analysis and addressing research question Q2, it is evident that age predominantly influences the frequency with which Statutory Auditors use certain AI technologies in audit work, with gender and educational qualifications also exerting an influence in specific instances.

Q3 - What are the Factors that Shape Statutory Auditors' Perception Regarding the Significance of AI in Performing the Audit Function?

For the Q3, respondents used a Likert scale to rate the importance of AI technologies for audit task performance. Table 5 ranks AI technologies according to the importance level attributed by respondents, with the mode of the responses determining the order. The percentages reflect the proportion of responses relative to the total number of respondents familiar with each technology.

Table 5. Importance given to AI technologies.

AI Technologies	Importance	Absolute Frequency	Relative Frequency
Robotic Process Automation	Very Important	20	35,1%
Image Recognition	Important	19	30,2%
Process Mining	Important	15	37,5%
Virtual Agents	Important	6	37,5%
Computer Vision	Important	8	34,5%
Machine Learning	Reasonably Important	19	37,3%
Deep Learning	Reasonably Important	8	38,1%
Voice Recognition	Slightly Important	19	27,5%
Natural Language Processing	Slightly Important	6	31,6%
Artificial Neural Networks	Not Important	7	31,8%

Similarly, the hypotheses considered when applying the Chi-Square test are:

H_0: Auditors' perception of the importance of AI technologies in the execution of audit work is not influenced by gender, age, educational qualifications, or the number of years of experience as a Statutory Auditor.
H_1: Auditors' perception of the importance of AI technologies in the execution of audit work is influenced by gender, age, educational qualifications, and the number of years of experience as a Statutory Auditor.

The results presented in Table 6 are also based on Fisher's exact test because the conditions necessary for the application of the Chi-Square test were not met.

Based on Table 6, since the significance value (Sig.) is mostly above 0.05, the null hypothesis (H_0) is not rejected in most cases. However, for the five instances where the Sig. value is less than or equal to 0.05, the null hypothesis (H_0) is rejected, indicating that the independent variable in question has a statistically significant influence on the dependent variable in those cases:

1. The auditors' perception of the importance of Process Mining (PM) in conducting audit work is influenced by the gender and the age of the Statutory Auditor.
2. The perception of auditors regarding the importance of NLP in performing audit tasks is influenced by the age of the Statutory Auditor.
3. The auditors' perception of the relevance of Image Recognition and Robotic Process Automation in audit task execution is influenced by the educational qualifications of the Statutory Auditor.

We conclude that the perception of auditors on the importance of AI technologies in the execution of audit work is, indeed, affected by the independent variables of gender, age, and educational qualifications for certain AI technologies.

Table 6. Chi-Square test statistic value (Sig.) on factors that can influence auditors' perception regarding the importance of AI technologies in carrying out audit work.

AI Technologies	Gender	Age	Qualifications	Experience as Auditor
Machine Learning	0,227	0,877	0,766	0,561
Deep Learning	0,333	0,174	0,328	0,559
Artificial Neural Networks	1,000	0,695	0,769	0,495
Process Mining	0,037	0,016	0,767	0,382
Natural Language Processing	0,591	0,030	0,825	0,558
Computer Vision	0,783	0,274	0,951	0,894
Voice Recognition	0,108	0,210	0,275	0,172
Image Recognition	0,070	0,898	0,016	0,187
Virtual Agents	0,600	0,061	0,295	0,391
Robotic Process Automation	0,847	0,117	0,015	0,374

5 Final Considerations

The research examines the determinants of AI adoption and perception among Statutory Auditors, highlighting the profession's need to adapt to technological advancements. Key findings indicate that age, gender, and educational background significantly influence auditors' usage frequency and perceived importance of AI in auditing, suggesting these demographics may hinder or enhance AI integration into auditing practices.

The survey of Portuguese Statutory Auditors provides insights into the adoption of AI in auditing, revealing a trend where younger and more educated auditors are more receptive to AI technologies. This suggests a generational shift towards increased efficiency and accuracy in auditing practices. The paper emphasizes the necessity for education and training to bridge the identified gaps, especially for older auditors and those less educated in AI, and highlights the need for continuous research into the ethical and societal impacts of AI in the auditing sector.

As AI technologies advance, auditors must also progress, using AI to improve risk assessment, fraud detection, and decision-making. Successful adoption requires addressing demographic factors influencing auditors' engagement with technology, ensuring equitable and thorough integration of AI within the profession.

The study's primary constraints stem from the small sample size and the exclusive focus on the Portuguese auditing landscape. Future research should incorporate qualitative data from interviews for a more in-depth analysis and broaden the scope to an international context with a larger sample to enhance the generalizability of the findings.

References

1. Zhang, C.: Intelligent process automation in audit. J. Emerg. Technol. Account. **16**, 69–88 (2019). https://doi.org/10.2308/JETA-52653
2. Couceiro, B.A. dos S.: Inteligência Artificial em Auditoria Interna: proposta de modelo de auditoria interna a projetos de inteligência artificial. (2021)
3. Canada, C.: The Data-Driven Audit: How Automation and AI are Changing the Audit and the Role of the Auditor (2020)
4. Russell, S., Norvig, P.: Artificial Intelligence: A Modern Approach. University of California, Berkeley (2021)
5. Lecun, Y., Bengio, Y., Hinton, G.: Deep learning. Nature **521**(7553), 436–444 (2015). https://doi.org/10.1038/nature14539
6. Silver, D., Singh, S., Precup, D., Sutton, R.S.: Reward is enough. Artif. Intell. **299**, 103535 (2021). https://doi.org/10.1016/J.ARTINT.2021.103535
7. Floridi, L., et al.: An ethical framework for a good AI society: opportunities, risks, principles, and recommendations. Philos. Stud. Ser. **144**, 19–39 (2021). https://doi.org/10.1007/978-3-030-81907-1_3/COVER
8. Marcus, G., Davis, E.: Rebooting AI: building artificial intelligence we can trust. **273**
9. Stahl, B.C., Brooks, L., Hatzakis, T., Santiago, N., Wright, D.: Exploring ethics and human rights in artificial intelligence – a Delphi study. Technol. Forecast. Soc. Change. **191**, 122502 (2023). https://doi.org/10.1016/J.TECHFORE.2023.122502
10. Stahl, B.C.: Embedding responsibility in intelligent systems: from AI ethics to responsible AI ecosystems. Sci. Rep. **13**(1), 1–8 (2023). https://doi.org/10.1038/s41598-023-34622-w
11. Manita, R., Elommal, N., Baudier, P., Hikkerova, L.: The digital transformation of external audit and its impact on corporate governance. Technol. Forecast. Soc. Change. **150**, 119751 (2020). https://doi.org/10.1016/J.TECHFORE.2019.119751
12. Puthukulam, G., Ravikumar, A., Sharma, R.V.K., Meesaala, K.M.: Auditors' perception on the impact of artificial intelligence on professional skepticism and judgment in Oman. Univ. J. Account. Financ. **9**, 1184–1190 (2021). https://doi.org/10.13189/UJAF.2021.090527
13. Munoko, I., Brown-Liburd, H.L., Vasarhelyi, M.: The ethical implications of using artificial intelligence in auditing. J. Bus. Ethics **167**, 209–234 (2020). https://doi.org/10.1007/s10551-019-04407-1
14. Searcy, D.L., Woodroof, J.B.: Continuous auditing: leveraging technology. CPA J. **73**, 17–27 (2003)
15. Rocha, O.: Contabilidade Assistida: O impacto das tecnologias de inteligência artificial na Contabilidade (2022)
16. Rao, A., Verweij, G., Cameron, E.: Sizing the prize: what's the real value of AI for your business and how can you capitalise? PwC. 2017 (2022)

Personal Data Protection and Public Disclosure of Data Relating to Taxpayers Debtors to the Portuguese Tax Authority

Sara Luís Dias[✉]

School of Management, Polytechnic University of Cávado and Ave, Barcelos, Portugal
sldias@ipca.pt

Abstract. To fulfill its duties related to the management and collection of taxes, the Portuguese Tax Authority has access to a variety of taxpayers' personal data of personal, financial, professional, and family nature, among others.

In order to protect the privacy of taxpayers, this entity is, according to the General Tax Law, obliged to keep confidential these data. This confidentiality obligation is in accordance with the legal rules of personal data's protection, provided by the General Regulation on the Protection of Personal Data and the Personal Data Protection Act in force in the Portuguese legal system.

However, the possibility of publicly disclosing the data of individual and collective taxpayers whose tax situation has not been regularized is currently determined in the law as a possibility that does not infringe the aforementioned obligation of confidentiality.

In this paper, we will try to understand if this possibility conflicts with the principles and values pursued by the rules on the protection of personal data, distinguishing between business and non-business taxpayers, and understanding the reasons of public interest, transparency and the safeguarding of commercial and competitive interests that underlie the publication of personal data of those taxpayers.

Keywords: Confidentiality · Personal Data · Tax Debtors

1 The Confidentiality Obligation of the Portuguese Tax Authority (Constitutional and Legal Obligations)

The Portuguese Tax Authority[1] has access to a different type of personal data (personal, family, professional and financial information) of taxpayers (natural and legal persons).

It is a consequence of the practice of its administrative acts as calculation of taxes, control of the tax situation of taxpayers, tax inspections, collection taxes, among others.

In fact, the possibility of carrying out tax inspections and audits, the cross-checking of information with other taxpayers and the access to bank and financial documents

---

[1] Hereinafter AT.

Á. Rocha et al. (Eds.): WorldCIST 2024, LNNS 990, pp. 317–324, 2024.
https://doi.org/10.1007/978-3-031-60328-0_32

facilitates the access to personal information which are processed, used and stored by this public entity [1].

In this context, it is needed to provide for a legal regime to protect taxpayers, ensuring confidentiality and secrecy on the part of AT in the processing of this kind of information[2]. Legally, it appears in the provisions of article 64 (1) of the General Tax Law[3], under which AT employees are obliged to keep confidential all accessed and collected personal data of taxpayers. In the context of the tax inspection procedure this obligation is also referred in article 22 of the Complementary Regime of the Tax and Customs Inspection Procedure, in order to reinforce the obligation of confidentiality imposed on inspectors.

It aims to ensure the right to privacy, constitutionally established in Article 26 (1) of the Constitution of the Portuguese Republic, respect for private and family life and the protection of their personal data [2], which must be processed carefully and fairly, for specific purposes and with the consent of the person concerned or on other legitimate grounds provided for by law (Articles 7 and 8 of the Charter of Fundamental Rights of the European Union).

The violation of this confidentiality's obligation by AT employees may represent the practice of a crime or an administrative offense, under the terms of articles 91 and 115 of the General Regime of Tax Infractions, respectively.

2 The Possibility of Disclosing Personal Data of Taxpayers in Debt

This obligation to maintain confidential personal data of taxpayers ceases in the situations provided for in paragraphs 2 and 4 of article 64 of LGT and in the other special situations provided for by law. These are essentially cases in which the taxpayer authorizes the disclosure of his/her tax information, situations of cooperation with other public entities and when AT is obliged to cooperate with the Courts.

2.1 Public Lists of Taxpayers in Debt

However, the legislator also introduced two other exceptions to this obligation of secrecy in topics a) and b) of paragraph 5 of article 64 of LGT: disclosure of the list of taxpayers whose tax situation has not been regularized (topic a)) and annual publication of incomes (topic b)).

We give special attention to paragraph a) as, in this case, it is possible to publish lists of taxpayers (both natural and legal persons) whose tax situation has not been regularized, including their full name and tax identification number[4]. According to paragraph 6 of this article, "regularized tax situation" must be defined as referred in article 177-A of the Tax Procedure Code[5]. The situation will be considered regularized when tax enforcement

---

[2] According to Portuguese Superior Courts, the personal tax number is also a confidential information (decision of *Tribunal Central Administrativo Sul* of 25/06/2020, process no 2796/19.7BELRS, www.dgsi.pt).

[3] Hereinafter LGT.

[4] In topic b) of paragraph 5 of article 64 of LGT, the public lists must contain generic information, without identifying the taxpayers to whom the published income is related.

[5] Hereinafter CPPT.

proceedings is suspended by the provision of a suitable guarantee to the entire debt, under the terms of article 169 of CPPT.

These lists of debtors are organized in a hierarchical form, with various scales that increase according to the value of the debt and are published on the Tax Authority's website and are publicly accessible[6].

2.2 Grounds to Justify the Publicity of These Debtor's Lists

There are many reasons to justify the publicity of these lists with personal data of taxpayers:

i) It could be said that such lists aim to guarantee the transparency and efficiency of the tax system, making generally known who the debtors are;
ii) Since the tax credit is a public credit that "belongs to everyone" - since the collection of revenue through the levying of taxes is generally aimed at satisfying the collective and social needs of citizens - it makes sense that the existence of these debts should be generally known;
iii) It is also relevant in commercial terms, as it allows economic agents to know the tax situation of those with whom they intend to do business, allowing them to get from this information ideas about their economic and financial situation (taking into consideration the nature of the debts and the harmful consequences they involve for debtors[7], these debts can signify a difficult economic situation).

Notwithstanding all these hypothetical reasons behind the publication of this data, it is important to understand whether, in view of the rules in force for the protection of personal data of natural persons, this possibility is legal, appropriated and justified[8].

3 The Protection of Taxpayers' Personal Data (Regulation (EU) 2016/679 of April 27)

The current rules on the protection of personal data of natural persons go further than the duty of secrecy laid down in Tax law[9], providing a stricter, stronger and broader degree of protection, so it is now important to analyze these rules.

---

[6] Available in https://static.portaldasfinancas.gov.pt/app/devedores_static/de-devedores.html.

[7] For example, the existence of debts of this nature is one of the facts on which the insolvency situation is based. In accordance with the provisions of Article 20(1)(g) of the Insolvency and Company Recovery Code, the generalized non-compliance with tax obligations for more than six months is an objective precondition for insolvency [8].

[8] We can also discuss this problem of access to personal data presented in the the Standard Audit File for Tax (SAFT): a standardized file in XML submitted by the contributors in the Tax Authority with the objective of allowing an easy export, and at any time, of a predefined set of accounting and invoicing records. This file often incorporates personal data and there is also a risk of violation of personal data protection rules. It is not the problem analyzed in this paper, but it is also important to mention it.

[9] Decision of *Tribunal da Relação do Porto* of 16/12/2015, process no 478/13.2TAAMT-A.P, www.dgsi.pt.

We highlight the Regulation (EU) 2016/679 of April 27 on the protection of natural persons with regard to the processing of personal data which came into force on May 25, 2018[10].

This Regulation has significantly changed the paradigm of legislation on this matter, expanding and defining more rigorously the concept of *personal data* relating to natural persons, providing for certain obligations in the collection, recording, processing, transmission and storage of this data and listing an extensive range of principles on this subject.

The concept of *personal data* includes all information relating to an identified or identifiable natural person, i.e. all elements which allow a person to be identified, directly or indirectly (Article 4 (1) of GDPR), covering all social, family, physical, mental, ideological, professional or other aspects relating to the person which, individually or in combination, allow them to be identified [3].

Following the entry into force of the aforementioned Regulation, in Portugal the Law no. 58/2019 (*Personal Data Protection Law*[11]) was published on August 8, 2019, implementing the GDPR.

4   The Application of Personal Data Protection Laws to the Portuguese Tax Authority

AT is also obliged to comply the provisions of the Personal Data Protection Act (article 2(1)), including the obligation of appoint a data protection officer (Article 12 of the LPDP and Article 37 of the GDPR), which was done by Order No. 10370/2017, published in *Diário da República* No. 230/2017, of November 29. The *Data Protection Officer* (DPO) must act autonomously and independently, and he is responsible for informing and advising the data's user, monitoring processing compliance, carrying out a data protection impact assessment (article 39 of the GDPR and article 11 of LPDP).

In this context, when this public entity applies certain type of processing data (applying, for example, new technologies) which can involve a high risk to the rights of natural taxpayers, it must carry out an "impact assessment" of the intended processing operations on the protection of personal data (article 35 of the GDPR) in order to describe the processing, understand and assess, its necessity and proportionality and contribute to the minimization of those risks, assessing them and determining the necessary procedures to be applied.

Failure to comply with these and other rules of this laws may result in the application of the fines mentioned in article 44 of the LPDP. It should be noted, however, that paragraph 2 of this legal provision allows public entities, upon duly substantiated request, to ask the *National Data Protection Commission* to waive the application of those fines for a period of three years from the entry into force of this law (ended in 2022).

---

[10] General Data Protection Regulation (hereinafter GDPR).

[11] Hereinafter LPDP.

5   The Disclosure of Data Relating to Companies' Taxpayers

The rules now in force on the GDPR do not apply to the companies but only to natural persons as, in accordance with its article 1 (2), it aims to protect the «fundamental rights and freedoms of natural persons, in particular their right to the protection of personal data».

However, even in the case of companies, it is also possible to find some rights and values of personal nature [4] such as the right to the name (trade name), other distinctive signs (trademarks, logos), honor, good image, secrecy (correspondence, trade secrets, know-how) but these rights are directly connected to the commercial activity they carry out, to the strict extent that they are considered necessary or convenient for the pursuit of the commercial purpose.

Therefore, here it is not in cause a true general right of personality, in the terms in which it is provided for in Article 70 (1) of the Civil Code [5]. This right is closely linked to natural persons, to their physical, intellectual and emotional existence, to the human personality [6]. In this sense, Article 160 of the Civil Code seems to expressly exclude rights and obligations «inseparable from the individual personality» from companies.

For this reason, we believe that regarding commercial companies, there is no special reason to avoid the publicity of information related with the existence of tax and social security debts. This information is, as we have explained, useful and relevant in the commercial circle, allowing economic agents to consider doing business with commercial companies that, being on the list, demonstrate that their tax situation is not regularized, i.e. they have debts of this nature which, even if claimed or contested, are not guaranteed. In addition, a lot of information of these companies are accessible publicly in the Commercial Registration.

This is information that, along with other published and publicly accessible information (such as the annual accounts of companies, in accordance with Article 3 (n) and 42 of the Commercial Registry Code, or other information on the securities markets and derivative financial instruments published by the *National Securities Commission*), promotes transparency and security in commercial relations.

It should be noted, however, that all of these purposes could be defeated if, as - in practice - seems to be the case, the AT is not careful and rigorous in constantly updating this information, keeping a record of taxpayers who no longer have tax debts or, if they do, they are already duly guaranteed, and their tax situation is regularized. This possibility raises many other problems of legality and responsibility, which, although we point them out, are not worth developing here.

6   The Disclosure of Data Relating to Taxpayers Who are Natural Persons Entrepreneurs

In our opinion, the publication of personal data of natural persons (non-business persons) in the public lists of debtors to the Tax Authorities and to Social Security clearly violates the rules of the GDPR and the LPDP. The disclosure of the full name and tax identification number of a natural person allows any other person to identify the person in question, knowing that they have tax debts, and it opens the door to the access of other type of data related with that person.

It seems to us that, in the case of natural persons who have personal tax debts (personal income tax debts, for example), often arising from family and intimate situations, there are no reasons of public interest that justify such exposure and risks. The ideas of public embarrassment, shame, harmed image to force these people to pay these debts are motives that cannot prevail here, nor do they seem to be sufficient to force these people to regularize their tax situation.

The *National Commission for the Protection of Personal Data* has already published some opinions (statements no 38/2005, 16/2006, 56/2017 e 54/2018[12]) about the publication of these lists of debtors, highlighting the impact that the disclosure of this data has on people's private lives. According to their statements, this possibility is excessive, with effects that do not end at the moment of publication but are reflected in the future, violating the principle of proportionality to which the legislator is bound (Articles 18 (2), 26 (1) and 35 (3) of the Constitution of Portuguese Republic). The provision and maintenance of this information facilitates access, collection, cross-checking and use of this data for the most different purposes, such as profiling, which can be used as a means of discrimination and unfair stigmatization of people. In this regard, this entity has been suggesting that the legislative impact of these rules that allows the publication of these lists of individual debtors must be measured, specifically analyzing the need to maintain this possibility.

In general, we agree with this position, considering that the publication of these lists is, as far as non-business individuals are concerned, a measure that restricts their rights, freedoms and guarantees, disproportionate and excessive in relation to the objectives it seems to be trying to achieve.

However, we need to consider the following scenario: There are some natural persons who are entrepreneurs, practicing commercial acts and executing an economical activity in the same way as a company. Natural persons who are individually engaged in a business activity and who, in this context, present tax debts.

In this case, we cannot defend the same position. We believe these data protection rules are also not applicable to natural persons who act in these conditions, presenting business debts.

In order to define the concept of "entrepreneurial natural persons", and considering the lack of a legal definition, we use the definition of entrepreneur referred in article 5 of the Insolvency Law, which define this person as someone who, in a non-corporate manner or under the guise of a commercial company, practice any economic activity with the aim of generating profits. [7] goes further in defining this concept, understanding that a company is «a productive organization or mediator of wealth, which carries out, in a stable manner, a certain economic activity according to the market to which it is directed», understanding that it does not differ from the entrepreneur who owns it, nor from the establishment that is part of it. In short, based on these positions, we believe that a natural person who is an entrepreneur is one who, individually but with productive or professional means, practice a commercial activity.

For these reasons, we cannot adopt the same view as for non-business individuals, particularly, for example, with regard to VAT[13] debts, which are related to the taxpayer's

---

[12] www.cnpd.pt.

[13] Value Added Tax.

business or commercial activity. Even in the case of personal income tax debts, these may be related to the professional activity, referring to an economic or business substrate owned by the natural person.

In addition, even in cases where debts of a private nature are involved, knowledge of those debts may be relevant in commercial and competitive terms. In the absence of a separation between personal and business assets, the financial situation of these people can be relevant information for anyone who project to do business with them.

We reiterate that, in relation to these people, the same reasons of public interest, transparency and security that we pointed out in relation to companies justify the publication of this type of information. By choosing to set up a commercial activity as an individual, the taxpayer is subject to sharing with other traders and/or entrepreneurs' information about the economic and financial situation that results or interfere with the exercise of their activity.

7 Conclusions

With this study, we can conclude that the Portuguese Tax Authority is also bounded by national and European legal rules relating to the protection of personal data.

In carrying out its functions of a public nature, this entity must ensure the protection of taxpayers' data (personal, family, professional and intimate information) which it accesses, processes and stores, adopting the necessary measures to fulfill this duty.

In fact, this obligation of adoption measures of security and secrecy is broader than the duty of confidentiality to which the Tax Authority's officials are subject and which is regulated by tax laws.

This raises even more reservations and questions regarding the possibility, provided for by law, of publicly disclosing personal data relating to individuals who do not have their tax situation regularized.

We believe that, when analyzing this issue, it is important to distinguish two different situations: the publication of such information regarding non-business individuals and the publication of data regarding business individuals.

In relation to the first group of persons such disclosure seems to us as inappropriate and disproportionate, since there are no commercial advantages or reasons of public interest that are sufficiently relevant to eliminate the negative impact that the disclosure of such information may cause on the lives of these people (providing and maintaining this information makes it easier to access, collect, cross-reference and use these people's personal data for the most diverse purposes, such as profiling, which can be used as a means of discrimination and unfair stigmatization of people).

Regarding natural persons who are entrepreneurs, other factors must be taken into account when they are registered in the list of public debtors to the Tax Authority. These people are practicing a commercial activity, dealing with other economic agents in a relationship that goes beyond their personal and private lives and their tax debts are often related to the exercise of these economic activities.

In these cases, sharing this data can be useful and relevant in commercial terms (as it is in relation to companies), allowing other economic agents to consider doing business with people who do not have a regularized tax situation. This information promotes freedom, transparency and security in commercial relations.

We believe that this is an issue that requires greater consideration and analysis on the part of our entities, as there are several questions that arise regarding the actions of public entities and the need to fulfill the public interest and the protection of taxpayers' personal data and privacy. We hope that, in the near future, more and better studies can look into this matter.

References

1. Amorim, J.C., Rocha, A.P.: O dever de confidencialidade na relação jurídico-fiscal – Notas sobre a detenção e a divulgação de dados pessoais dos contribuintes por parte da Administração Tributária. In: Cadernos de Justiça Tributária (21), pp. 25–35 (2018)
2. Campos, D.L., Rodrigues, B., Sousa, J.L.: De: Lei Geral Tributária Anotada e Comentada, 4th edn. Encontro da Escritora Editora, Lisboa (2012)
3. AAVV. Comentário ao Regulamento Geral de Proteção de Dados e à Lei n.º 58/2019 (coord. Cordeiro, A. B. M.). Almedina, Coimbra (2021)
4. Azevedo, M.A.: A problemática da extensão dos direitos de personalidade às pessoas colectivas, maxime, às sociedades comerciais. Rev. Direito das Soc. **II**, 123–144 (2010)
5. Sousa, R.V.A.C.: O Direito Geral de Personalidade. Coimbra Editora, Coimbra (2011)
6. Ascensão, J.O.: Teoria Geral do Direito Civil, vol. I, 2nd edn. Coimbra Editora, Coimbra (2000)
7. Cunha, P.O.: Direito das Sociedades Comerciais, 7th edn. Almedina, Coimbra (2019)
8. Dias, S.L.: O crédito tributário no processo de insolvência e nos processos judiciais de recuperação. Almedina, Coimbra (2021)
9. Abreu, J.M.C.: Curso de Direito Comercial, vol. II, 7th edn. Das sociedades, Almedina, Coimbra (2021);
10. Cordeiro, A.M.: Tratado de Direito Civil Português. I - Parte Geral, Tomo III, Almedina, Coimbra (2012)
11. Rodrigues, J.N., Teves, D.M.: A Proteção de Dados Pessoais e a Administração Pública – O novo paradigma jurídico. AAFDL Editora, Lisboa (2020)

Beyond Labels and Barriers: Women's Ongoing Journey in the Auditing Profession

Silvia Bernardo[1](✉), Isabel Pedrosa[1,2](✉) (iD), and Daniela Monteiro[3] (iD)

[1] Polytechnic Institute of Coimbra, Coimbra Business School, Quinta Agrícola - Bencanta, 3045-231 Coimbra, Portugal
iscac15351@alumni.iscac.pt, ipedrosa@iscac.pt
[2] Polytechnic Institute of Porto, CEOS.PP, Porto, Portugal
[3] ISEG, University of Lisbon, Lisbon, Portugal

Abstract. This paper aims to synthesize the primary metaphorical barriers representing the difficulties women face in advancing their careers over time: glass ceiling, glass cliff, labyrinth, leaky pipeline and queen bee. It also seeks to assess the validity of these metaphors in relation to actual realities. The methodology involves analyzing, synthesizing, and compiling various articles and studies on the topic. This analysis revealed that over time, these barriers evolve, adapting to the present yet continuing to exist. Some barriers are more pertinent than others, as evidenced by critiques of the justifications for their definitions. This paper underscores the need for actions to combat gender inequality, specially within the auditing profession, not only within organizations but also at the societal level.

Keywords: Metaphors · Barriers · Women' careers · Gender inequality · Auditing

1 Introduction

We can say that women have had, over time, and continue to have, barriers that make it difficult for them to advance in their careers. As a way of describing these difficulties imposed on women, metaphors emerged that evolved over the years. The use of metaphors has become notorious, as it allows us to structure the way we understand and visualize underrepresentation at the level of leadership, also influencing what is not visible. However, since the focus of these metaphors is gender inequality that causes female underrepresentation in various circumstances, the most consensual way of exposing it is through explaining "the less prototypical group becomes the effect to be explained" [8].

Metaphors regarding gender stereotypes are numerous, however, we chose to analyze those that are more directed towards female representation in organizational leadership, to understand the evolution of barriers over time and its validity today. Namely, the glass ceiling [8, 26], the glass cliff [25], the labyrinth [12, 15], the leaky pipeline [8] e queen bee [16].

Á. Rocha et al. (Eds.): WorldCIST 2024, LNNS 990, pp. 325–334, 2024.
https://doi.org/10.1007/978-3-031-60328-0_33

The method used in this paper involved the analysis of several articles and studies carried out within the topic, its synthesis and compilation.

The present paper is structured as follows: the second section encompasses metaphors for female advancement in their careers, criticisms of these metaphors and suggestions for possible actions for change; the third section about female leadership in auditing profession points several considerations on how women's leadership and women in profession may have positive outcomes concerning the efficiency and efficacy of the work; last section compiles the conclusions.

2 Metaphors for Female Advancement, Criticism, and Suggestions for Change

2.1 Glass Ceiling

The glass ceiling metaphor was the first response to emerge regarding the enormous gender inequality in existing positions of power and authority [15]. Portrays a transparent and subtle barrier, stronger enough to block women's advancement to the highest levels of work hierarchies [26]. Bruckmüller et al. [8] summarize it as the phenomenon that traps women so that they are unable to reach senior or executive positions, as these positions appear to be exclusively restricted to men. The analysis carried out by Steil [26] includes two aspects that help to understand this phenomenon: the social perspective and the perspective of power. The first suggests that the reduced number of women occupying high hierarchical positions may be related to the fact that men, as a group, try to preserve the self-esteem attributed to them for occupying leadership positions, thus differentiating themselves from other groups. The second, corroborating the first, suggests that men tend to make decisions in accordance with their self-interest in order to remain in a position of power. Both aspects strengthen the study's conclusion that the barrier exists due to the values of male groups that influence the organizational structure, since they are the ones who have the power to do so.

Bruckmüller et al. [8] point factors to the glass ceiling: gender stereotypes and differentiated access to informal support networks and the possibility of mentoring, with stereotypes being more aimed at reiterating what leaders should be like. Bendl and Schmidt [5] criticize the glass ceiling for not including the variable time, since it treats discrimination as a static and stable phenomenon, which cannot experience changes, but can only be moved. They also refer as a failure the fact that it does not refer to who has the power to control the firmness of this barrier, concluding that by targeting structural aspects of organizations, it ends up representing the concept of "existence of discrimination". Carli and Eagly [12] also criticize the fact that the barrier exists only at the penultimate level of the career, at the level immediately below women reaching top positions and that until then the challenges to be faced are reduced. They denote that because it is made of glass, it conveys the image that, until the moment women are prevented from moving forward, this obstacle is invisible and undetectable. Another issue raised by Eagly and Carli [15] is the fact that there have already been women breaking this barrier, which contradicts the essence of the metaphor. However, by portraying a unique and invariable obstacle to women's advancement, the glass ceiling's most notable flaw is that it does not encompass the complexity and variety of challenges that women are forced to face on their path to leadership [15].

Cook and Glass [14] suggest that one possibility of overcoming barriers such as the glass ceiling will be through the diversification of institutional boards. In their analysis, authors gather evidence that the integration of women in management has a positive impact on the increase in nominations of other women to CEOs and the length of their mandates. The same study also shows that gender diversity among those who make decisions contributes to the increase in women's promotions in senior positions and that boards made up solely or mostly of men favor the probability of nominating men to CEOs. In this way, authors concluded that by adopting this measure, women will also have more opportunities to demonstrate their leadership capabilities, and female presence in these positions will be more observable and will allow access to professional support networks made up of women at the same level, culminating in higher satisfaction and performance.

2.2 Glass Cliff

Ryan et al. [25] described the glass cliff as a reality to be explained regarding the attribution of leadership positions to women associated with the organization's poor performance during the period of their leading. To them this is a pattern observed in crisis situations and not a mere isolated occurrence. The study of this phenomenon brings another perspective regarding the leadership positions held by women and the circumstances that underlie them. In this sense, Ryan et al. [25] observed the type of positions that women occupy, with no intention of highlighting or blaming them for the low results obtained during their leadership period. Authors concluded that the glass cliff is the product of a diverse set of factors: organizational strategy, stereotypical beliefs, and partial selections, tending to be more evident in organizational crisis environments, with the exception that the type of crisis influences the occasions in which the phenomenon may or may not arise. Bruckmuller and Branscombe [7] consider two reasons for the development of this phenomenon: 1) the maintenance of the status quo related to male predominance in leadership and 2) the stereotypes associated with gender and leadership. Bruckmuller and Branscombe [7] name men's attributes of competitiveness and self-confidence as strong associations with leadership positions and women's interpersonal attributes such as intuitive ability and consideration for others' feelings as indispensable in business crisis situations. Therefore, men adapt more easily to stereotypical patterns related to leaders, while women are particularly better able to bring about the change that is necessary in times of crisis [8].

Bruckmuller and Branscombe [7] clarify that, normally, this phenomenon is analyzed from the perspective that women have more opportune and necessary characteristics in times of crisis. Although it can also be concluded that men do not have such characteristics, which makes them less capable of playing the role of leader on these occasions, so what puts women at the top of the glass cliff is men's lack of capacity for the demands of these positions.

Ryan et al. [25] also highlight that the biggest problem appears to be more related to the comfortable leadership positions granted to men, than to the problems that women face as leaders. They suggest that to save women from the inevitability of the glass cliff is to change the direction of attention to the privileged place of men in accessing more comfortable leadership positions (glass cushion). Bruckmuller and Branscombe [7] obtained evidence that the phenomenon is completely extinct in contexts with a history of female leadership, so they reinforce the conclusions of Ryan et al. [25] when highlighting that the development of this phenomenon is related to the history and continuity of the structuring of organizations around male leaders, which results in the reinforcement of the stereotypical relationship between leadership and masculinity.

Cook and Glass [14] also mention the way performance is assessed as a contributing factor to this phenomenon. They found that, when companies adopt subjective evaluation measures, it is less likely that glass cliff patterns will be observed, while those that use measures focused on financial objectives tend to develop these patterns.

2.3 Labyrinth

Eagly and Carli [15] portray another metaphor that arises because they believe that the glass ceiling no longer captures in the best way the experience lived by women in reaching top positions. The authors represent labyrinth as the path that women have ahead of them throughout their career, full of curves and counter curves (various barriers, as opposed to the single barrier of the glass ceiling) that may be both expected and unexpected, but which in some way will lead to an achievable goal. Therefore, recognizing that obstacles are present on this path, but they are not discouraging – the same path that women face to rise to leadership positions. They also reinforce that passing through a labyrinth "requires persistence, awareness of one's progress, and careful analysis of the puzzles that lie ahead" [15]. In a more comprehensive analysis, Carli and Eagly [12] consider that the metaphor they propose is optimistic and realistic, in that it conveys confidence that women can successfully achieve leadership and uncertainty associated with this achievable success, respectively.

The objective of the metaphor is not to blame women or the situation for the reduced progress, but to reinforce that female success is manifested in the interaction between their skills and the motivation and challenges that make up the context in which they operate [12]. Thus, Eagly and Carli [15] also define the objective of joining forces to combat more effectively the various barriers that make up the labyrinth, through understanding them and how some women manage to successfully overcome them. They point out barriers like: the existence of traces of gender prejudices that favor men; resistance to female leadership based on the attitudes taken by the leader; issues related to leadership styles as there are styles that are more effective than others, and women's way of leading tends to be more effective and also more participatory and collaborative; responsibilities of family life, which they consider to be the "worst curve in the labyrinth" for women, as it has the greatest impact on their careers; reduced socialization, considering the previous barrier, women tend to invest less in social life due to lack of time and because it is not their priority.

2.4 Leaky Pipeline

There is also another metaphor that is not directed so much towards the barriers themselves, focusing mainly on the consequences that arise from them, even without specifying which barrier causes them – leaky pipeline [8]. It represents the effect of the higher dropout rates of women at the academic level, in Science, Technology, Engineering and Mathematics, STEM [6, 8]. The dropout by women in certain professions, even when there is no gender gap (as the academic career), reveal that they are gradually less present, by choosing to follow alternative careers [6, 8]. One of the interesting aspects of this metaphor is related to the portrait it makes of the various moments in which abandonment occurs [8], which implies that part of its solution involves reforming this academic area, in this specific case [6].

2.5 Queen Bee

The queen bee phenomenon, although it is not exactly a metaphor, demonstrates the effect that women themselves have on the career ladder of other women. Faniko et al. [16] define it as the behavior adopted by women who have been successful in predominantly male-dominated organizations, which tends to contribute to preventing other women from having the same opportunities to advance in their careers, as it does not favor the professional development of the women who report to them.

However, despite the image that this type of behavior transpires, there are studies demonstrating that it is not a matter of malice among women in a situation of competition at work [18]. In addition to the tendency to separate themselves from other women in the work context, it is not a natural attribute of women, and must be perceived as a strategic behavior that gives them adaptability to the idealization, even in masculine terms, of success in organizations [17]. In other words, this strategy involves the need to move away from group stereotypes associated with women, to overcome sexist expectations that exist in organizations [18].

Faniko et al. [16] point out the queen bee as a consequence of the discrimination experienced by women during their career construction, so it is not a factor in professional competitiveness among women in general. Faniko et al. [18] concluded that in organizational contexts where women face less difficulties in achieving their professional ambitions, the tendency to adopt the queen bee posture is also lower. This phenomenon should not be understood as a cause of discrimination between women, but as a consequence of the discrimination they experienced before [17].

Faniko et al. [16] demonstrate that women in senior positions try to distinguish themselves from subgroups of women, namely those at the initial stage of their careers and who tend to support the implementation of gender quotas if they are aimed at women at the same hierarchical level as them, but not if they are aimed at those who are in the initial phase. Authors suggest that what gives this phenomenon a negative connotation is that the behavior associated with it is not what is expected from women. This contrasts with the gender identity and modesty behavior revealed by women in lower hierarchical positions. Faniko et al. [16] mention that women at higher levels show solidarity and create a supportive environment for colleagues who position themselves at the same level as them, revealing themselves to be harsher in their assessments of those who report to them. Thus, authors conclude that the biggest challenge to increasing gender diversity in management positions will also be related to managing the low career expectations of women in lower positions, so that they are able to value more and improve their commitment to their professional career. Faniko et al. [18] reinforce their previous conclusions, based on the perception that women in top positions have of the sacrifices they made for the success of their career, they are not willing to support and promote opportunities for other women, if they are junior professionals, which may reveal that, for them, it is a question of justice. Therefore, the authors consider that the differences observed between these hierarchical positions (top and bottom) constitute a barrier to success, the possibility of passing which depends on the willingness to make sacrifices for one's career.

In a different context, Bruckmüller et al. [8], when analyzing the maternal wall metaphor, they also draw attention to the materialization of the injustice that is for women who are mothers to see their career decisions being based on work rules that make family/personal life incompatible with work commitments and on models of successful careers that are designed around the male life and body. This aspect is related with the sacrifices that women must make to assert themselves in successful careers.

2.6 General Criticism of Metaphors

Despite the deserved recognition that these metaphors should have for their common objective of giving visibility and combating gender inequality, they are also subject to criticism. As an example, Bruckmüller et al. [8] criticize metaphors in general for leading the focus to be directed towards numbers and the description of the problem of female underrepresentation (covered here, the glass ceiling) or towards identifying the moment and places in which this happens (mentioned here, leaky pipeline and maternal wall), suggesting that the focus should highlight women's experience in the workplace (mentioned here, labyrinth and glass cliff).

Another issue raised by Bruckmüller et al. [8] is the tendency that metaphors follow when treating women as the "effect to be explained", leaving a gap in terms of male stereotypes and forcing the idea that it is women who must rethink their behavior to lead to change that will reduce gender inequality. The authors consider that it is questionable that the focus of the metaphors remains on women and emphasize that keeping the focus on the stereotypes associated with them also has the consequence of legitimizing and reinforcing the advantageous and privileged status of men, contrary to what is intended. Another aspect that the authors point out as counterproductive is related to focusing on women's particularities so as not to reach the top, which distracts attention from companies' responsibilities in creating equal opportunities. In the same way, it encourages the idea that only individual women who make an effort can reach the top and ensures that they gather strength and move towards joint actions.

2.7 Possible Measures to Change the Unequal Reality

Since the purpose of metaphors is also to encourage action to change the current unequal reality, some authors name some of the actions to be taken that they consider most important and productive.

There are several barriers that keep women in less important positions, however, the responsibilities are not only associated with men: are also related to women. Therefore, it is imperative that the work to change the observed inequalities is carried out by both genders and companies [19].

As proposals for actions to be adopted, there are procedures at the level of organizations by promoting more structured informal support networks, mentoring and work-life balance programs and policies for the enjoyment of employees in general, without revealing or pointing to gender inequality as a reason for this [8]. The authors also suggest that the approach is not so associated with metaphors and focuses more on the privileges and advantages that men enjoy and the reasons why this is still the case. Faniko et al. [17] suggest that the focus should not only be on increasing female representation but also on making institutions more inclusive through the recognition of heterogeneous models of success, to also encourage changes in the basis of the ideology of success that is still based in masculine qualities. On the other hand, Archer and Kam [4] state that the title associated with a top position also has an influence on the way people will perceive their occupation in terms of gender, naturally tending to believe that it will be held by a man: the linguistic terms associated with these positions also constitute a barrier to female leadership, so it would be favorable to adopt neutral language for the designations of top positions.

The increase in visibility of women in leadership positions also has a positive effect on other women who aspire to build a career in that direction, namely by inspiring and motivating them [19, 23]. Another perspective of change to be adopted is suggested at the level of policies and legislation issued by some countries regarding quotas in business staff in countries with weak child social support services: countries should implement, beyond gender quotas, policies that promote those services, to allow women to carry out their professional tasks during their children's childhood [1].

3 Female Leadership and the Audit Profession

The auditing profession is related to ensuring the reliability of financial information [13], as it is a job that, normally, is carried out by a team with different skills and experiences [3]. Therefore, auditor's work includes three aspects: relationship with the audited entity, work management and carrying out detailed tasks [11].

Despite the different perceptions stated before, in a profession that still has some associated stereotypes, there are attributes that women have that favor both their presence in the audit profession and in auditing teams' leadership. Diversity that women may add to the profession has a positive impact on the quality of the work they carry out [2, 11, 20] and in the work carried out by their teams [20], their professional judgment and the decisions they make as they are more risk averse [22]. Their leadership also influences the effectiveness of their companies' internal control by reducing the auditor's effort and risk [21]. It also has a beneficial impact on organizations and employees [27], both through its leadership's encouraging practices which include better communication, more collaboration environments, collective participation [19, 24], stronger and safer physically and psychologically [19], in which they form and guide more efficient teams [9] and united around a common objective [27].

4 Practical Implications

The analysis allowed us to understand and visualize the barriers that have been imposed on women over time in professions in general, but specifically in the auditing profession. This also allowed us to understand how to direct actions to change the current inequality in the auditing profession, considering the type of barriers women face. Therefore, we consider that the barriers with greatest applicability in the context of auditing are those defined by the leaky pipeline and queen bee metaphors, since this profession is associated with a superior intellectual authority figure and this attribute is more easily conceived of for men, but also due to the demands of the profession that can imply greater difficulties for women.

5 Conclusions

We can state that, over the years, organizational contexts and structures have evolved and imposed adaptation of existing metaphors, and also lead to new ones. Therefore, in the work environment, women do not encounter a linear challenge (glass ceiling) that allows them to grow professionally and individually in a simple way. On the contrary, they face barriers for which no one prepares them, and it is difficult to have understanding so they can move forward (labyrinth). Still, the few women who manage to reach leadership positions continue to have a puzzle of expectations ahead of them (queen bee) and a predisposition to deal with crisis or problematic situations (glass cliff). This leads not only to a higher resistance to female leadership, but also to the increasing in dropout among women throughout their careers (leaky pipeline).

In this sense, based on the analysis of the metaphors, we consider the need to act in the face of female underrepresentation to change mentalities and behaviors, through corresponding measures at organizational level, to be notable. Well, around forty years have passed since the definition of the first metaphorical barrier emerged and, despite some progress being made, the core of the metaphors remains current. Despite the conditions that favors women in auditing, they are still underrepresented, in special in leadership.

As future work we suggest carrying out studies in female leadership environments that allow us to understand the structuring of female careers in which there have been no influences from metaphorical barriers. Audit profession is appropriated to continue this research: in individual professional contexts, specifically in audit, we suggest the development of studies that allow conclusions to be drawn regarding the queen bee and leaky pipeline phenomenon, as we consider these to be the metaphors with the greatest applicability in this area given the demands of the profession.

Acknowledgements. This work received financial support from the Polytechnic Institute of Coimbra within the scope of Regulamento de Apoio à Publicação Científica dos Professores e Investigadores do Instituto Politécnico de Coimbra (Despacho n.º 12598/2020).

References

1. Adams, R.B.: Women on boards: the superheroes of tomorrow? Leadersh. Q. **27**(3), 371–386 (2016). https://doi.org/10.1016/j.leaqua.2015.11.001
2. Amondarain, J., Aldazabal, M.E., Espinosa-Pike, M.: Gender differences in the auditing stereotype and their influence on the intention to enter the profession. J. Behav. Exp. Finan. **37** (2023). https://doi.org/10.1016/j.jbef.2022.100784
3. Amyar, F., Hidayah, N.N., Lowe, A., Woods, M.: Investigating the backstage of audit engagements: the paradox of team diversity. Account. Audit. Accountability J. **32**(2), 378–400 (2019). https://doi.org/10.1108/AAAJ-08-2016-2666
4. Archer, A.M.N., Kam, C.D.: She is the chair(man): Gender, language, and leadership. Leadersh. Quart. **33**(6) (2022). https://doi.org/10.1016/j.leaqua.2022.101610
5. Bendl, R., Schmidt, A.: From 'Glass ceilings' to 'Firewalls'— different metaphors for describing discrimination [Article]. Gend. Work Organ. **17**(5), 612–634 (2010). https://doi.org/10.1111/j.1468-0432.2010.00520.x
6. Blickenstaff, J.C.: Women and science careers: leaky pipeline or gender filter? Gender Educ. **17**(4), 369–386 (2005). https://www.researchgate.net/publication/228384649_Women_and_Science_Careers_Leaky_Pipeline_or_Gender_Filter
7. Bruckmuller, S., Branscombe, N.R.: The glass cliff: when and why women are selected as leaders in crisis contexts. Br. J. Soc. Psychol. **49**(Pt 3), 433–451 (2010). https://doi.org/10.1348/014466609X466594
8. Bruckmüller, S., Ryan, M. K., Haslam, S.A., Peters, K.: Ceilings, cliffs, and labyrinths: exploring metaphors for workplace gender discrimination. In: The SAGE Handbook of Gender and Psychology, vol. 27, pp. 450–464 (2013). https://doi.org/10.4135/9781446269930.n27
9. Bustos-Contell, E., Porcuna-Enguix, L., Serrano-Madrid, J., Labatut-Serer, G.: Female audit team leaders and audit effort. J. Bus. Res. **140**, 324–331 (2022). https://doi.org/10.1016/j.jbusres.2021.11.003
10. Byrnes, J.P., Miller, D.C., Schafer, W.D.: Gender differences in risk taking: a meta-analysis. Psychol. Bull. **125**(3), 367–383 (1999). https://doi.org/10.1037/0033-2909.125.3.367
11. Cameran, M., Ditillo, A., Pettinicchio, A.: Audit team attributes matter: how diversity affects audit quality. Eur. Account. Rev. **27**(4), 595–621 (2017). https://doi.org/10.1080/09638180.2017.1307131
12. Carli, L.L., Eagly, A.H.: Women face a labyrinth: an examination of metaphors for women leaders [Article]. Gender Manage. **31**(8), 514–527 (2016). https://doi.org/10.1108/gm-02-2015-0007
13. Chung, J., Monroe, G.S.: A research note on the effects of gender and task complexity on an audit judgment. Behav. Res. Account. **13**(1), 111–125 (2001). https://doi.org/10.2308/bria.2001.13.1.111
14. Cook, A., Glass, C.: Women and top leadership positions: towards an institutional analysis. Gend. Work Organ. **21**(1), 91–103 (2014). https://doi.org/10.1111/gwao.12018
15. Eagly, A.H., Carli, L.L.: Women and the Labyrinth of Leadership [Article]. Harvard Bus. Rev. **85**(9), 63–71 (2007). https://search.ebscohost.com/login.aspx?direct=true&AuthType=sso&db=bth&AN=26128729&site=ehost-live&custid=ns063492
16. Faniko, K., Ellemers, N., Derks, B.: Queen Bees and alpha males: are successful women more competitive than successful men? Eur. J. Soc. Psychol. **46**(7), 903–913 (2016). https://doi.org/10.1002/ejsp.2198
17. Faniko, K., Ellemers, N., Derks, B.: The Queen bee phenomenon in academia 15 years after: does it still exist, and if so, why? Br. J. Soc. Psychol. **60**(2), 383–399 (2021). https://doi.org/10.1111/bjso.12408

18. Faniko, K., Ellemers, N., Derks, B., Lorenzi-Cioldi, F.: Nothing changes, really: why women who break through the glass ceiling end up reinforcing It. Pers. Soc. Psychol. Bull. **43**(5), 638–651 (2017). https://doi.org/10.1177/0146167217695551
19. Franczak, J., Margolis, J.: Women and great places to work: gender diversity in leadership and how to get there. Organ. Dyn. **51**(4) (2022). https://doi.org/10.1016/j.orgdyn.2022.100913
20. Garcia-Blandon, J., Argilés-Bosch, J.M., Ravenda, D.: Is there a gender effect on the quality of audit services? J. Bus. Res. **96**, 238–249 (2019). https://doi.org/10.1016/j.jbusres.2018.11.024
21. Garcia-Blandon, J., Argilés-Bosch, J.M., Ravenda, D., Castillo-Merino, D.: Women leading the audit process and audit fees: A European study. Eur. Res. Manage. Bus. Econ. **29**(1) (2023). https://doi.org/10.1016/j.iedeen.2022.100206
22. Hardies, K., Breesch, D., Branson, J.: Are female auditors still women? Analyzing the sex differences affecting audit quality. SSRN Electron. J. (2010). https://doi.org/10.2139/ssrn.1409964
23. Hoyt, C.L., Murphy, S.E.: Managing to clear the air: stereotype threat, women, and leadership. Leadersh. Q. **27**(3), 387–399 (2016). https://doi.org/10.1016/j.leaqua.2015.11.002
24. Lipkin, N.: Why women are natural born leaders. Forbes (2019). https://www.forbes.com/sites/nicolelipkin/2019/11/19/why-women-are-natural-born-leaders/
25. Ryan, M.K., Haslam, S.A., Morgenroth, T., Rink, F., Stoker, J., Peters, K.: Getting on top of the glass cliff: reviewing a decade of evidence, explanations, and impact. Leadersh. Q. **27**(3), 446–455 (2016). https://doi.org/10.1016/j.leaqua.2015.10.008
26. Steil, A.: Organizações, gênero e posição hierárquica - compreendendo o fenômeno do teto de vidro. Revista De Administracao Da Universidade De Sao Paulo **32**, 62–69 (1997). https://www.academia.edu/32692793/Organizações_gênero_e_posição_hierárquica_compreendendo_o_fenômeno_do_teto_de_vidro
27. Trinidad, C., Normore, A.H.: Leadership and gender: a dangerous liaison? Leadersh. Org. Dev. J. **26**(7), 574–590 (2005). https://doi.org/10.1108/01437730510624601

Promoting Fiscal Incentives for Urban Regeneration: Local Government Digital Presence

Ana Arromba Dinis[✉]

Research Centre On Accounting and Taxation (CICF), School of Management, Polytechnic University of Cávado and Ave, Barcelos, Portugal
adinis@ipca.pt

Abstract. This research aims to analyse urban regeneration in Portugal in the context of environmental sustainability. Urban regeneration is much more than an information technology problem, as it increasingly requires social, economic, political, management and communication skills. Urban regeneration cannot be reduced to a mere real estate business, as it is seen as a public policy that treats the city as a collective good and aims to deal with the cities of the future. It is therefore important to highlight the role of environmental taxation as an extra-fiscal policy that can influence taxpayers' behaviour (in this case, towards the use of more sustainable materials in urban regeneration). This study seeks to show how Portuguese municipalities have a digital presence and how this presence can be relevant in the process of spreading fiscal incentives (namely to companies), in the context of their local significance. Thus, based on an exploratory analysis of the municipalities of the Portuguese Northern Region, this research aims to present a portrait of the digital presence of municipalities and how it is linked to fiscal incentives for urban regeneration. From the author's point of view, this is a groundbreaking study because it puts the issue of environmental taxation in the context of urban regeneration on the table.

Keywords: Urban rehabilitation · fiscal incentives · digital strategies

1 Introduction

This study explores the sustainability standards that allow buildings to be more energy efficient, through fiscal incentives.

Urban regeneration has a fundamental role to play, as it is seen as an opportunity for cities and municipalities to regenerate the future of buildings, always using materials that do not pollute the environment and are sustainable and low carbon [14].

Urban regeneration is much more than an information technology problem, as it increasingly requires social, economic, political, management and communication skills [19]. Therefore, urban regeneration cannot be reduced to a mere real estate business, as it is seen as a public policy that treats the city as a collective good and aims to deal with the cities of the future [17].

Á. Rocha et al. (Eds.): WorldCIST 2024, LNNS 990, pp. 335–343, 2024.
https://doi.org/10.1007/978-3-031-60328-0_34

It is therefore important to highlight the role of environmental taxation as an extra fiscal policy that can influence taxpayers' behaviour [6], this reason, towards the use of more sustainable materials in urban regeneration.

So, the starting point for this research is the fiscal benefits that are granted to companies in Portugal as one of the solutions for sustainable urban regeneration in the national housing strategy, and the digital presence of municipalities is analysed.

This research is exploratory, and it is developed to assess how municipalities in Northern Portugal behave in this regard, particularly in terms of publishing fiscal incentives for urban regeneration on the municipality's website and what kind of digital presence they have on other digital platforms. Administratively, the North of Portugal consists of 86 municipalities, representing almost 28 per cent of the municipalities of Portugal. This justifies the importance of this study.

The research objectives of this work are, firstly, to present the global phenomenon of the environmental impact of buildings; secondly, to discuss the importance of fiscal policies and their impact on environmental sustainability; thirdly, to discuss fiscal incentives for building renovation as a local policy to protect the environment and natural resources. Fourthly, to discuss the granting of fiscal incentives for building regeneration in Portugal, in the local context in 2015 to 2020. The fifth and last is to analyse the digital presence of the municipalities in the North of Portugal where the companies benefiting from these fiscal incentives are located.

To achieve the initial objectives, four research questions are outlined that it is sought to answer: 1) What role can local building renovation policies play in guiding Portuguese municipalities towards a more sustainable future? 2) How do Portuguese municipalities communicate fiscal incentives for building renovation? 3) How is the digital presence of the Portuguese municipalities where the companies that take advantage of the fiscal benefits for urban regeneration are based?

This research is divided into five sections. The first is this Introduction with the explanation of the topic addressed, as well as the methodology of the research and its structure. The second section addresses urban rehabilitation in the context of environmental sustainability, being a literature review presented on the concept of urban rehabilitation, some of the challenges listed in the strategic plan and how the fiscal benefits for urban rehabilitation are seen in the Portuguese tax system. In section three, it is addressed the theme of digital presence, in terms of concept, the advantages and disadvantages that may arise, and its strategies in connection with the theme under study. In the fourth section, the digital presence of the Municipalities of the Portuguese Northern Region is analysed. Finally, section five concludes, and presents the research limitations, and some future research perspectives.

2 Literature Review

2.1 Urban Rehabilitation in the Context of Environmental Sustainability

The protection and enhancement of the cultural heritage of populations, the defence of nature and the environment, the conservation of natural resources and the proper planning of the territory are fundamental tasks of the State [11].

Regeneration has therefore become an integral part of urban policy, setting objectives for revitalisation and sustainable intervention in the built environment [4].

For [9], urban regeneration is a form of integrated intervention in the existing urban fabric, preserving and modernising all or a significant part of the urban and property heritage, by carrying out works to transform or improve urban infrastructure systems, facilities and urban or green spaces for collective use, as well as construction, reconstruction, extension, alteration, conservation, or demolition of buildings.

It should be noted that [10] point out that, buildings are responsible for almost forty per cent of carbon emissions, mainly due to the energy consumed for heating and cooling. Also, urban areas are facing major challenges due to rapid urbanisation, including the demand for natural resources and the impact of climate change [2].

Well, the approval of financial incentives for green buildings plays an important role in promoting sustainable development and carbon emission reduction strategies [12]. In this way, the approval of financial incentives for green buildings plays an important role in promoting sustainable development and carbon reduction strategies [3].

Likewise, local policies for the renovation of buildings can help to understand the orientation of municipalities towards a more sustainable development [16] and local policies on building refurbishment can help to understand the orientation of communities towards a more sustainable future society [1].

In addition, environmental taxation is seen as one of the supports for a more green future, through the application of fiscal incentives that can steer taxpayers' behaviour towards environmental sustainability [5].

In this sense, about the rehabilitation of buildings, Portuguese tax legislation establishes that properties completed more than thirty years ago or located in urban regeneration areas benefit from fiscal incentives, namely exemptions from local taxes.

2.2 The Fiscal Benefits of Urban Rehabilitation in the Portuguese Tax System: Brief Reference

In the Portuguese context, it can be said that the legal framework for urban regeneration is an important instrument for the materialisation and institutionalisation of public decisions regarding the regeneration of buildings.

For example, in Portugal, it is up to the municipalities to define the urban regeneration areas and to communicate and promote the use of the financial incentives authorised by law and by the State. So, Portugal's urban regeneration tax policy reinforces the values of sustainability (in financial, social, and environmental terms) and equity (fair distribution of tax burdens and benefits). It is worth noting that the scheme approved in two thousand and seventeen broadened the concept and scope of urban regeneration areas and introduced the need to develop solutions that consider functional, economic, social, cultural, and environmental aspects (following on from the Portuguese green tax reform of two thousand and fourteen).

The fiscal incentive consists, for example, of a 3-year exemption from the local property tax (with the possibility of renewal for a further 5 years) granted to owners of urban buildings or autonomous fractions completed more than 30 years ago or located in urban rehabilitation areas. The following cumulative conditions must be met: the building must have been rehabilitated under the Portuguese Urban Rehabilitation Regime because of the planned intervention, its state of conservation must be two levels higher than that previously attributed and at least good, and the energy efficiency and thermal quality requirements must also be met. Taxpayers who purchase property for redevelopment purposes may also be exempt from municipal property transfer tax. Furthermore, under the current law, fiscal incentives are targeted at property owners and, as far as businesses are concerned (which is the focus of this study), micro and small businesses account for over ninety-nine per cent of Portuguese businesses.

Table 1 exhibits the fiscal incentives for urban rehabilitation that were granted in Portugal to companies from 2015 to 2020 and broken down by municipalities of the Northern Region.

Table 1. Fiscal incentives for urban regeneration (by municipalities and companies)

Municipalities of the Northern Region	Fiscal incentives granted 2015–2020 (€)	Number of beneficiaries (Companies)
Porto	191,746.71	26
Viana do Castelo	60,568.14	5
Braga	51,604.13	1
Matosinhos	48,704.80	3
Maia	17,449.39	2
Vila Real	10,521.43	1
Sao João da Madeira	10,235.58	2
Penafiel	9,688.07	2
Chaves	6,835.32	1
Santa Maria da Feira	6,809.38	1
Barcelos	4,876.19	2
Arcos de Valdevez	4,577.41	1
Vila Nova de Gaia	3,598.17	3
Ponte de Lima	2,527.32	1
Oliveira de Azeméis	2,513.52	1
Valongo	1,827.07	1
Gondomar	1,533.91	1
Esposende	1,172.33	1
Mirandela	116.41	1
Espinho	106.56	1

Source: Adapted from the Portuguese Tax Agency. Accessed 17 Nov. 2023

Table 1 shows that is the municipality where most of the companies that have benefited from urban regeneration are located, with this amount reaching a value of €191,746.71, and the number of beneficiary entities of 26 entities. The municipalities of Viana do Castelo and Braga are next, where companies that received benefits of €60,568.14 (5 companies) and 51,604.13 (1 company), respectively, are located. The municipality where the companies were based received the lowest amount of tax breaks was Espinho. It's important to note that companies can reduce their tax burden by adopting sustainable behaviour, which is linked to their willingness to renovate buildings. Nevertheless, the number of beneficiary companies is still very small. This also supports the relevance of this study. So, the aim is to understand how municipalities make public this type of fiscal incentive.

2.3 Digital Presence

For [8] digital presence is the set of strategies used to attract new customers, create new relationships with the public, increase notoriety and build customer loyalty. Therefore, to improve public services and promote public objectives, government social media has been integrated as part of the government's management tools [20].

Government Digital presence strategies are mixed. These can be done through attending to the creation of internet content, a blog or a website, mail-marketing; ads, and social media [18].

Some authors highlight that social network is the modality of communication and virtual interaction that is more dynamic and interactive, since it allows its elements to have permanent communication, anytime and anywhere [15].

For [13], the way of communicating is in increasing transformation, and municipalities realizing that the internet could be an advantage for the dissemination of their products/services they need to invest in online social networks to be able to interact more easily with their consumers/customers, to instantly detect their satisfaction or dissatisfaction with the service provided, among others. But there's still some way off [7].

3   Research Methodology

3.1   Collection of Data on Municipalities with a Digital Presence

Starting with companies that have benefited from fiscal incentives for urban regeneration (information collected from the Portuguese Tax and Customs Authority's Finance Portal), this study sought to analyse the digital presence of the municipalities in Northern Portugal where they are located, considering the information on their websites.

Of the 86 municipalities in Northern Portugal, twenty are the base of companies that have taken advantage of fiscal incentives for urban regeneration. Table 2 presents information on fiscal incentives for urban regeneration on the websites of the municipalities where these companies are located.

The analysis shows that all municipalities have information on fiscal incentives for urban regeneration on their websites. In a way, this seems to show the importance of making this information available so that companies are aware of the fiscal incentive under investigation. At the same time, the aim was to understand the digital presence of these municipalities, since, as mentioned above, this can be a factor of proximity to

Table 2. Fiscal incentives for urban regeneration information (municipalities website)

Municipalities	Fiscal incentives for urban regeneration information (website)	Information on urban regeneration areas
Arcos de Valdevez	yes	yes
Barcelos	yes	yes
Braga	yes	yes
Chaves	yes	yes
Espinho	yes	yes
Esposende	yes	yes
Gondomar	yes	yes
Maia	yes	yes
Matosinhos	yes	yes
Mirandela	yes	yes
Oliveira de Azeméis	yes	yes
Penafiel	yes	yes
Ponte de Lima	yes	yes
Porto	yes	yes
Santa Maria da Feira	yes	yes
Sao João da Madeira	yes	yes
Valongo	yes	yes
Viana do Castelo	yes	yes
Vila Nova de Gaia	yes	yes
Vila Real	yes	yes

Accessed 17 Nov. 2023

the public and, regarding the fiscal incentives under study, this study can be relevant in explaining the wealth with which companies can access and benefit from the information.

The analysis of this study continued with research on the information on the websites of the municipalities. The aim was to find out what kind of digital presence the municipalities have.

4 Results

Considering the objectives of this study and analysing each of the websites of the municipalities where the companies benefiting from the urban renewal fiscal incentive are located, Table 3 shows the digital presence of each municipality.

According to Table 3, all twenty municipalities analysed have a very active presence on the social network Facebook. Only four municipalities have WhatsApp contact available on its website. Of the twenty municipalities that were analysed, only three of the

Table 3. Digital presence (municipalities website)

Municipalities	Facebook	WhatsApp	Instagram	LinkedIn	YouTube	Twitter	RSS
Arcos de Valdevez	s	x	s	s	s	s	s
Barcelos	s	x	s	x	s	s	x
Braga	s	x	s	x	s	s	x
Chaves	s	x	x	x	s	x	x
Espinho	s	x	s	s	s	x	x
Esposende	s	x	s	x	s	x	s
Gondomar	s	x	s	s	s	s	s
Maia	s	s	s	x	s	s	x
Matosinhos	s	x	s	s	s	s	x
Mirandela	s	x	x	x	x	s	s
Oliveira de Azeméis	s	s	s	x	s	s	s
Penafiel	s	x	s	x	s	s	x
Ponte de Lima	s	x	s	s	s	s	s
Porto	s	s	s	x	s	x	x
Santa Maria da Feira	s	x	s	x	s	s	x
Sao João da Madeira	s	x	s	x	s	s	x
Valongo	s	s	s	s	s	s	x
Viana do Castelo	s	x	x	x	s	x	x
Vila Nova de Gaia	s	x	s	x	s	s	x
Vila Real	s	x	s	x	s	x	x

Municipalities	LinkTree	Issuu	Vimeo	Pinterest	Flickr	Google+	Telegram	Pocket
Arcos de Valdevez	x	x	x	x	x	x	x	x
Barcelos	x	x	x	x	x	x	x	x
Braga	x	x	x	x	x	x	x	x
Chaves	x	x	x	x	x	x	x	x
Espinho	s	x	x	x	x	x	x	x
Esposende	x	s	x	x	x	x	x	x
Gondomar	x	x	s	x	x	x	x	x
Maia	x	x	x	x	x	x	s	s
Matosinhos	x	x	x	s	x	x	x	x
Mirandela	x	x	x	x	x	x	x	x
Oliveira de Azeméis	x	x	x	x	x	x	x	x
Penafiel	x	x	x	x	x	x	x	x
Ponte de Lima	x	s	x	s	s	s	x	x
Porto	x	x	x	x	x	x	x	x
Santa Maria da Feira	x	x	x	x	x	x	x	x
Sao João da Madeira	x	s	x	x	x	x	x	x
Valongo	x	x	x	x	x	x	x	x
Viana do Castelo	x	x	x	x	x	x	x	x
Vila Nova de Gaia	x	s	x	x	x	x	x	x
Vila Real	x	x	x	x	x	x	x	x

Accessed 17 Nov. 2023. Encoding: s-yes –; x- no

Portuguese municipalities in the north of Portugal don´t have an official Instagram page. Since LinkedIn is a business social network, it would make sense for all the municipalities analysed in the table to have an official page on this social network, but this was not the case, because when analysing the data, only six of the analysed municipalities have an official page on LinkedIn. About the presence that the municipalities have on the Web, and especially on the social networks of YouTube and Twitter, almost all municipalities are present in these two social networks, although less frequently than on Facebook. More residual, some municipalities are present on LinkTree, Issuu, Vimeo, Pinterest, Flickr, Google +, Telegram and Pocket.

Although this is an exploratory study, it is interesting to note that Porto is the municipality where most of the beneficiary companies are located, and it is also the municipality with the largest digital presence. This seems to indicate, in line with the literature review, that the proximity shaped by a digital presence on several channels is important in terms of contact with citizens/companies, and that through these channels they can also access relevant information (in this case, access to fiscal benefits linked to environmental sustainability).

5 Conclusions, Limitations, and Future Research Perspectives

This study aims to show how Portuguese municipalities have a digital presence and how this presence can be relevant in the process of disseminating fiscal incentives (especially to companies) in the context of their local importance. Thus, based on an exploratory analysis of the municipalities of the Portuguese Northern Region, this research aims to present a portrait of the digital presence of municipalities and how it is linked to fiscal incentives for urban regeneration.

It was possible to analyse that the municipality of Porto is the municipality where most of the beneficiary companies are located, and it is also the municipality with the largest digital presence; it was also found that all twenty municipalities analysed have a very active presence on the social network Facebook.

However, only municipalities in the north of Portugal were analysed. As this is an exploratory study, it would be important to extend the analysis to the 308 Portuguese municipalities and see if there is a correlation between publicity and companies' access to this type of fiscal incentive. The methodology used is exploratory and therefore does not identify all the variables that influence the understanding of tax incentives for urban regeneration. Future studies should consider including factors such as the scale of economic activity, the size of the population and the dimensionality of the housing stock. This dimensionality could explain why the municipality of Porto is considered to offer the most significant tax incentives, regardless of its digital presence.

Acknowledgement. This work is financed by national funds through the FCT - Foundation for Science and Technology, I.P., under project UIDB/04043/2020.

References

1. Abreu, M.I., de Oliveira, R.A.F., Lopes, J.: Local governance and network-based policies for housing energy-related renovations: insights from a Portuguese case. WIT Trans. Ecol. Environ. **254**, 163–175 (2021)
2. Chan, F.K.S., Chan, H.K.: Recent research and challenges in sustainable urbanisation. Resour. Conserv. Recycl. **184**, 106346 (2022)
3. Chen, L., et al.: Green construction for low-carbon cities: a review. Environ. Chem. Lett. **21**(3), 1627–1657 (2023)
4. Ciampa, F.: A creative approach for the architectural technology: using the ExtrArtis model to regenerate the built environment. Sustainability (Switzerland) **15**(11), 9124 (2023)

5. Dias, S.L., Ribeiro, J.S.: Taxation policies as an environmental protection instrument: the portuguese case. In: Dinis, A.A., David, F., Pereira, L., Dias, S.L. (eds.), Taking on Climate Change Through Green Taxation, pp. 265–281. IGI Global, Hershey (2023)
6. Dinis, A.A., Carvalho, A., Dias, S.L.: Sustainable taxi fleet: an approach to the contribution of green taxation. In: Taking on Climate Change Through Green Taxation, pp. 241–264. IGI Global, Hershey (2023)
7. Duygan, M., Fischer, M., Ingold, K.: Assessing the readiness of municipalities for digital process innovation. Technol. Soc. **72**, 102179 (2023)
8. Dwivedi, Y.K., et al.: Setting the future of digital and social media marketing research: perspectives and research propositions. Int. J. Inf. Manage. **59**, 102168 (2021)
9. Fernandes, J., Chamusca, P., Pinto, J., Tenreiro, J., Figueiredo, P.: Urban rehabilitation and tourism: lessons from Porto (2010–2020). Sustainability (Switzerland) **15**(8), 6581 (2023)
10. González-Torres, M., Pérez-Lombard, L., Coronel, J.F., Maestre, I.R., Yan, D.: A review on buildings energy information: trends, end-uses, fuels and drivers. Energy Rep. **8**, 626–637 (2022)
11. Gordon, J.E.: Geoconservation principles and protected area management. Int. J. Geoheritage Parks **7**(4), 199–210 (2019)
12. He, W., et al.: Promoting green-building development in sustainable development strategy : a multi-player quantum game approach. Exp. Syst. Appl. **240**, 122218 (2024)
13. Lafioune, N., Desmarest, A., Poirier, É. A., St-Jacques, M.: Digital transformation in municipalities for the planning, delivery, use and management of infrastructure assets: strategic and organizational framework. Sustain. Futures **6**, 100119 (2023)
14. Liu, Y., Sang, M., Xu, X., Shen, L., Bao, H.: How can urban regeneration reduce carbon emissions? A Bibliometric Review. Land **12**(7), 1–19 (2023)
15. Ma, X., Fan, X.: A review of the studies on social media images from the perspective of information interaction. Data Inf. Manage. **6**(1), 100004 (2022)
16. Mavi, R.K., Gengatharen, D., Mavi, N.K., Hughes, R., Campbell, A., Yates, R.: Sustainability in construction projects: a systematic literature review. Sustainability (Switzerland) **13**(4), 1–24 (2021)
17. Mykhnenko, V.: Smart shrinkage solutions? The future of present-day urban regeneration on the inner peripheries of Europe. Appl. Geogr. **157**, 103018 (2023)
18. Ponzoa, J.M., Gómez, A., Mas, J.M.: EU27 and USA institutions in the digital ecosystem: proposal for a digital presence measurement index. J. Bus. Res. **154**, 113354 (2023)
19. Xie, F., Liu, G., Zhuang, T.: A comprehensive review of urban regeneration governance for developing appropriate governance arrangements. Land **10**(5), 545 (2021)
20. Yuan, Y.P., et al.: Government digital transformation: understanding the role of government social media. Gov. Inf. Q. **40**(1), 101775 (2023)

**2nd Workshop on Data Mining
and Machine Learning in Smart Cities**

Deep Learning Approaches for Socially Contextualized Acoustic Event Detection in Social Media Posts

Vahid Hajihashemi[1] , Abdorreza Alavi Gharahbagh[1] ,
Marta Campos Ferreira[2] , José J. M. Machado[2] ,
and João Manuel R. S. Tavares[2]([⊠])

[1] Faculdade de Engenharia, Universidade do Porto, Rua Dr. Roberto Frias, s/n,
4200-465 Porto, Portugal
[2] Departamento de Engenharia Mecânica, Faculdade de Engenharia, Universidade do
Porto, Rua Dr. Roberto Frias, s/n, 4200-465 Porto, Portugal
tavares@fe.up.pt

Abstract. In recent years, social media platforms have become an essential source of information. Therefore, with their increasing popularity, there is a growing need for effective methods for detecting and analyzing their content in real time. Deep learning is a machine learning technique that teaches computers to understand complex patterns. Deep learning techniques are promising for analyzing acoustic signals from social media posts. In this article, a novel deep learning approach is proposed for socially contextualized event detection based on acoustic signals. The approach integrates the power of deep learning and meaningful features such as Mel frequency cepstral coefficients. To evaluate the effectiveness of the proposed method, it was applied to a real dataset collected from social protests in Iran. The results show that the proposed system can find a protester's clip with an accuracy of approximately 82.57%. Thus, the proposed approach has the potential to significantly improve the accuracy of systems for filtering social media posts.

Keywords: Machine Learning · Acoustic Signals · Instagram · Twitter

1 Introduction

Social media platforms have become increasingly used to post and share multimedia content. This presents an opportunity to use machine learning techniques, especially deep learning models, to perform contextualized event detection by analyzing text, images, videos, and audio in social media posts. Recent research has shown that sound analysis can improve the understanding of social media posts by providing additional contextual data. Models can better contextualize events by combining textual and video analysis with acoustic features. Researchers have trained deep neural networks using text, images, videos, social relations, and audio clips from social media posts. Audio data help models learn

Á. Rocha et al. (Eds.): WorldCIST 2024, LNNS 990, pp. 347–358, 2024.
https://doi.org/10.1007/978-3-031-60328-0_35

the relations between sounds and social content. This multimodal approach combining audio signals and social context can provide robust contextualized event detection. Ongoing challenges include collecting labelled data, reducing noise, and generalizing data across diverse audios. Overall, deep learning combined with sound analysis offers a semantic system to gain a contextualized understanding of events discussed on social media posts by integrating multiple data modalities.

Souza et al. [1] proposed a methodology to overcome two key challenges in using social media posts to monitor real-world events at a local scale: data scarcity and noise. Their approach addresses these issues using an evolutionary optimization strategy. Specifically, they applied their methodology to monitor dengue fever outbreaks in Brazil using Twitter data. Dengue monitoring from social media posts poses difficulties due to the sparseness of location-specific data and noise from irrelevant tweets. Their evolutionary approach identified optimized location-specific models that filter out noise and maximize the signal for improved event monitoring. It outperformed the baseline models by significantly increasing the correlation between dengue cases and Twitter data in some locations, almost doubling it. This demonstrates how evolutionary optimization can help develop localized models tailored to social media data challenges.

The authors concluded that their methodology enables more effective real-time monitoring of disease outbreaks and potentially other events using social media posts. Noise and scarcity are challenges in using social media posts as an event monitoring tool. Based on the above, this study aimed to extract audio information from social posts to classify them into normal, inappropriate, or offensive groups. Classical approaches to this problem involve Mel Frequency Cepstral Coefficients (MFCCs), frequency spectrum and hand-crafting features from sound clips based on fixed or variable size time frames, and training machine learning models, mainly deep learning models. The difficulty is that selecting the appropriate features requires deep expertise in the field. Deep learning methods, such as recurrent neural networks (RNNs), have recently provided promising results for challenging sound-processing tasks.

This study proposes a method for processing sound data from social media clips. The proposed method uses basic features like features obtained from comments and video processing to determine the candidate content. Subsequently, MFCCs and Fourier transform are used for feature extraction. Finally, the proposed method uses a Long short-term memory (LSTM) model for semantic analysis. Finally, the results are analyzed to identify patterns.

2   Related Work

Social media posts provide a continuous stream of user-generated content, sometimes with sound, that can offer insights into unfolding real-world events. Liang et al. [2] proposed an iterative mining algorithm to detect events in Twitter data. Their key innovation was viewing tweets as spatial-temporal signals and applying signal processing techniques like noise filtering to improve event detection. Specifically, they considered the number of occurrences in time and location as noisy signals. Aiello et al. [3] provided valuable insights into utilizing social media audio content for understanding real-world events and social contexts. Their work constructing urban sound maps from social media tags. Deep neural networks could be trained to detect these audio characteristics automatically. They also offer a new insight for urban planners to understand soundscapes, events, and communities. Overall, the pioneering work by Aiello et al. motivated future research into socially contextualized event intelligence using social media audio and deep learning. He et al. [4] introduced a study on the effectiveness of self-organized social media channels in providing actionable information during a disaster. They analyzed the #PorteOuverte hashtag during the 2015 Paris attacks. They demonstrated the potential of deep learning to detect audio cues to improve awareness and disaster response automatically.

Marques et al. [5] highlighted the challenges of automatically generating high-quality audio descriptions of online videos, which are crucial for increasing automatic understanding of social media content. This motivated further research into deep neural networks that better capture context and details. A study conducted by Callcut et al. [6] explored how social media can be used for early notification of multiple casualty events in hospitals. They studied two incidents: the Asiana Airlines crash in San Francisco and the Boston Marathon bombing. They revealed that social media posts began emerging within minutes of these events, showing promise for early event detection. A major problem was distinguishing meaningful information from noise in the data. To address this challenge, they employed a dual approach based on keywords and location of relevant posts.

A study by Tindall and Robinson [7] focused on understanding how social networks contribute to environmental activism in Clayoquot Sound, a pristine rainforest in British Columbia, Canada. Using a mix of interviews, the researchers examined the social connections among activists involved in protests of Clayoquot Sound. Their findings highlighted the pivotal role of these social networks in uniting people for the cause of preserving the rainforest. Gasco et al. [8] studied how social media posts can gather people's reactions to noise. They introduced a method that combines noise with social media data to better understand how noise affects individuals. They applied their method to Twitter data to uncover citizens' opinions about noise in Bilbao, Spain.Their primary challenge was sorting out relevant tweets from irrelevant ones, focusing on noise-related content. To address this challenge, they employed keyword filtering and machine learning techniques. Purwins et al. [9] presented an overview of cutting-edge deep learning methods in audio signal processing. Their review encompassed speech, music, and environmental sound processing, highlighting shared traits and distinctions across these domains. Kumar and Mukesh [10] aimed to automate event detection from social posts, focusing on platforms like Twitter and Facebook. Their approach had three phases. First, they identified meaningful word relationships using Microblog Word Co-occurrence Networks (WCN). Second, they used k-bridge decomposition and topological sorting to extract event-related keywords.

Lastly, they ranked these keywords using the Analytical Hierarchical Process (AHP) based on attributes like Edge Strength Density and Phrase Degree.

Ye et al. [11] studied Instagram's potential as a tool for improving communication during and after the COVID-19 pandemic. They introduced a method that combines information sharing with Instagram to gain a deeper insight into how information impacts people. A challenge was filtering out irrelevant content and identifying posts relevant to COVID-19. To tackle this, they employed a combination of keyword filtering and machine learning techniques. The study's approach, examining the impact of information on social media posts during an event like the pandemic, resonates with the exploration of deep learning methods for socially contextualized event detection. Belcastro et al. [12] studied social media posts to detect sub-events within disasters. They proposed combining machine learning techniques with social media analysis to identify disaster-related sub-events. Their challenge was to sort relevant disaster-related posts from irrelevant ones, and they used keyword filtering, named entity recognition, and clustering techniques to address this.

Verlin [13] examined how abbreviations are formed on Instagram. Data was collected by browsing Instagram for relevant hashtags and accounts. The study faced challenges in analyzing the informal Instagram text. The author identified patterns in abbreviation formation, categorizing processes as "old" conventions or "new" unconventional methods. Bahuguna et al. [14] conducted a comprehensive review of recent methods for detecting events on Twitter, utilizing both single and multimodal data. They highlighted graph-based models, Convolutional Neural Networks (CNNs), and neural networks. They classified these techniques as correlation-based, graph-based, matrix decomposition, neural network-based, and topic-based.

Li et al. [15] studied techniques for detecting events in Twitter data, such as clustering-based, term-based, and neural network-based methods. A notable challenge was the growing diversity of tweet content, including text, images, videos, and more. They reviewed techniques, datasets, evaluation metrics, and research opportunities, providing a comprehensive snapshot of the ongoing state of event detection within social media streams. Kolajo et al. [16] introduced the SMAFED framework as a solution to enhance event detection in social media posts. They focused on managing informal language elements such as slang, abbreviations, and acronyms (SAB), which can pose challenges. To address this, they developed an integrated knowledge base (IKB) containing SAB terms and the SABDA algorithm to clarify ambiguous SAB terms using context drawn from tweets.

Mredula et al. [17] reviewed 67 research articles published between 2009 and 2021, all focused on event detection through social media data. The reviewed articles offered a comprehensive insight into the methods, tools, technologies, datasets, and metrics utilized in this field. The authors systematically categorized these articles based on methods, such as shallow machine learning, e.g., Support Vector Machine (SVM), random forest, deep learning, such as LSTM and CNN, rule-based, and other diverse approaches. Singh et al. [18] proposed an innovative technique for real-time event detection on Twitter, employing a specialized community detection algorithm. The core of their methodology relied on a graph-based community detection algorithm. Lasri et al. [19] devised a real-time system to assess sentiment towards Moroccan universities via Twitter. They tested random forest, Naive Bayes, logistic regression, decision tree, support vector machine, and XGBoost classifiers.

According to the above enlightenments, the main goal of this study was to propose a socially contextualized acoustic event detection (AED) approach in social media posts. The proposed approach fuses text, images, and other information to analyze abnormal events in social media posts.

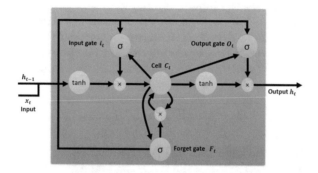

Fig. 1. Block diagram of an LSTM.

3   LSTM

In the domain of acoustic signal analysis for socially contextualized event detection within social media posts, utilizing LSTM networks is a pivotal breakthrough. LSTMs represent a specialized category of RNNs distinguished by their remarkable capacity to capture temporal dependencies in sequential data, rendering them exceptionally suited for processing acoustic signals. RNNs, at their core, exhibit proficiency in modelling temporal sequences and long-time dependencies. Nevertheless, conventional RNNs face issues like vanishing gradients when processing extended sequences. LSTMs have been ingeniously engineered to tackle these limitations by introducing memory cells and forget gate mechanisms into the architecture. The amalgamation of a CNN with an LSTM, creating a hybrid CNN-LSTM model, harnesses the synergistic strengths of both architectures for effective feature extraction and sequential modeling. Within this framework, the CNN layers are responsible for localized feature extraction, capturing spatial characteristics, while the LSTM layers are used in modelling temporal dependencies that evolve over time. In the context of audio input, the raw waveform undergoes multiple convolutional layers processing to extract low-level acoustic patterns. The pivotal constituents of the LSTM consist of the cell state, c_t, hidden state, g_t, and the essential gates named input, i_t, forget, f_t, and output, o_t, Fig. 1. The cell state serves as a memory repository that keeps information over time, while the input and forget gates control the flow of information into and out of the cell. Subsequently, the output gate determines how much the cell state influences the next output in the sequence. LSTM Gate equations are:

$$
\begin{aligned}
i_t &= \sigma \left(W_i \left[h_{t-1}, v_t \right] + b_i \right), \\
f_t &= \sigma \left(W_f \left[h_{t-1}, v_t \right] + b_f \right), \\
o_t &= \sigma \left(W_o \left[h_{t-1}, v_t \right] + b_o \right),
\end{aligned}
\tag{1}
$$

where i_t, f_t, o_t represent the input, forget, and output gates, respectively, h_{t-1} is the previous hidden state, v_t signifies the input at the current time step, W_i, W_f and W_o correspond to the weight matrices for each gate, b_i, b_f and b_o are the bias terms, and σ denotes the sigmoid activation function, constraining values within the range of 0 (zero) to 1 (one). The LSTM Cell Update can be written as:

$$
\begin{aligned}
g_t &= \tanh(W_g[h_{t-1}, v_t] + b_g), \\
c_t &= f_t * c_{t-1} + i_t * g_t,
\end{aligned}
\tag{2}
$$

where g_t represents the candidate cell state generated by applying the hyperbolic tangent, $tanh$, c_{t-1} denotes the cell state from the previous time step, i_t, f_t dictates the extent to which each element is updated, and c_t is subsequently passed through the output gate, o_t, to yield h_t. In event detection in social media posts, audio signals offer invaluable temporal context that text cannot provide. LSTMs innate sequential modeling capabilities make them exceptionally suited for identifying acoustic patterns evolving over time, which may hold predictive significance for detecting events. Furthermore, to enrich contextual cues, such as location and user metadata, attention mechanisms can be added to the model. For instance, an attention vector, a_t, can assign weightage to the relative significance of audio features, v_{at}, and contextual features, v_{ct}, at each time step:

$$
\begin{aligned}
a_t &= \text{softmax} \left(W_a \left[v_{at}, v_{ct} \right] \right), \\
v_t &= a_t * v_{at} + (1 - a_t) * v_{ct},
\end{aligned}
\tag{3}
$$

where v_{at} represents the audio feature input, v_{ct} corresponds to the contextual feature input, W_a denotes the weight matrices that assess the alignment between v_{at} and v_{ct}, and the Softmax function normalizes the scores to sum to 1 (one), effectively rendering them as probabilities. The used approach makes a socially aware representation, v_t, that the LSTM layer can effectively utilize. The sequential modelling capability of the LSTM facilitates the identification of audio patterns, while the attention vector weighting mechanism incorporates pertinent social context. This hybrid CNN-LSTM architecture combined the strengths of CNNs and LSTMs, including local feature extraction through convolution, sequential modelling via recurrence, and socially informed representation learning. It adeptly detects events in social media posts as a holistic, end-to-end model.

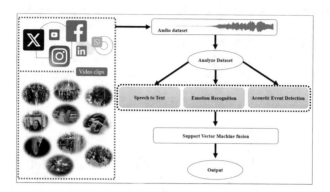

Fig. 2. Block diagram of the proposed method.

4   Proposed Method

The proposed method is an intelligent system based on LSTM and CNN models for hybrid AED in Instagram and Telegram clips, which consists of three main parallel steps. Each step was related to a specific task. The first step is "an API for converting Speech to Text" designed to detect a limited number of keywords from "Protest slogans". The second step involves emotion recognition, focusing on detecting Anger, Squeeze, and Excitation emotions. The third step is an AED system that identifies chanting, screaming, or shooting sounds. The results of these steps are fused in an SVM block to obtain the final output. A block diagram of the proposed method is shown in Fig. 2.

4.1   Speech-to-Text API

The Speech-to-Text API can accurately transcribe spoken text from 125 languages worldwide. However, in this research, the focus was just on Farsi and English languages. Consequently, these two languages were exclusively employed when using it. Once installed, the plugin can be invoked to convert English and Farsi speech into text with remarkable precision. Nonetheless, the primary drawback of API is its susceptibility to ambient noise and extraneous sound. An attempt was made to enhance the API's performance by filtering urban sounds, such as HAM signals, and minimizing background noise. Another limitation of this API is its inability to handle multi-speaker scenarios, which is common in crowded environments. Multiple speakers introduce inherent noise into audio samples that cannot be easily eliminated. Two parallel emotion recognition and AED networks were here employed to mitigate the potential errors and weaknesses.

4.2   Emotion Recognition

A deep neural network of three parallel components was employed to detect sound emotions. This approach was based on the method for emotion recognition based on acoustic signals proposed in [20]. As depicted in Fig. 3, the approach utilizes three distinct LSTM networks. Each network employs different features, including MFCC, spectrogram, and raw audio signals. Then, the outputs of these three networks are combined using a co-attention module to produce the result of the multimodal approach.

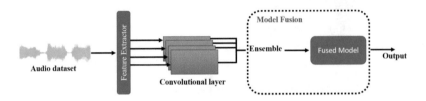

Fig. 3. Diagram of the AED block.

4.3   AED

The AED system used the urban sound tagging system proposed by Bai et al. [21]. These authors developed a multimodal urban sound tagging system using audio features and spatiotemporal context. In this study, only the audio features were used. The authors extracted four spectrogram representations from audio clips: log-Mel, log-linear, harmonic, and percussive spectrograms. The spectrograms provided complementary time-frequency information about the input sounds. Then, the authors fed the spectrograms into CNNs and residual CNNs to classify the audio data. These features were also tested with the models proposed in [22,23] for sound classification tasks. The model proposed by Bai et al. [21] yielded better results than the ones proposed in [22,23]. The diagram of this block is shown in Fig. 3.

4.4   Fusion

The fusion step comprises a binary SVM, which was trained to separate normal audio from important parts. As the classifier does not need to know the events in each frame, all frames are tested to include any event or have no event, and a final score is assigned to the total audio. The effect of different SVM kernel functions on the system accuracy was evaluated, and the best kernel was selected.

Table 1. Result of each step on test data.

	Speech to Text		Emotion Recognition		AED	
Usual	554	292	559	287	603	243
Protests	756	1398	708	1446	564	1590
Accuracy (%)	65.07		66.83		73.10	

5 Result

To evaluate the effectiveness of the proposed method, two top social media platforms in Iran: Telegram and Instagram, which are known to host a significant number of Persian news channels and audiences, were used. Based on available statistics from Persian news Telegram channels, 330 million social content items were posted during the first 100 d of the Mahsa Amini protests, i.e., from 16th September to 24th December 2022. Simultaneously, Persian-language Instagram social channels shared seven and a half million posts, garnering 4 billion views. Given the distinct characteristics of these two platforms, almost all Instagram posts contained audio, while only 40% of Telegram posts contained audio. For this study, 10,000 posts, including audio, were selected from both Instagram and Telegram. Although the sound quality varied considerably among these posts, they all had stereo sounds. Approximately 72% of the selected posts pertained to protests, with the remaining 28% containing other content.

The objective of the developed approach was to differentiate between posts related to protests and general posts based on their audio content. To assess the proposed method comprehensively, sports and public event clips, including screams and emotional sounds, were added to the dataset. The voice content of all clips was extracted using the Moviepy library in Python software, and all audio was resampled to a 16 kHz frequency. The average of the stereo sound channels was used as the input audio signal. In all simulations, 70% of the samples were used as train samples, and the remaining 30% were used as test samples. Table 1 presents the results of each step separately for the test samples. The training and test samples for the three steps were the same due to the proposed system structure.

The scores assigned to each class in the three steps were used as input for the fusion step, so the fusion step had six inputs. The fusion step was trained using the training data scores to determine the correct class in the output. The effect of three SVM kernel functions: Gaussian, linear, and polynomial, on the output was tested. The results are presented in Table 2, which indicates that the best results were obtained by using a polynomial kernel and, therefore, it was selected as the optimal kernel for the SVM classifier. According to the results obtained, one can conclude that the proposed system can accurately identify social events within social media posts based on their sounds. Therefore, the proposed system can be integrated into automatic social media content-processing systems to enhance efficiency.

Table 2. Result of fusion step on test data.

	Gaussian		Linear		Polynomial	
Usual	566	286	653	199	701	151
Protests	722	1426	492	1656	372	1776
Accuracy (%)	66.40		76.97		82.57	

6   Conclusion

This article proposed a social media content detection system based on analyzing audio content from posted media. The proposed system employs deep learning, audio features, and a fusion classifier that utilizes a binary SVM to enhance performance. The suggested approach combines three different systems: speech-to-text, emotion recognition, and AED. These systems are integrated into a fusion system that can accurately determine the social domain of the posted content. It can also identify whether a post is related to public discourse or social protests. To evaluate the effectiveness of the proposed method, it was tested using a real dataset collected from social protests in Iran in 2022 achieving very promising results.

Acknowledgements. This article is partially a result of the project Sensitive Industry, co-funded by the European Regional Development Fund (ERDF), through the Operational Programme for Competitiveness and Internationalization (COMPETE 2020), under the PORTUGAL 2020 Partnership Agreement. The first author would like to thank "Fundação para a Ciência e Tecnologia" (FCT) for his Ph.D. grant with reference 2021.08660.BD.

References

1. Souza, R.C.S.N.P., de Brito, D.E.F., Cardoso, R.L., de Oliveira, D.M., Meira, W., Pappa, G.L.: An evolutionary methodology for handling data scarcity and noise in monitoring real events from social media data. In: Bazzan, A.L.C., Pichara, K. (eds.) IBERAMIA 2014. LNCS (LNAI), vol. 8864, pp. 295–306. Springer, Cham (2014). https://doi.org/10.1007/978-3-319-12027-0_24
2. Liang, Y., Caverlee, J., Cao, C.: A noise-filtering approach for spatio-temporal event detection in social media. In: Hanbury, A., Kazai, G., Rauber, A., Fuhr, N. (eds.) ECIR 2015. LNCS, vol. 9022, pp. 233–244. Springer, Cham (2015). https://doi.org/10.1007/978-3-319-16354-3_25
3. Aiello, L.M., Schifanella, R., Quercia, D., Aletta, F.: Chatty maps: constructing sound maps of urban areas from social media data. Royal Soc. Open Sci. **3**(3), 150690 (2016). https://doi.org/10.1098/rsos.150690
4. He, X., Lu, D., Margolin, D., Wang, M., Idrissi, S.E., Lin, Y.-R.: The signals and noise: actionable information in improvised social media channels during a disaster. In: Proceedings of the 2017 ACM on Web Science Conference, pp. 33–42 (2017). https://doi.org/10.1145/3091478.3091501

5. dos Santos Marques, J.M., Valente, L.F.G., Ferreira, S.B.L., Cappelli, C., Salgado, L.: Audio description on Instagram: evaluating and comparing two ways of describing images for visually impaired. In: ICEIS, issue 3, pp. 29–40 (2017). https://doi.org/10.5220/0006282500290040

6. Callcut, R.A., Moore, S., Wakam, G., Hubbard, A.E., Cohen, M.J.: Finding the signal in the noise: could social media be utilized for early hospital notification of multiple casualty events? PLOS one **12**(10), e0186118 (2017). https://doi.org/10.1371/journal.pone.0186118

7. Tindall, D.B., Robinson, J.L.: Collective action to save the ancient temperate rainforest: social networks and environmental activism in Clayoquot sound. Ecol. Soc. **22**(1) (2017). https://doi.org/10.5751/ES-09042-220140

8. Gasco, L., Clavel, C., Asensio, C., de Arcas, G.: Beyond sound level monitoring: exploitation of social media to gather citizens subjective response to noise. Sci. Total Environ. **658**, 69–79 (2019). https://doi.org/10.1016/j.scitotenv.2018.12.071

9. Purwins, H., Li, B., Virtanen, T., Schlüter, J., Chang, S.-Y., Sainath, T.: Deep learning for audio signal processing. In: IEEE Journal of Selected Topics in Signal Processing, vol. 13, no. 2, pp. 206–219 (2019). https://doi.org/10.1109/JSTSP.2019.2908700

10. Kumar, M., et al.: An event detection technique using social media data (2019). http://hdl.handle.net/10603/285467

11. Ye, S., et al.: Turning information dissipation into dissemination: Instagram as a communication enhancing tool during the Covid-19 pandemic and beyond. J. Chem. Educ. **97**(9), 3217–3222 (2020). https://doi.org/10.1021/acs.jchemed.0c00724

12. Belcastro, L., et al.: Using social media for sub-event detection during disasters. J. Big Data **8**(1), 1–22 (2021). https://doi.org/10.1186/s40537-021-00467-1

13. Verlin, S.: Abbreviation establishment in Instagram social media. ETDC: Indonesian J. Res. Educ. Rev. **1**(4), 588–598 (2022). https://doi.org/10.51574/ijrer.v1i4.753

14. Bahuguna, R., Nisha Chandran, S., Gangodkar, D.: Recent trends in event detection from twitter using multimodal data. In: AIP Conference Proceedings, vol. 2481, no. 1. AIP Publishing (2022). https://doi.org/10.1063/5.0104560

15. Li, Q., Chao, Y., Li, D., Lu, Y., Zhang, C.: Event detection from social media stream: methods, datasets and opportunities. In: 2022 IEEE International Conference on Big Data (Big Data), pp. 3509–3516. IEEE (2022). https://doi.org/10.1109/BigData55660.2022.10020411

16. Kolajo, T., Daramola, O., Adebiyi, A.A.: Real-time event detection in social media streams through semantic analysis of noisy terms. J. Big Data **9**(1), 1–36 (2022). https://doi.org/10.1186/s40537-022-00642-y

17. Mredula, M.S., Dey, N., Rahman, M.S., Mahmud, I., Cho, Y.-Z.: A review on the trends in event detection by analyzing social media platforms' data. Sensors **22**(12), 4531 (2022). https://doi.org/10.3390/s22124531

18. Singh, J., Pandey, D., Singh, A.K.: Event detection from real-time twitter streaming data using community detection algorithm. Multimedia Tools Appl., 1–28 (2023). https://doi.org/10.1007/s11042-023-16263-3

19. Lasri, I., Riadsolh, A., Elbelkacemi, M.: Real-time twitter sentiment analysis for Moroccan universities using machine learning and big data technologies. Int. J. Emerging Technol. Learn. **18**(5), (2023). https://doi.org/10.3991/ijet.v18i05.35959

20. Zou, H., Si, Y., Chen, C., Rajan, D., Chng, E.S.: Speech emotion recognition with co-attention based multi-level acoustic information. In: ICASSP 2022-2022 IEEE International Conference on Acoustics, Speech and Signal Processing (ICASSP), pp. 7367–7371. IEEE (2022). https://doi.org/10.1109/ICASSP43922.2022.9747095
21. Bai, J., Chen, J., Wang, M.: Multimodal urban sound tagging with spatiotemporal context. IEEE Trans. Cogn. Dev. Syst., 555–565 (2022). https://doi.org/10.1109/TCDS.2022.3160168
22. Hajihashemi, V., Alavigharahbagh, A., Oliveira, H.S., Cruz, P.M., Tavares, J.M.R.S.: Novel time-frequency based scheme for detecting sound events from sound background in audio segments. In: Tavares, J.M.R.S., Papa, J.P., González Hidalgo, M. (eds.) CIARP 2021. LNCS, vol. 12702, pp. 402–416. Springer, Cham (2021). https://doi.org/10.1007/978-3-030-93420-0_38
23. Hajihashemi, V., Gharahbagh, A.A., Cruz, P.M., Ferreira, M.C., Machado, J.J., Tavares, J.M.R.: Binaural acoustic scene classification using wavelet scattering, parallel ensemble classifiers and nonlinear fusion. Sensors **22**(4), 1535 (2022). https://doi.org/10.3390/s22041535

Abnormal Action Recognition in Social Media Clips Using Deep Learning to Analyze Behavioral Change

Abdorreza Alavi Gharahbagh[1], Vahid Hajihashemi[1],
Marta Campos Ferreira[2], José J. M. Machado[2],
and João Manuel R. S. Tavares[2]([✉])

[1] Faculdade de Engenharia, Universidade do Porto, Rua Dr. Roberto Frias, s/n,
4200-465 Porto, Portugal
[2] Departamento de Engenharia Mecânica, Faculdade de Engenharia, Universidade do
Porto, Rua Dr. Roberto Frias, s/n, 4200-465 Porto, Portugal
`tavares@fe.up.pt`

Abstract. With the increasing popularity of social media platforms like
Instagram, there is a growing need for effective methods to detect and
analyze abnormal actions in user-generated content. Deep learning is
part of a broader family of machine learning methods based on artificial
neural networks with representation learning that can learn complex patterns. This article proposes a novel deep learning approach for detecting
abnormal actions in social media clips, focusing on behavioural change
analysis. The approach uses a combination of Deep Learning and textural, statistical, and edge features for semantic action detection in video
clips. The local gradient of video frames, time difference, and Sobel and
Canny edge detectors are among the operators used in the proposed
method. The method was evaluated on a large dataset of Instagram and
Telegram clips and demonstrated its effectiveness in detecting abnormal
actions with about 86% of accuracy. The results demonstrate the applicability of deep learning-based systems in detecting abnormal actions in
social media clips.

Keywords: Social Media · User-Generated Content · Temporal
Analysis · Instagram · Twitter

1 Introduction

The proliferation of short video clips on social media platforms such as Instagram has opened new possibilities for studying human behavior and actions
through computer techniques. One of these techniques is deep learning to analyze behavioral changes in media video clips. This approach uses the vast amount
of data available on social media posts to train machine learning models that
recognize and classify different actions and behaviors. By analyzing patterns

Á. Rocha et al. (Eds.): WorldCIST 2024, LNNS 990, pp. 359–370, 2024.
https://doi.org/10.1007/978-3-031-60328-0_36

and changes in behavior over time, it is possible to gain insights into how people interact with each other and their environment. In web archiving, Schefbeck et al. [1] introduced a fascinating paradigm shift by highlighting the archiving challenges posed by the dynamic Social Web. They demonstrated the role of the social Web as a democratic platform for public expression about social events. They also analyzed the contextualization of the social web, extracted insights about social events, subjects, and entities, and created comprehensive and socially contextualized digital archives [1]. Batrinca et al. [2] analyzed these clips to autonomously detect abnormal behaviors that might indicate disorders, risk factors, or behavioural changes. Recent advances in cutting-edge deep learning techniques, such as Convolutional Neural Networks (CNNs), are helping governments, customers, and digital companies to detect abnormal actions in media video clips semantically. These systems require initial training on a labelled database to train patterns of normal and abnormal actions. Automatic content filtering reviews content to restrict or block nasty clips or users. This could include material that compromises security or violates content policies based on international rules. In an ideal, innovative abnormal activity detection system, users can define inappropriate or offensive content for the media platform system, which will hide inappropriate or offensive posts, stories, and live videos.

In a human action recognition (HAR) system, the main problem is classifying a sequence of frames in a video clip recorded by specialized harnesses or smartphones into known, well-defined movements. In abnormal action detection, the main problem is classifying a clip into an acceptable or unacceptable group. Given the large number of clips and media produced and uploaded every second, the temporal nature of the clips, and the lack of an easy way to define inappropriate or offensive clips, this is a difficult problem. Based on the above explanations, this work aimed to classify social posts into normal, inappropriate, or offensive groups. Classical approaches to this problem involve edge, textural, frequency, and hand-crafting features from the frame image sequence or video based on fixed or variable size windows and training machine learning models, such as ensembles of decision trees. The difficulty is that this feature engineering requires deep expertise in the field. Deep learning methods such as CNN and recurrent neural networks have recently provided promising results on challenging activity recognition tasks with little or no data feature engineering.

This article proposes a method for abnormal activity detection in social media posts, which uses some basic features to find the region of interest. Then, the local histogram and co-occurrence matrix are used for feature extraction. Finally, the proposed method uses a CNN classifier to classify clips. The last step is to analyze the results to find a pattern of behavior change.

2   Related Work

Murthy et al. [3] presented a study on interpreting Instagram images shared on Twitter during Hurricane Sandy. This research focused on the dynamic nature of social media platforms such as Twitter and Instagram, which provide quick and free information to a wide audience, proving particularly valuable for crisis communication. The authors showed that the visual data and tweets generated during the hurricane provided important insights for understanding the social impact of such events [3]. Alam et al. [4] presented a comprehensive image processing system for social media posts called Image4Act. The main goal of this system is to capture, enhance, and categorize visual content shared on social media platforms. The system uses a deep neural network and perceptual hashing techniques to effectively deal with the inherent noise prevalent in social media image posts [4]. Sherchan et al. [5] presented a tool known as the Australian Crisis Tracker (ACT) to improve access to important information accessible through social media platforms, particularly for individuals and organizations related to natural disasters. They split each tweet into some parts within a pipeline, including filtering, metadata extraction, and image retrieval. The tweets were finally divided into coherent clusters. Geboers and Van De Wiele [6] proposed a method for automatically ranking images in social posts using image processing, hashtags, and retweet counts. They used hashtag frequencies versus time, retweets, and images as input and extracted patterns using co-occurrence analysis, a network of hashtags, object annotation, and retweet statistics.

Gul et al. [7] presented an online abnormal human activity detection technique for patient monitoring. Their approach uses a You Only Look Once (YOLO) CNN. Graf et al. [8] conducted a comprehensive study of posts shared on Twitter and Instagram related to orthodontics. Over 30 d, they collected 361 posts about orthodontics on Twitter and Instagram and ranked them using a combination of qualitative content analysis. They examined the differences between Twitter and Instagram posts and showed that the latter platform had a higher frequency of posts classified as positive in sentiment. Appel, Gil et al. [9] analyzed the challenge of social media models in future marketing. They analyzed challenges such as omni-social presence, the emergence of new forms of social influence and influencers, privacy concerns, loneliness and isolation, and customer care. Jacobsen and Barnes [10] evaluated the impact of social media on Gen Z consumer behavior. Their research highlights that social media increases intrinsic pressure on Generation Z, affecting their moral compass and sometimes leading to atypical behavior in the marketplace.

Zhou and Deng [11] studied strategies for detecting abnormal activities involving multiple individuals in video surveillance scenarios. The studied methods include various techniques, such as spatiotemporal features for tracking motion, optical flow, statistical models to encapsulate unusual patterns and sparse coding methods wherein dictionaries of typical patterns are learned, enabling the detection of anomalies as deviations from the norm. Kınlı et al. [12] introduced the methods and outcomes of the AIM 2022 challenge on Instagram Filter Removal. The main objective of this challenge was to enhance filtering accuracy on images that deceive filters. Two preliminary studies on this task were included to establish a comparative baseline. A review of video-based HAR using deep learning across various video datasets is given in [13]. They reviewed deep architectures such as CNN, Recurrent Neural Networks (RNNs), and Deep belief networks (DBNs). Yeo et al. [14] presented an innovative multiscale color attention network called CAIR to remove Instagram filters. Their network was a nonlinear, activation function free architecture known as NAFNet. Additionally, CAIR contains a color attention module that effectively extracts color information to improve accuracy.

Gongane et al. [15] present a strategy for identifying harmful content in social media posts through Artificial Intelligence (AI) techniques. Their approach integrates AI methods, particularly Natural Language Processing (NLP), with Machine Learning (ML) algorithms and Deep Neural Networks. Kushwaha et al. [16] presented an approach to human activity recognition by fusing dynamic motion patterns, Optical Flow (OF), and Histogram of Oriented Gradients (HOG) to produce a robust and perspective-independent feature descriptor. This work addresses significant challenges associated with human activity recognition, such as interference from camera motion, irregular human motion patterns, fluctuating illumination conditions, complicated backgrounds, and variations in the shape and size of human subjects across video clips of the same activity category.

Pogadadanda et al. [17] presented a comprehensive study of abnormal activity detection in the context of surveillance. The article discussed the key merits, challenges, and noteworthy contributions of HAR algorithms and their applications across various scenarios. In the face of rising crime rates, ensuring public safety and security has become a primary societal concern. In response, security cameras have become constant watchdogs that keep a close eye on the behavior of the public [17]. According to the above, it is possible to analyze the changes in the videos of users from different geographic regions by using HAR and abnormal action detection algorithms to detect behavioral changes. Based on the review, none of the previous researchers has used this approach to analyze the change in social behavior over time.

3   CNN

In the proposed method of analyzing abnormal actions in Instagram clips using a CNN. CNNs are specialized algorithms for complex pattern recognition, such as in image and video, designed to process input data through multiple layers with filters that extract various features. These features are then processed using nonlinear activation functions like ReLU. The core of a CNN is the convolutional layer, which employs filters to create feature maps by convolving with input data. During training, these filters adaptively learn to minimize the difference between predicted and actual outputs.

CNNs are highly effective because they can learn hierarchical data representations. They start by recognizing basic features like edges and corners in early layers and progress to understanding more complex attributes like object components in later layers. This hierarchical learning enables CNNs to achieve remarkable accuracy in challenging recognition tasks. In the context of social media posts, CNNs can be used to recognize abnormal actions and analyze behavioral changes in user posts. The convolution operation in CNNs can be mathematically expressed as:

$$I(x, y) = \sum_{(i=1)}^{H} \sum_{(j=1)}^{W} K(i, j). \ P(x - i, y - j),$$ (1)

where $I(x, y)$ represents the output of the convolution operation, H and W are the dimensions of the kernel, $K(i, j)$ is the kernel value at position (i, j), and $P(x - i, y - j)$ is the input pixel at position $(x - i, y - j)$. In CNNs, filters are applied to input data in a hierarchical manner. This hierarchical learning enables CNNs to recognize intricate patterns and achieve remarkable accuracy. CNNs are particularly effective for HAR in videos by leveraging convolutional filters to extract discriminative spatiotemporal features. In video action recognition, CNNs process sequences of frames as input, and typically, the architecture begins with convolutional layers, which apply 3D or 2D filters to capture spatio-temporal information. The convolutional operation can be written as:

$$y_i^l = f(\sum_j w_{ij}^l * x_j^{l-1} + b_i^l),$$ (2)

where y_i^l is the uutput feature map, f the non-linear activation function, w_{ij}^l the filter kernel, x_j^{l-1} the input, and b_i^l the bias. Pooling layers follow convolutional layers to downsample feature maps and reduce computational requirements. The architecture culminates in fully connected layers for action classification. Advanced designs incorporate techniques like 3D convolutional filters, recurrent connections, and attention mechanisms to enhance spatio-temporal modeling. Key advantages of using CNNs for video action recognition include:

- Automatic feature extraction from raw pixel inputs;
- Hierarchical feature learning at multiple levels of abstraction;
- Weight sharing across space and time for parameter efficiency;
- Access to large labeled video datasets for representation learning.

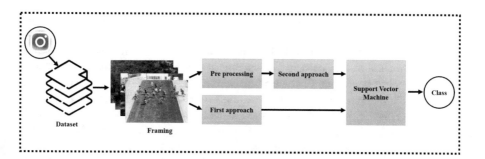

Fig. 1. Block diagram of the proposed method.

4   Proposed Method

This article proposes a method that uses two distinct approaches to detect events in social video posts. Figure 1 shows the block diagram of the proposed method. The first approach is based on the complex action recognition system proposed in [18]. Within this deep neural network event detection system, the main layer is the Timeception layer, as illustrated in Fig. 2, which consists of a combination of one-dimensional (1D) and two-dimensional (2D) convolution layers, along with a concatenation layer. This deep neural network structure enables the model to capture long-term and short-term temporal dependencies like an LSTM structure.

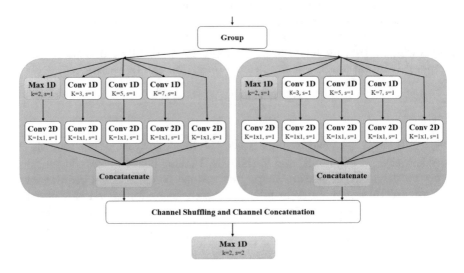

Fig. 2. Block diagram of the Timeception layer [18].

One notable advantage of the Timeception layer is its proficiency in handling fragmented information, which is processed concurrently through the 1D or 2D convolution blocks. Moreover, its simplified structure and reduced computational overhead when updating the weight parameters compared to more complex structures like LSTM make it particularly interesting for many applications. In the final model used in the current study, four Timeception layers were leveraged as the backbone of the system, alongside the use of ResNet-152 [19] as the used CNN. Each video input must contain a minimum of 128 frames; thus, given the typical quality of social media posts, the minimum duration should exceed 6 s, a criterion all the training clips used in this study meet. The output of ResNet-152 serves as the input for the Timeception layers and is then applied to some standard layers, such as BatchNorm and ReLU. This network proves highly efficient in terms of computational speed and effectiveness for action detection tasks. Other parameters used in this step are aligned with those detailed in [18].

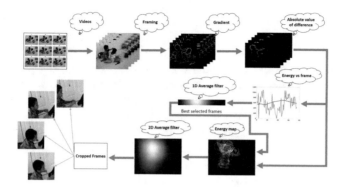

Fig. 3. The pre-processing step of the second HAR approach.

In the second approach, the method proposed by Gharahbagh et al. [20], which was designed to aid HAR systems, was used, which employs gradient and edge operators to find regions of interest within video frames exhibiting notable visual changes. Figure 3 illustrates the block diagram of this technique. The resultant frames are then used to train a HAR system.

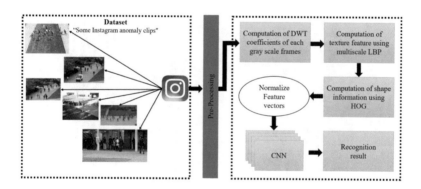

Fig. 4. Block diagram of the second HAR approach.

The HAR methods introduced in [21–25] were tested in the current study. Due to the obtained results, the method proposed in [24] was modified and a CNN-based system was used with its features. This model is based on the network architecture described in [24], which used the correlation between spatial and temporal features. Figure 4 shows the block diagram of the second approach. The last step is a binary SVM, which was trained to assign normal or abnormal labels to an input social video post. In this step, the classifier judges based on two scores assigned by the two classifier approaches, and the final score is assigned to the input clip. The effect of different SVM kernel functions on the system accuracy was studied, and the best kernel was selected.

5   Result

To evaluate the accuracy of the proposed abnormal event detection system, some clips were collected from two social media platforms in Iran: Telegram and Instagram. These platforms are known to host many Persian news, social, marketing, and political channels. Based on available statistics from social media posts, Telegram channels shared 330 million social posts during the first 100 d of the Mahsa Amini protests, which occurred from 16th September to 24th December 2022). Persian-language Instagram influencers shared seven and a half million posts during the same period. Given the different characteristics of these two platforms, almost all Instagram posts are clips, while only 40% of the Telegram posts contain video. For the current study, 10,000 clips were selected from both Instagram and Telegram platforms. Approximately 67% of the selected clips include abnormal actions, with the remaining 33% containing normal content. The implemented approach aimed to differentiate between posts related to abnormal and normal actions. To assess the proposed method comprehensively, sports and public event clips, including running and normal emotional actions, were added to the experimental dataset. The quality of the video clips under study is different, and they were all resized to 128×128 to overcome this problem.

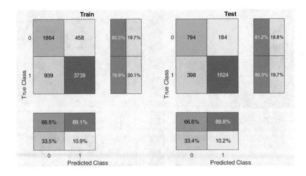

Fig. 5. Confusion matrix obtained for the first approach.

In all simulations, 70% of the samples were used as training data, and the remaining were used as testing data. Figures 5 and 6 present the results of each approach separately for the training and testing datasets. The used training and testing datasets were the same for the two approaches. The data of Fig. 5 indicates that the accuracy achieved by the first approach was about 75% in the training dataset and a few better in the testing dataset. Using the second approach, the obtained result was better: about 80% in the training dataset and 80.8% in the testing dataset, Fig. 6. This can result from the added preprocessing step.

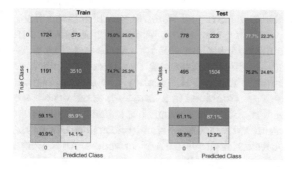

Fig. 6. Confusion matrix obtained for the second approach.

The scores assigned to each class in the two parallel classifier steps were used as input for the binary SVM final classifier. The SVM was trained using the training data scores to determine the correct final class. The effect of three SVM kernel functions: Gaussian, linear, and polynomial, was tested on the final output of the proposed system, Table 1. As the best results were obtained using the linear kernel, it was selected as the optimal kernel for the SVM classifier. According to the results of Table 1, the proposed system can accurately separate abnormal actions from normal within social media video posts using a combination of two parallel deep learning based approaches. Therefore, the proposed system seems interesting to be used as a smart filter in social media posts.

Table 1. Confusion Matrix obtained for the SVM final classifier.

	Training Set		Test Set	
Linear	1974	292	891	143
	634	4100	263	1703
Gaussian	1882	389	852	177
	791	3938	297	1674
Polynomial	1879	417	839	165
	798	3906	344	1652

6 Conclusion

This research introduces a system to separate abnormal actions from normal ones in social media video posts. The proposed system uses two deep learning based approaches and an extra pre-processing step. Using the added pre-processing step, training one of the deep neural networks is conducted using the spatiotemporal features extracted from regions of interest from the input video clips. This pre-processing step enhances the performance of the HAR method compared to the method without this pre-processing. The core emphasis of the proposed method lies in deep learning techniques, employing added layers to extract spatiotemporal features. Ultimately, the outcomes from these two deep learning techniques are combined in a final step to enhance the overall efficiency. Experimental results obtained through implementing the proposed method on real datasets demonstrated that the simultaneous use of two HAR methods can significantly improve the final system's effectiveness. An important feature of the proposed system is its adaptability to low-quality videos, which reduces computational requirements and allows for online implementation. In conclusion, the proposed method can be used as an intelligent and trainable semantic filter for social media video posts.

Acknowledgements. This article is partially a result of the project Sensitive Industry, co-funded by the European Regional Development Fund (ERDF), through the Operational Programme for Competitiveness and Internationalization (COMPETE 2020), under the PORTUGAL 2020 Partnership Agreement. The second author would like to thank "Fundação para a Ciência e Tecnologia" (FCT) for his Ph.D. grant with reference 2021.08660.BD.

References

1. Schefbeck, G., Spiliotopoulos, D., Risse, T.: The recent challenge in web archiving: archiving the social web. Context **7**, 9 (2012)
2. Batrinca, B., Treleaven, P.C.: Social media analytics: a survey of techniques, tools and platforms. Ai Soc. **30**, 89–116 (2015). https://doi.org/10.1007/s00146-014-0549-4
3. Murthy, D., Gross, A., McGarry, M.: Visual social media and big data. interpreting Instagram images posted on twitter. Digit. Culture Soc. **2**(2), 113–134 (2016). https://doi.org/10.14361/dcs-2016-0208
4. Alam, F., Imran, M., Ofli, F.: Image4Act: online social media image processing for disaster response. In: Proceedings of the 2017 IEEE/ACM International Conference on Advances in Social Networks Analysis and Mining 2017, pp. 601–604 (2017). https://doi.org/10.1145/3110025.3110164
5. Sherchan, W., Pervin, S., Butler, C.J., Lai, J.C., Ghahremanlou, L., Han, B.: Harnessing Twitter and Instagram for disaster management. IBM J. Res. Dev. **61**(6), 8:1-8:12 (2017). https://doi.org/10.1147/JRD.2017.2729238
6. Geboers, M.A., Van De Wiele, C.T.: Machine vision and social media images: why hashtags matter. Soc. Media+ Soc. **6**(2), 2056305120928485 (2020). https://doi.org/10.1177/2056305120928485
7. Gul, M.A., Yousaf, M.H., Nawaz, S., Ur Rehman, Z., Kim, H.: Patient monitoring by abnormal human activity recognition based on CNN architecture. Electronics **9**(12), 1993 (2020). https://doi.org/10.3390/electronics9121993
8. Graf, I., Gerwing, H., Hoefer, K., Ehlebracht, D., Christ, H., Braumann, B.: Social media and orthodontics: a mixed-methods analysis of orthodontic-related posts on Twitter and Instagram. Am. J. Orthod. Dentofac. Orthop. **158**(2), 221–228 (2020). https://doi.org/10.1016/j.ajodo.2019.08.012

9. Appel, G., Grewal, L., Hadi, R., Stephen, A.T.: The future of social media in marketing. J. Acad. Market Sci. **48**(1), 79–95 (2020). https://doi.org/10.1007/s11747-019-00695-1

10. Jacobsen, S.L., Barnes, N.G.: Social Media, Gen Z and consumer misbehavior: Instagram made me do it. J. Market. Dev. Competitiveness **14**(3), 51–58 (2020). https://doi.org/10.33423/jmdc.v14i3.3062

11. Zhou, Y., Deng, M.: A review of multiple-person abnormal activity recognition. J. Image Graph. **9**(2), 55–60 (2021). https://doi.org/10.18178/joig.9.2.55-60

12. Kinli, F., et al.: Aim 2022 challenge on Instagram filter removal: methods and results. In: European Conference on Computer Vision, pp. 27–43. Springer (2022). https://doi.org/10.1007/978-3-031-25066-8_2

13. Pham, H.H., Khoudour, L., Crouzil, A., Zegers, P., Velastin, S.A.: Video-based human action recognition using deep learning: a review (2022). arXiv preprint arXiv:2208.03775

14. Yeo, W.-H., Oh, W.-T., Kang, K.-S., Kim, Y.-I., Ryu, H.-C.: CAIR: fast and lightweight multi-scale color attention network for Instagram filter removal. In: European Conference on Computer Vision, pp. 714–728. Springer (2022). https://doi.org/10.1007/978-3-031-25063-7_45

15. Gongane, V.U., Munot, M.V., Anuse, A.D.: Detection and moderation of detrimental content on social media platforms: current status and future directions. Soc. Netw. Anal. Min. **12**(1), 129 (2022). https://doi.org/10.1007/s13278-022-00951-3

16. Kushwaha, A., Khare, A., Prakash, O.: Human activity recognition algorithm in video sequences based on the fusion of multiple features for realistic and multi-view environment. Multimedia Tools Appl. 1–22 (2023). https://doi.org/10.1007/s11042-023-16364-z

17. Pogadadanda, V., Shaik, S., Neeraj, G.V.S., Siralam, H.V., Rao, K.B., et al.: Abnormal activity recognition on surveillance: a review. In: 2023 Third International Conference on Artificial Intelligence and Smart Energy (ICAIS), pp. 1072–1077. IEEE (2023). https://doi.org/10.1109/ICAIS56108.2023.10073703

18. Hussein, N., Gavves, E., Smeulders, A.W.: Timeception for complex action recognition. In: Proceedings of the IEEE/CVF Conference on Computer Vision and Pattern Recognition, pp. 254–263 (2019)

19. He, K., Zhang, X., Ren, S., Sun, J.: Deep residual learning for image recognition. In: Proceedings of the IEEE Conference on Computer Vision and Pattern Recognition, pp. 770–778 (2016)

20. Gharahbagh, A.A., Hajihashemi, V., Ferreira, M.C., Machado, J.J., Tavares, J.M.R.: Best frame selection to enhance training step efficiency in video-based human action recognition. Appl. Sci. **12**(4), 1830 (2022). https://doi.org/10.3390/app12041830

21. Carreira, J., Zisserman, A.: Quo vadis, action recognition? a new model and the kinetics dataset. In: Proceedings of the IEEE Conference on Computer Vision and Pattern Recognition, pp. 6299–6308 (2017)

22. Zheng, Z., An, G., Ruan, Q.: Motion guided feature-augmented network for action recognition. In: 2020 15th IEEE International Conference on Signal Processing (ICSP), vol. 1, pp. 391–394. IEEE (2020). https://doi.org/10.1109/ICSP48669.2020.9321026

23. Chen, E., Bai, X., Gao, L., Tinega, H.C., Ding, Y.: A spatiotemporal heterogeneous two-stream network for action recognition. IEEE Access **7**, 57:267-57:275 (2019). https://doi.org/10.1109/ACCESS.2019.2910604

24. Yudistira, N., Kurita, T.: Correlation net: spatiotemporal multimodal deep learning for action recognition. Signal Process. Image Commun. **82**, 115731 (2020). https://doi.org/10.1016/j.image.2019.115731

25. Alavigharahbagh, A., Hajihashemi, V., Machado, J.J., Tavares, J.M.R.: Deep learning approach for human action recognition using a time saliency map based on motion features considering camera movement and shot in video image sequences. Information **14**(11), 616 (2023). https://doi.org/10.3390/info14110616

2nd Workshop on Enabling Software Engineering Practices Via Last Development Trends

Exploring Software Quality Through Data-Driven Approaches and Knowledge Graphs

Raheela Chand[1]([⊠]), Saif Ur Rehman Khan[2], Shahid Hussain[3], Wen-Li Wang[3], Mei-Huei Tang[4], and Naseem Ibrahim[3]

[1] Department of Computer Science, COMSATS University Islamabad (CUI), Islamabad, Pakistan
raheela.chand@yahoo.com
[2] Department of Computing, Shifa Tameer-e-Millat University, Park Road Campus, Islamabad, Pakistan
[3] Department of Computer Science and Software Engineering, School of Engineering, Penn State University, Behrend, USA
[4] Computer and Information Science, Gannon University, Pennsylvania, USA

Abstract. Context: The quality of software systems has always been a crucial task and has led to the establishment of various reputable software quality models. However, the automation trends in Software Engineering have challenged the traditional notion of quality assurance, motivating the development of a new paradigm with advanced AI-based quality standards.

Objective: The goal of this paper is to bridge the gap between theoretical frameworks and practical implementations on the aspects of software quality.

Methodology: This study involved an extensive literature review of software quality models, including McCall, Boehm, Dromey, FURPS, and ISO/IEC 25010. The detailed information about quality attributes from each model was systematically synthesized and organized into datasets, data frames, and Python dictionaries. The resulting resources were then shared and made accessible through a public GitHub repository.

Results: In brief, this research provides (i) a comprehensive dataset on software quality containing catalogs of quality models and attributes, (ii) a Python dictionary encapsulating the quality models and their associated characteristics for convenient empirical experimentation, (iii) the application of advanced knowledge graph techniques for the analysis and visualization of software quality parameters, and (iv) the complete construction steps and resources for download, ensuring easy integration and accessibility.

Conclusion: This study builds a foundational step towards the standardization of automating software quality modeling to enhance not just quality but also efficiency for software development. For our future work, there will be a concentration on the practical utilization of the dataset in real-world software development contexts.

© The Author(s), under exclusive license to Springer Nature Switzerland AG 2024
Á. Rocha et al. (Eds.): WorldCIST 2024, LNNS 990, pp. 373–382, 2024.
https://doi.org/10.1007/978-3-031-60328-0_37

Keywords: Software Quality Models · Unified Framework · Data-Driven Paradigm · Decision Support Systems · Knowledge Graphs · Empirical Study · Datasets · Python Libraries

1   Introduction

Quality-oriented software development is paramount in the software industry, exerting a direct influence on customer satisfaction [1], organizational reputation [2], and operational efficiency[3]. This has led to persistent efforts by practitioners and academia to advance quality standards, resulting in the introduction of specialized frameworks such as software vulnerability frameworks [4] and those for malware detection [5,6]. Additionally, various structured approaches, like the Quality Attributes Extractor in Agile-driven software development [7] and the Quality Features Extractor (QFE) for Android Apps [8], have been proposed. Comprehensive models such as McCall [9,10], Boehm, Dromey, FURPS, and ISO/IEC 25010 [11] have been established to evaluate different dimensions of software quality.

In parallel, traditional information sources like user stories, user reviews [12], and domain experts [13] have played a vital role in shaping the understanding of quality attributes. However, the software development landscape is undergoing significant transformations, particularly with the integration of advanced artificial intelligence (AI) based quality standards [14]. This shift challenges conventional notions of quality assurance, prompting a re-evaluation of existing practices. The literature review conducted in this study revealed a noteworthy gap-a lack of studies offering a comprehensive repository for quality parameters [15–17]. To address this gap, our prior work presented a dataset for user reviews [18]. This study extends that study and aims to reduce biases introduced by domain experts and promote a more universal understanding of knowledge within the quality domain.

To summarize, the contributions of this research are as follows:

1. Synthesis of a comprehensive list of software quality attributes from five prominent quality models.
2. Development of knowledge graphs depicting the complex relationships between quality attributes and their defining characteristics.
3. Provision of downloadable datasets and Python notebooks for accessibility and utilization by researchers, practitioners, and stakeholders.
4. Presentation of a unified and cohesive perspective on software quality attributes, fostering a holistic understanding for effective software development and risk management.

The rest of the paper is organized as follows. Section 2 explains the related work of this study. Section 3 describes the steps for constructing the dataset and Sect. 4 provides the results and compiled data. Section 5 provides research implications and Sect. 6 examines potential threats to validity. Finally, Sect. 7 concludes the paper.

2   Related Work

This section investigates studies in quality-oriented software development, spanning practical applications, empirical studies, and conceptual research proposals. Ahmed et al. [7] proposed a conceptual framework for automatic extraction and prioritization of quality attributes in agile-based software development, using natural language processing. Tyang et al. [3] contributed with a three-stage framework for prioritizing perceived quality attributes using social media data and text mining. Chand et al. [8] proposed an automatic Quality Feature Extraction (QFE) framework for Android apps, utilizing Natural Language Processing (NLP) [19] and Topic Modeling [20] to extract quality attributes from user reviews. Croft et al. [3] explored data quality problems in software vulnerability datasets, revealing issues in the accuracy and duplication of vulnerability labels. Aparna et al. [1] presented a fault prediction model for identifying components susceptible to faults. Several studies on single-quality attributes such as usability, security, malware detection, and reliability have provided valuable insights. However, existing studies lack an empirically comprehensive understanding of quality attributes. This study aims to fill this gap by contributing structured datasets and visual representations for a comprehensive understanding of software quality.

3   Construction of Quality Attributes Dataset

In this section, we describe the systematic process employed to construct the dataset of quality attributes, as depicted in Fig. 1. The following steps outline the comprehensive methodology applied:

Fig. 1. Dataset Construction Steps

3.1   Step 1: Models Selection

The first step involved selecting the quality standard models that collectively provide a comprehensive view of software quality. The chosen models, the McCall Model, Dormey Model, Boehm Model, FURPS Model, and ISO/IEC 25010, were based on their prominence and ability to cover diverse dimensions of quality aspects.

3.2   Step 2: Attributes Identification

The second step involved identifying and extracting the quality attributes of each quality model. For this process, a detailed examination of model documentation and the literature was reviewed, ensuring a thorough understanding of the attributes encapsulated by each model. This review is explained in detail in the results and data compilation section of this study.

3.3   Step 3: Data Unification and Categorization

The third step involved understanding the quality models and constructing the categorization mechanism of the dataset. To compile a cohesive dataset, quality attributes were gathered and standardized from various models. The standardization process started with aligning similar attributes across models to ensure uniformity. Next, the attributes were organized into common themes, i.e., simplifying the dataset for enhanced clarity and usability.

3.4   Step 4: Definition Refinement

This step involved refining attribute definitions for precision. Synthesizing information from the literature, we crafted comprehensive definitions, integrating insights on purpose, benefits, and evaluation criteria. Data collected through the literature review informed nuanced definitions, with manual corrections for technical precision and alignment with industry terminology.

3.5   Step 5: Characteristics Extraction

The fifth step covered the extraction of characteristics associated with each attribute. These characteristics offer supplementary context and insights into the nature of the attributes, thereby enriching the dataset with detailed information. Given the complexity of attributes and their specific meanings in software development, we resorted to a literature review to identify and extract three key characteristics defined for each quality attribute.

3.6   Step 6: Dataset Compilation

The final step involved compiling the refined quality attributes, their definitions, and characteristics into a structured tabular format. To complement these tables, detailed annotations for each attribute are provided to facilitate a comprehensive understanding of the empirical studies. Please refer to Sect. 4 for annotations and the downloadable resource files.

4   Results and Discussion

This section presents a compilation of the final results and information about the dataset usage.

4.1   Data Compilation

The resulting dataset presents a synthesized table of 14 distinctive quality attributes, derived from comprehensive analyses of prominent industry models, including McCall, Dromey, Boehm, FURPS, and ISO/IEC 25010. Each quality attribute is succinctly described by three defining characteristics in the synthesized table. The final list is provided in Table 1. Due to page limitations, detailed tables for each model are available for download in the provided link. The downloadable resource serves as a valuable reference for developers and researchers, facilitating informed decision-making in software development and risk management.

4.2   Data Annotation

This section provides comprehensive resources for utilizing the synthesized list of quality attributes. The information is organized into separate Python data frames, dictionaries, .csv files, and knowledge graphs.

Python DataFrames for Quality Models. To provide accessibility and usability of the quality attributes synthesized from various quality models, Python DataFrames are created. The provided structured representations can be seamlessly integrated into programming environments, analyses, and research projects. These data frames encapsulate the quality attributes, their definitions, and three defining characteristics for each attribute.

Dictionary Resources. Two distinct dictionaries were thoughtfully created for each quality model, catering to the diverse needs of researchers. The *Quality Attribute with Definition Dictionary* pairs each attribute with a precise definition, offering a concise and accurate resource for clarity. Complementing this, the *Quality Attribute with Characteristics Dictionary* delves into defining features, providing a comprehensive breakdown for researchers seeking a deeper exploration of attribute attributes. These dictionaries stand as essential references, facilitating a nuanced understanding and practical application of quality attributes.

GitHub Repository: .csv Files. The comprehensive dataset of quality attributes for each model, along with corresponding information, is archived in the GitHub repository. Available in a structured .csv format, researchers can easily download and integrate this dataset into their studies or analytical workflows for various research purposes.

Knowledge Graphs. We explored knowledge graphs as visual representations to intuitively depict interactions between different quality attributes [21,22]. Utilizing the NetworkX library in Python, directed graphs were constructed, representing quality attributes and their defining characteristics. The systematic steps, illustrated in Fig. 2, to construct knowledge graphs for diverse software quality models are detailed below:

Table 1. Merged Quality Attributes from Various Models

Quality Attribute	Definition	Characteristic 1	Characteristic 2	Characteristic 3
Compatibility	The degree to which a system can operate with other systems or software	Platform independence	Hardware independence	Operating System independence
Correctness	The accuracy and reliability of the system in delivering the intended results	Number of defects	User requirements	System requirements
Compliance	Adherence to relevant standards, regulations, and customer requirements	Customer Support	User Feedback	Response Time
Efficiency	Optimal use of resources to achieve the desired system performance	Resource usage	Memory usage	Device usage
Functionality	The ability of the system to deliver the intended features and services	Service	Malfunction	Performance
Flexibility	The ease with which the system can be adapted or extended to meet changing requirements	Customizability	Configurability	Extensibility
Interoperability	The ability of the system to work seamlessly with other systems	Protocol compatibility	Data format compatibility	API compatibility
Maintainability	The ease with which the system can be maintained, updated, and repaired	Code readability	Modularity	Comment density
Portability	The ability of the system to run on different platforms or environments	Audio	Graphics	Video
Reliability	The system's ability to perform consistently and predictably under various conditions	MTTF (Mean Time to Failure)	MTTR (Mean Time to Repair)	Failure rate
Security	The protection of the system and its data against unauthorized access and attacks	Encryption strength	Authentication methods	Access control
Supportability	The ease with which the system can be supported, serviced, and maintained	Accessibility	User Assistance	Troubleshooting support
Testability	The ease with which the system can be tested to ensure its correctness and reliability	Test coverage	Test case count	Test automation
Usability	The user-friendliness and ease of use of the system	User satisfaction	Learnability	Ease of Use

Graph Construction: Directed graphs featured quality attribute nodes (colored in 'light blue') and characteristic nodes (colored in 'light coral'). Edge connections showcased relationships between attributes and characteristics. Layout optimization used the Kamada-Kawai algorithm for an aesthetically pleasing and informative arrangement. To exemplify, Fig. 3 provides a visual depiction of the Knowledge Graph for the McCall Model.

Visualization and Accessibility: Graphs were visualized using Matplotlib, incorporating node colors, sizes, fonts, and edge styles. The Kamada-Kawai layout ensured clear representation. The resulting visual representations were systematically saved in PNG format for easy sharing and integration into scholarly documents and presentations.

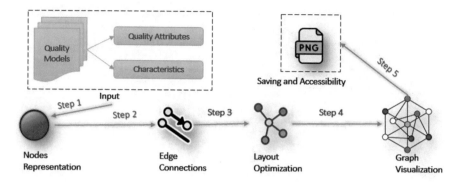

Fig. 2. Construction Steps of Knowledge Graphs

Utility and Accessibility: The Python notebook containing code for generating these graphs is available for download, empowering researchers, practitioners, and stakeholders to leverage these visual representations for analysis, research, and educational activities.

4.3 Downloadable Link

Note: For convenience, all relevant files and resources related to this research are available in the openly accessible GitHub repository [23].

5   Research Implications

The study holds multifaceted implications for software engineering. It establishes a foundational groundwork for automated tools and algorithms, impacting software quality assessment and prediction. The provided datasets and dictionaries are crucial resources, contributing to advancements in machine learning and natural language processing. Additionally, the synthesized quality attributes serve as educational assets, enriching software engineering curricula. The study offers practical guidance for software development through detailed attribute definitions. Furthermore, it reveals prospects for integrating neural networks and advanced algorithms in software development, paving the way for sophisticated automation. The research significantly enhances predictive modeling, fostering innovation in computational approaches and proactive quality management throughout the software development life cycle.

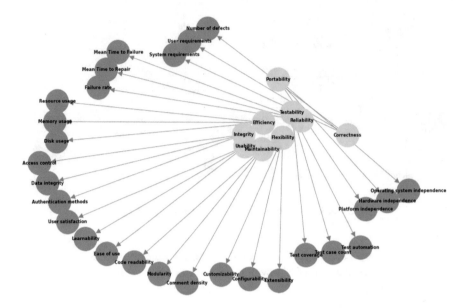

Fig. 3. Knowledge Graph of the McCall Model

6   Threats to Validity

In ensuring the reliability of our research findings, we acknowledge potential threats to both internal and external validity. The intricacies of dataset construction pose a risk to internal validity, despite rigorous efforts to standardize attribute definitions. Additionally, the external validity of our results is constrained by the specific focus on five prominent quality models, limiting their seamless extension to alternative software quality models and restricting generalization to a broader range of quality attributes and models within diverse software development practices. Our commitment to addressing these threats is reflected in the detailed measures taken to minimize their impacts throughout the study.

7   Conclusion

This research paper introduces a dataset of software quality attributes and employs state-of-the-art knowledge graph techniques to enhance data visualization. It represents a pioneering effort in laying the groundwork for the automation and universal application of attributes in software development research. While previous studies on attribute-based analyses often relied on expert opinions, introducing potential biases, our data-driven approach mitigates such biases. The provision of a downloadable public repository, Python codes, and detailed construction steps with pseudocode ensures ease of reproducibility and reliability. Future work aims to apply this dataset in real-world scenarios, mapping attributes to specific software contexts for further exploration and application.

References

1. Mohapatra, A., Pattnaik, S., Pattanayak, B.K., Patnaik, S., Laha, S.R.: Software quality prediction using machine learning, pp. 137–146 (2022)
2. Sugumaran, E.C.V.: Classification and security assessment of android apps. Discover Internet of Things **3**, 1–17 (2023)
3. Croft, R., Babar, M.A., Kholoosi, M.M.: Data quality for software vulnerability datasets, pp. 121–133 (2023)
4. Cui, J., Wang, L., Zhao, X., Zhang, H.: Towards a predictive analysis of android vulnerability using statistical codes and machine learning for IoT applications. Comput. Commun. **155**, 125–131 (2020)
5. Cagatay, C., Görkem, G., Bedi, T.: Applications of deep learning for mobile malware detection: a systematic literature review. Neural Comput. Appl. **34**, 1007–1032 (2021)
6. Koushki, M.M., AbuAlhaol, I., Raju, A.D., Zhou, Y., Giagone, R.S., Shengqiang, H.: On building machine learning pipelines for android malware detection: a procedural survey of practices, challenges, and opportunities. Cybersecurity **5**, 1–37 (2022)
7. Ahmed, M., Khan, S.U.R., Alam, K.A.: An NLP-based quality attributes extraction and prioritization framework in agile-driven software development. Autom. Softw. Eng. **30**, 1–7 (2023)
8. Chand, R., Khan, S.U.R., Hussain, S., Wang, W.L.: An ML-based quality features extraction (QFE) framework for android apps (2023)
9. Alanazi, S., Akour, M., Anbar, M.: Enterprise resource planning quality model ERPQM. In: First International Conference of Intelligent Computing and Engineering (ICOICE), pp. 1–5 (2019)
10. Elberkawi, E.K., Dakhil, M.I., Almukhtari, A.A., Abdelsalam, Maatuk, M.: Assessing quality standards in electronic gates of educational institutions: a field study, pp. 1–6 (2020)
11. Moumane, K., Idri, A., Abran, A.: Usability evaluation of mobile applications using ISO 9241 and ISO 25062 standards. Springerplus **5**, 1801–2193 (2016)
12. Dehkordi, M.R., Seifzadeh, H., Beydoun, G., Nadimi-Shahraki, M.H.: Success prediction of android applications in a novel repository using neural networks. Complex Intell. Syst. **6**, 573–590 (2020)
13. Jain, P., Sharma, A., Aggarwal, P.K.: Key attributes for a quality mobile application, pp. 50–54 (2020)
14. Maia, V., Rocha, A.R., Gonçalves, T.G.: Identification of quality characteristics in mobile applications, pp. 1–119 (2020)
15. Mishra, A., Otaiwi, Z.: DevOps and software quality: a systematic mapping. Comput. Sci. Rev. **38**, 100308 (2020)
16. Politowski, C., Petrillo, F., Ullmann, G.C., de Andrade Werly, J., Gu, Y.-G.: Dataset of video game development problems, pp. 553–557 (2020)
17. Motogna, S., Lupsa, D., Ciuciu, I.: A NLP approach to software quality models evaluation, pp. 207–217 (2019)
18. Chand, R., Khan, S.U.R., Hussain, S., Wang, W.L.: TTAG+R: a dataset of google play store's top trending android games and user reviews, pp. 580–586 (2022)
19. Abdallah, M., Jaber, K.M., Salah, M., Jawad, M.A.: An e-learning portal quality model: from al-zaytoonah university students' perspective, pp. 553–557 (2021)
20. Aljarallah, S., Lock, R.: A comparison of software quality characteristics and software sustainability characteristics, pp. 1–11 (2020)

21. Tamašauskaitė, G., Groth, P.: Defining a knowledge graph development process through a systematic review. ACM Trans. Softw. Eng. Methodol. **32**, 40 (2023)
22. Peng, C., Xia, F., Naseriparsa, M., Osborne, F.: Knowledge graphs: opportunities and challenges. Artif. Intell. Rev. **56**, 13:071-13:102 (2023)
23. Chand, R.: Dataset of software quality attributes. Github Repository (2023). https://github.com/AndroidGamesResearch/Dataset-of-Software-Quality-Attributes.git

Author Index

Á. Rocha et al. (Eds.): WorldCIST 2024, LNNS 990, pp. 383–384, 2024.
https://doi.org/10.1007/978-3-031-60328-0

Printed in the United States
by Baker & Taylor Publisher Services